de Gruyter Studies in Mathematics 8

Editors: Heinz Bauer · Peter Gabriel

Tammo tom Dieck

Transformation Groups

Walter de Gruyter
Berlin · New York 1987

Author

Dr. Tammo tom Dieck
Professor of Mathematics
Georg-August-Universität
Göttingen

Library of Congress Cataloging in Publication Data

Dieck, Tammo tom
Transformation groups.
(De Gruyter studies in mathematics ; 8)
Bibliography: p.
Includes index.
1. Topological transformation groups. 2. Lie groups.
J. Title. II. Series.
QA613.7.D53 1987 514'.3 87–5450
ISBN 0-89925-029-7 (U.S.)

CIP-Kurztitelaufnahme der Deutschen Bibliothek

TomDieck, Tammo:
Transformation groups / Tammo tom Dieck. –
Berlin ; New York : de Gruyter, 1987.
(De Gruyter studies in mathematics ; 8)
ISBN 3-11-009745-1 ISBN 978-3-11-009745-0
NE: GT

Cover design: Rudolf Hübler, Berlin. Typesetting and Printing: Tutte Druckerei GmbH, Salzweg-Passau. Binding: Dieter Mikolai, Berlin.

Ich habe den Bau eingerichtet und er scheint wohlgelungen. Von außen ist eigentlich nur ein großes Loch sichtbar, dieses führt aber in Wirklichkeit nirgends hin, schon nach ein paar Schritten stößt man auf natürliches festes Gestein.

Freilich manche List ist so fein, daß sie sich selbst umbringt, das weiß ich besser als irgendwer sonst und es ist gewiß auch kühn, durch dieses Loch überhaupt auf die Möglichkeit aufmerksam zu machen, daß hier etwas Nachforschungswertes vorhanden ist.

K.

Preface

This book introduces the reader to the theory of compact transformation groups. The theory of transformation groups deals with symmetries of mathematical objects. This viewpoint obviously needs some restriction. Thus I have chosen to concentrate on the algebraic topology of Lie transformation groups.

An introduction is (by definition) not a complete presentation of a theory. Rather, methods and ideas are explained. Even for the purpose of an elementary introduction to the theory of transformation groups, at least two more books are necessary: Transformation groups on topological and smooth manifolds. Geometric representation theory.

The topics of this book are organized into four Chapters. Chapter I presents the basic language. Chapter II is concerned with cell-complexes, homotopy theory and (co-)homology theory. Chapter III explains localization methods in equivariant cohomology. Chapter IV is centered around the Burnside ring and additive invariants. Each Chapter has its own introduction.

A reference of the form III(6.2) refers to Theorem (Definition, etc.) (6.2) of Chapter III. The Roman numeral is omitted whenever the reference concers the chapter where it appears. References to the bibliography at the end of the book have the form P. A. Smith [1938a], referring to author and year of publication, and, if necessary, are distinguished by additional letters a, b, and c.

I have tried not to repeat material which is readily available in other textbooks: See the section "Further reading" at the end of this book. Since the theory of transformation groups is not a "primary theory", the reader is expected to be acquainted with basic algebraic topology. Occasionally, some knowledge about topological groups or representation theory might be helpful. There are numerous references to the literature to provide the reader with background material.

The presentation varies in difficulty. Basic notions and results are developed in detail while more special applications may rely on references to the bibliography. Moreover, I have often mentioned topics which are related to the text, thus giving a guide to further reading. The bibliography is by no means complete. It contains only those items which I have used or referred to in the text.

The numerous exercises vary in character: They ask the reader to verify statements used in the text. They give further information, sometimes with references to the literature. They present examples. They are genuine exercises.

Stefan Bauer and Peter Hauck read the manuscript. They suggested many mathematical and linguistical improvements. Wolfgang Lück helped with the proof-reading. I thank them for their generous and accurate work.

Naturally, I would have liked to write the book in my mother tongue. Jedoch ...

Der Knecht wirft beide Arm empor,
als wollt er sagen: „Laß doch, laß!"

Göttingen, December 1986 Tammo tom Dieck

Contents

Chapter I: Foundations

The first chapter comprises the following sections.
1. The basic terminology related to a transformation group: Group action, orbit space, fixed point set, and equivariant map. The problem of turning the homeomorphism group of a space into a topological group.
2. A collection of examples from various branches of mathematics to indicate the role of symmetry considerations. Basic concepts such as representations and homogeneous spaces, which are used throughout the book.
3. Point set topology of transformation groups. Proper actions of locally compact groups.
4. Elementary material about changing the group: Restriction, additive and multiplicative induction.
5. Differentiable manifolds as a tool to prove the basic slice theorem. Qualitative results about differentiable manifolds, such as the principal orbit type theorem or finiteness of orbit types.
6. Classifying spaces for families of subgroups. They generalize the classifying spaces of bundle theory and are basic for the method of induction over orbit bundles.
7. Equivariant maps as non-equivariant sections. Inductive construction of equivariant maps. Equivariant fibrations.
8. Principal bundles with automorphism group and their classifying spaces.
9. Equivariant vector bundles: A rather short list of elementary properties.
10. Introduction to the combinatorial aspects of a transformation group. The universal covering of a transformation group as a functor.
11. Generalization of classical representation theory of modules over group rings to algebra in functor categories. Explanation of this viewpoint by describing projective modules and a few notions from algebraic K-theory.

Although in some of the earlier sections we consider general groups acting on arbitrary topological spaces, the intention of the book is to study compact Lie group actions. So, if not obvious from the text or otherwise stated, the reader should assume that the group acting is a compact Lie group. Subgroups are supposed to be closed. The transformation groups, if necessary, should have closed isotropy groups and closed fixed point sets. Usually, spaces are assumed to be Hausdorff spaces.

1. Basic notions.

Transformation groups describe (continuous) symmetries of geometric objects. This section introduces the basic notions and notations that will be used throughout the text. The next section describes examples to illustrate these concepts.

Let G be a topological group and X a topological space. A **left action of G on X** (also, a **left operation of G on X**) is a continuous map

(1.1) $\varrho\colon G \times X \to X$

such that

(1.2)
(i) $\varrho(g, \varrho(h, x)) = \varrho(gh, x)$ for $g, h \in G,\ x \in X$
(ii) $\varrho(e, x) = x$ for $x \in X,\ e \in G$ unit.

A **left G-space** (also, a **transformation group**) is a pair (X, ϱ) consisting of a space X together with a left action ϱ of G on X. We shall usually denote the G-space (X, ϱ) just by its underlying topological space X. It is convenient to denote $\varrho(g, x)$ by gx. The rules (1.2) then take the familiar form $g(hx) = (gh)x$ and $ex = x$. Occasionally, we call any set-mapping (1.1) satisfying (1.2) a left action and specify continuous action if necessary.

A **right action** is a map $X \times G \to X$, $(x, g) \mapsto xg$ satisfying $(xh)g = x(hg)$ and $xe = x$. If $(x, g) \mapsto xg$ is a right action, then $(x, g) \mapsto xg^{-1}$ is a left action. Usually, we work with left G-spaces and omit the word left.

The **left translation** $L_g\colon X \to X$, $x \mapsto gx$ **by** g is a homeomorphism of X with inverse $L_{g^{-1}}$. This follows from the rules $L_g L_h = L_{gh}$, $L_e = \mathrm{id}(X)$, which are just reformulations of (1.2). The map $g \mapsto L_g$ is a homomorphism of G into the group of homeomorphisms of X. The action is called **effective** if the kernel of $g \mapsto L_g$ is $\{e\}$. It is called **trivial** if the kernel is G itself. An action is called **free** if $gx = x$ always implies $g = e$. Occasionally, we denote the left translation by l_g.

(1.3) Example. Let H be a subgroup of G. The group multiplication $H \times G \to G$, $(h, g) \mapsto hg$ is a free left H-action. There is a similar right action. A group also acts on itself by conjugation $G \times G \to G$, $(gh) \mapsto ghg^{-1}$. If G is a topological group, then these actions are continuous.

Let X be a G-space. Then $R = \{(x, gx) \mid x \in X, g \in G\}$ is an equivalence relation on X. The set of equivalence classes $X \bmod R$ is denoted by X/G. The quotient map $q\colon X \to X/G$ is used to provide X/G with the quotient topology. This space is called the **orbit space** of the G-space X. The equivalence class of $x \in X$ is called the **orbit** Gx **through** x. A more systematic notation would be $G\backslash X$ for the orbit space of a left action and X/G for the orbit space of a right action. There are a few situations where we use this notation. An action of G on X is called **transitive** if X consists of a single orbit.

(1.4) Example. The orbit space of the right action $G \times H \to G$ is G/H, the space of right cosets gH. The space of left cosets should be $H\backslash G$. The map

$$G \times G/H \to G/H, \ (g', gH) \mapsto g'gH$$

is easily checked to be continuous. Any G-space G/H with this action is called a **homogeneous space**. If $K \subset G$ is another subgroup, then we can again form a left coset space $K\backslash(G/H)$, the **space of double cosets**.

A subset $F \subset X$ of a G-space X is called a **fundamental domain** of this G-space if $F \subset X \to X/G$ is bijective. A fundamental domain contains exactly one point from each orbit. Usually, there are many different fundamental domains, and the problem then is to choose one with particularly nice geometric properties.

For each $x \in X$, the set $G_x = \{g \in G \mid gx = x\}$ is a subgroup of G. This subgroup is called the **isotropy group** of X or of the G-space X at x. We let $\mathrm{Iso}(X)$ denote the set of isotropy groups of X. From $G_{gx} = gG_xg^{-1}$ it follows that $\mathrm{Iso}(X)$ consists of complete conjugacy classes of subgroups. Recall that subgroups H and K of G are called **conjugate** in G (notation $H \sim K$) if and only if there exists $g \in G$ such that $H = gKg^{-1}$. The **conjugacy class** of H is denoted by (H). We call H **subconjugate** to K if H is conjugate to a subgroup of K. Subconjugation defines a partial order on the set of conjugacy classes of subgroups. We write $(H) < (K)$ if H is subconjugate to K.

(1.5) Example. Let $S(G)$ be the set of subgroups of G. We have an action (without continuity) $G \times S(G) \to S(G), (g, H) \mapsto gHg^{-1}$. The orbit of H is the conjugacy class (H).

(1.6) Proposition. *Let X be a G-space and $x \in X$. The map $G \to X$, $g \mapsto gx$ is constant on cosets gG_x and induces an injective map $q_x\colon G/G_x \to X$ whose image is the orbit through x.* □

We leave the verification to the reader. In general, q_x is not a homeomorphism onto its image (compare section 3).

We specify the following subsets of a G-space X

(1.7) $\quad X_H = \{x \in X \mid G_x = H\}$,

(1.8) $\quad X_{(H)} = \{x \in X \mid G_x \sim H\}$.

$X_{(H)}$ is called the **(H)-orbit bundle** of X. The space X is the disjoint union of its orbit bundles. One of the problems in transformation group theory is to analyse in which way spaces are built from their orbit bundles.

If H is a subgroup of G, then

(1.9) $\quad X^H = \{x \in X \mid hx = x \quad \text{for all} \quad h \in H\}$

is called the ***H*-fixed point set** of X. Points in X^G are sometimes called **stationary points** of the G-space. One has $X^H \supset X_H$. The complement is denoted by

(1.10) $X^{>H} = X^H \backslash X_H$.

The reason for this notation is the equality

(1.11) $X^{>H} = \bigcup_K X^K$, all $K \supset H$, $K \neq H$

as the reader should check.

If X is a G-space and $Y \subset X$ a subspace, we call Y **G-invariant** or a **G-subspace** if for all $g \in G$ and $y \in Y$ the relation $gy \in Y$ holds. Then we have an induced continuous action $G \times Y \to Y$, $(g, y) \mapsto gy$ and Y becomes a G-space. The orbits are the smallest G-subsets of a G-space; each G-subspace is a union of orbits. The orbit bundles are G-subspaces. Of particular importance is the orbit bundle $X_{\{e\}}$, the largest subspace of X where the action is free. The complement $X \backslash X_{\{e\}}$ is often called the **singular set** X_s of X. It is the union of the fixed point sets X^H, $H \neq \{e\}$.

Suppose X and Y are G-spaces. A map $f\colon X \to Y$ is called a **G-map** or a **G-equivariant map** if for all $g \in G$ and $x \in X$ the relation $f(gx) = gf(x)$ holds. When dealing with G-spaces, G-maps are usually supposed to be continuous. A G-map $f\colon X \to Y$ induces a map

(1.12) $f/G : X/G \to Y/G$, $Gx \mapsto Gf(x)$

between orbit spaces. If f is continuous, then f/G is continuous. If $f\colon X \to Y$ is equivariant, then, for all $x \in X$,

(1.13) $G_x \subset G_{fx}$.

If in (1.13) equality holds for all $x \in X$, then f is called **isovariant**. Of course, G-spaces and G-maps form a category. As usual in categories, we have the notion of an isomorphism between objects; the term **G-homeomorphism** is also used in this context.

If X is a G-space and Y a K-space for topological groups G and K, then $X \times Y$ becomes a $G \times K$-space with action

$$((g, k), (x, y)) \mapsto (gx, ky).$$

Similarly, this definition can be extended to an arbitrary number of factors.

The category of G-spaces has products: Suppose $(X_j \mid j \in J)$ is a family of G-spaces. We define a G-action on the topological product $\prod_{j \in J} X_j$ by

$$(g, (x_j \mid j \in J)) \mapsto (gx_j \mid j \in J),$$

which is also called the **diagonal action**. The reader may check that this gives the categorical product.

G-maps $f_0, f_1\colon X \to Y$ are called **G-homotopic** if there exists a continuous G-map, a **G-homotopy from f_0 to f_1**,

$$F : X \times [0,1] \to Y$$

such that $F(x,0) = f_0(x)$, $F(x,1) = f_1(x)$; here, $[0,1]$ has the trivial G-action and $X \times [0,1]$ the diagonal action. Each map $f_t : x \mapsto F(x,t)$ is then a G-map. As usual, one shows that being G-homotopic is an equivalence relation and that one has a category of G-spaces and G-homotopy classes of mappings. We use the symbol $[X, Y]_G$ for the set of G-homotopy classes of G-maps $X \to Y$. Furthermore, we write $f_0 \simeq_G f_1$ if f_0 and f_1 are G-homotopic.

If K is a subgroup of G and $\varrho: G \times X \to X$ a G-action, then we can **restrict** this action to K, thus obtaining the K-space $\operatorname{res}^G_K X = \operatorname{res}_K X$ which is X together with the action $\varrho | K \times X$. More generally, if $\varphi: G' \to G$ is a continuous homomorphism, then $\varrho(\varphi \times \mathrm{id}_X) = \varrho'$ is a G'-action on X.

We say that a left action $G \times X \to X$, $(g,x) \mapsto gx$ and a right action $X \times K \to X$, $(x,k) \mapsto xk$ **commute** if for all $g \in G$, $x \in X$, $k \in K$ the identity $g(xk) = (gx)k$ holds.

If G is a Lie group, M a smooth ($= C^\infty$ differentiable) manifold, then a **smooth action** of G on M is a smooth map $G \times M \to M$, $(g,m) \mapsto gm$ which is an action. A smooth manifold together with a smooth action is called a **smooth** or **differentiable G-manifold**.

Occasionally, we need the purely algebraic aspect of a transformation group. Let G be a group. It can be considered as a topological group with discrete topology and as such it is called a **discrete group**. A **G-set** X is a space X with discrete topology together with an action of the discrete group G on X or, equivalently, a set X together with a map $G \times X \to X$, $(g,x) \mapsto gx$ such that $(gh)x = g(hx)$, $ex = x$. Any G-space may be considered as a G-set by forgetting the topology on G and X. Giving a G-set X amounts to specifying a homomorphism from G into the permutation group of the set X.

(1.14) Proposition. *Let H and K be subgroups of G.*

(i) *There exists a G-map $G/H \to G/K$ if and only if H is conjugate to a subgroup of K.*

(ii) *If $a \in G$, $a^{-1}Ha \subset K$, then we obtain a G-map* $R_a: G/H \to G/K$, $gH \mapsto gaK$.

(iii) *Each G-map $G/H \to G/K$ has the form R_a for suitable $a \in G$ with $a^{-1}Ha \subset K$.*

(iv) *$R_a = R_b$ if and only if $ab^{-1} \in K$.*

Proof. Let $f: G/H \to G/K$ be equivariant. Choose $a \in G$ such that $f(eH) = aK$, $e \in G$ the identity. Equivariance yields $aK = f(eH) = f(hH) = hf(eH) = haK$ for all $h \in H$, hence $a^{-1}Ha \subset K$.

Suppose $a^{-1}Ha \subset K$. If $g_1H = g_2H$, then $g_1 = g_2h$ for some $h \in H$ and $ha = ak$ for some $k \in K$. Therefore,

$$g_1aK = g_2haK = g_2akK = g_2aK;$$

hence R_a is well-defined. It is a G-map by construction. We have proved (i) and (ii) and leave the verification of (iii) and (iv) to the reader. □

Let H be a subgroup of G and NH or $N_G H$ the **normalizer** of H in G, i.e. $NH = \{n \in G \mid n^{-1}Hn = H\}$. The group H is a normal subgroup of NH and NH/H will generally be denoted by WH. A G-selfmap of G/H has the form $R_n\colon G/H \to G/H$, $gH \mapsto gnH$ for $n^{-1}Hn \subset H$. In general, such an element n need not be contained in NH. It belongs to NH precisely when R_n is a G-automorphism of G/H. We thus have an isomorphism

$$WH \to \operatorname{Aut}_G(G/H), \ n^{-1}H \mapsto R_n$$

of WH onto the group of G-automorphisms of G/H. The following is easily verified.

(1.15) Proposition. *The action*

$$G/H \times WH \to G/H, \quad (gH, nH) \mapsto gnH$$

is a free action. □

Let $\varrho\colon G \times X \to X$ be a continuous action of G on X. Let $\mathscr{H} = \operatorname{Homeo}(X)$ denote the group of homeomorphisms of X. The adjoint map $\bar{\varrho}\colon G \to \mathscr{H}$, $g \mapsto L_g$ is a homomorphism and $\varepsilon\colon \mathscr{H} \times X \to X$, $(f, x) \mapsto f(x)$ is an action. The question naturally arises whether there exist topologies on the set $\mathscr{H}$ such that assertions of the following type are true:

(i) $\bar{\varrho}$ is continuous.
(ii) $\mathscr{H}$ is a topological group.
(iii) ε is a continuous action.

There exist different suggestions for defining topologies on $\mathscr{H}$ (Arens [1946]). We shall only consider the compact-open-topology of Fox (see e.g. Kelley [1955], Ch. 7; Bourbaki [1961], Ch. 10, § 3; Maunder [1970], Ch. 6). We recall the definition and a few properties.

Let X and Y be topological spaces and Y^X the set of continuous maps $X \to Y$. Let $K \subset X$ be compact (not necessarily Hausdorff) and $U \subset Y$ open. We put $W(K, U) = \{f \in Y^X \mid f(K) \subset U\}$. The sets of the form $W(K, U)$ constitute a subbasis of the **compact-open-topology** (CO-topology) on Y^X.

If $f\colon X \times Y \to Z$ is continuous, we have the adjoint map $\bar{f}\colon X \to Z^Y$, $\bar{f}(x)(y) = f(x, y)$.

(1.16) If Z^Y carries the CO-topology, then $\bar{f}$ is continuous. We therefore have an induced map

(1.17) $\alpha\colon Z^{X \times Y} \to (Z^Y)^X, \ f \mapsto \bar{f}.$

This map has the following properties:

(1.18) (i) If X is a Hausdorff space, then α is continuous.
(ii) If Y is locally compact, then α is surjective.
(iii) If X and Y are Hausdorff spaces, then α is an embedding.
(iv) If X and Y are Hausdorff spaces and Y is locally compact, then α is a homeomorphism.

It is also important to consider the evaluation map $\varepsilon\colon\ Y^X \times X \to Y$, $(f, x) \mapsto f(x)$. To this end we recall that a space is called locally compact if each neighbourhood of a point contains a compact neighbourhood. From (1.18, ii) we obtain:

(1.19) If X is locally compact, then ε is continuous. Composition of maps is continuous in the following case.

(1.20) Let X and Y be locally compact. Then $Z^Y \times Y^X \to Z^X$, $(g,f) \mapsto g \circ f$ is continuous.

If we specialize to $\varrho\colon\ G \times X \to X$ and define $\mathscr{H} = \mathrm{Homeo}(X) \subset X^X$ as a subspace, then we obtain

(1.21) Proposition. (i) *$\bar{\varrho}$ is continuous. If X is locally compact and $\bar{\varrho}$ continuous, then ϱ is continuous.*
(ii) *If X is locally compact, then $\mathscr{H} \times \mathscr{H} \to \mathscr{H}$, $(f, g) \mapsto f \circ g$ and $\mathscr{H} \times X \to X$, $(f, x) \mapsto f(x)$ are continuous.* □

There remains the problem under which conditions $\mathscr{H} \to \mathscr{H}$, $f \mapsto f^{-1}$ is continuous for a locally compact X, because then $\mathscr{H}$ will be a topological group in the CO-topology.

(1.22) Proposition. *Let X be a compact Hausdorff space. Then* Homeo(X) *is a topological group in the CO-topology.*

The proof of the continuity of $f \mapsto f^{-1}$ uses a reformulation of the CO-topology. For simplicity, let X be a compact Hausdorff space and Y a uniform space with the induced topology being Hausdorff. Recall that a uniform structure is given by specifying a suitable collection $\mathfrak{U}$ of sets $U \subset X \times X$ containing the diagonal. For $K \subset X$ compact, $U \in \mathfrak{U}$, and $f_0 \in Y^X$, let

$$N(f_0, K, U) = \{f \in Y^X \mid (f(x), f_0(x)) \in U \text{ for all } x \in K\}.$$

Then:

(1.23) The sets $N(f_0, K, U)$ form a neighbourhood basis of f_0 in the *CO*-topology.

For a proof see Bourbaki [1961], p. 43.

If Y is a compact Hausdorff space, there exists a unique uniform structure $\mathfrak{U}$ inducing the given topology: The $U \in \mathfrak{U}$ are the neighbourhoods of the diagonal in $X \times X$. Using (1.23) and this uniform structure, a proof of (1.22) is now easy.

Proof of (1.22). If $f_0 \in \mathscr{H}$, then f_0 is uniformly continuous. Thus, given $V \in \mathfrak{U}$ with $(x, y) \in V \Rightarrow (y, x) \in V$, there exists W such that $(x, y) \in W$ implies $(f_0^{-1}(x), f_0^{-1}(y)) \in V$. Let $f \in \mathscr{H}$ satisfy $(f_0(x), f(x)) \in W$ for all $x \in X$. Then $(x, f_0^{-1}f(x)) \in V$ for all $x \in X$ and, since f is bijective, $(f^{-1}(x), f_0^{-1}(x)) \in V$ for all $x \in X$. Using (1.23), this shows continuity of $f \mapsto f^{-1}$ at f_0. □

There exist locally compact spaces X such that $\mathscr{H} \to \mathscr{H}$, $f \mapsto f^{-1}$ is not continuous; however, one can show:

(1.24) Proposition. *Let X be a locally compact and locally connected Hausdorff space. Then $\mathscr{H}$, equipped with the CO-topology, is a topological group.*

See Bourbaki [1961], p. 74, Ex. 17. See also (2.11). □

References. We assume that the reader has some knowledge of topological group theory. We mention at this point the following monographs about various aspects of topological groups and transformation groups:
Borel [1960], Bourbaki [1960], Bredon [1972], Bröcker-tom Dieck [1985], Chevalley [1946], Hewitt-Ross [1963], Hochschild [1965], Montgomery-Zippin [1955], Pontrjagin [1957].

(1.25) Exercises.

1. Show: $G \to G/H$ is open, $\mathrm{id} \times q\colon G \times G \to G \times G/H$ is a quotient map, the action of (1.4) is continuous. Prove or disprove: $K\backslash(G/H)$ and $(K\backslash G)/H$ are homeomorphic.
2. What are the isotropy groups in example (1.5)?
3. Prove (1.6), (1.11), continuity of (1.12), (1.15).
4. What are the isotropy groups of the G-space G/H?
5. When is the action on G/H effective? (Example (1.4))

We assume that the reader has met topological groups before. He may check his understanding of topological groups in dealing with the following exercises.

6. Let H be a subgroup of the topological group G. Then the closure $\bar{H}$ is a subgroup. If H is normal in G, then $\bar{H}$ is normal. If H is commutative and G

Hausdorff, then $\bar{H}$ is commutative. (Is Hausdorff necessary?) The connected component G_0 of e in G is a closed normal subgroup.

7. If H is closed in G, then the normalizer $NH = \{g \in G \mid gHg^{-1} = H\}$ is closed in G.
8. Let G be connected and V a neighbourhood of e. Then $G = \bigcup_{n \geq 1} V^n$. Here, $V^n = \{g = g_1 g_2 \ldots g_n \mid g_i \in V\}$.
9. Let H be an open subgroup of G. Then H is closed in G.
10. Let U be a neighbourhood of e in G. Then there exists a symmetric neighbourhood V, i.e. $V = V^{-1}$, of e such that $V^2 \subset U$. For such V one has $\bar{V} \subset U$. Use this to show: If G is a T_0-space, then G is regular (and in particular Hausdorff).
11. Let H be a subgroup of G and U a neighbourhood of e in G. If $\bar{U} \cap H$ is closed in G, then H is closed in G.
12. Let G be a T_0-group and H a subgroup which is locally compact with respect to its induced topology. Then H is closed in G.
13. Let $\lambda \in \mathbb{R}$. Let $f_\lambda\colon \mathbb{Z} \to \mathbb{C}$, $n \mapsto \exp(2\pi i \lambda n)$. Let $\mathcal{O}(\lambda)$ be the coarsest topology such that f_λ is continuous. Show that $(\mathbb{Z}, \mathcal{O}(\lambda))$ is a topological group. When are $\mathcal{O}(\lambda_1)$ and $\mathcal{O}(\lambda_2)$ equal? Use sufficiently many f_λ in order to find a topological group $(\mathbb{Z}, \mathcal{O})$ without countable base for the topology. (See Hewitt-Ross [1963], II.4.22.)
14. Let X and Y be G-spaces. There is a G-action on the set $\mathrm{Map}(X, Y)$ of mappings from X to Y defined by $(g, f) \mapsto (x \mapsto gf(g^{-1}x))$. The fixed point set consists of the G-equivariant maps.
    If Y^X carries the CO-topology and X is locally compact, then this action $G \times Y^X \to Y^X$ is continuous.
15. (Chain condition for compact Lie groups.)
    If $H_1 \supset H_2 \supset H_3 \supset \ldots$ is a sequence of closed subgroups of the compact Lie group G, then there exists k such that $H_n = H_k$ for $n > k$.
    The partial order of subconjugacy on the set of conjugacy classes of closed subgroups of the compact Lie group can be refined to a well-ordering.

2. General remarks. Examples.

Groups are intended to describe symmetries of mathematical objects, i.e. groups are meant to be transformation groups. Generally spoken, the theory of transformation groups deals with symmetries (= automorphisms) of mathematical objects, thus with most of mathematics. It is clear that in this generality no satisfactory **theory** can be developed. But nevertheless, a **viewpoint** is obtained. The automorphism group of an object is part of its internal structure and should be regarded as important information about the object. We mention at this point that **Galois theory** makes paradigmatic use of this aspect; as the reader knows, fixed point sets of field automorphisms are important.

(2.1) Category theory. We still retain a viewpoint of great generality. In dealing with transformation groups, it is necessary or at least useful to have at hand a systematic language. The process of obtaining such a language is expressed by the slogan: Make categorical notions "equivariant".

Let $\mathfrak{C}$ be any category. We can consider the associated category of $\mathfrak{C}$-objects with symmetry, Sym $\mathfrak{C}$. The objects are triples (G, ϱ, X), G a group, $X \in \mathrm{Ob}(\mathfrak{C})$, and $\varrho\colon G \to \mathrm{Aut}_{\mathfrak{C}}(X)$ a homomorphism of G into the group of $\mathfrak{C}$-automorphisms of X. A morphism from (G_1, ϱ_1, X_1) to (G_2, ϱ_2, X_2) is a pair (φ, f) consisting of a homomorphism $\varphi\colon G_1 \to G_2$ and a morphism $f\colon X_1 \to X_2$ in $\mathfrak{C}$ such that, for each $g \in G_1$, commutativity $f \circ \varrho_1(g) = \varrho_2(\varphi(g)) \circ f$ holds.

Composition of morphisms is defined in the obvious manner $(\varphi_1, f_1) \circ (\varphi_2, f_2) := (\varphi_1 \varphi_2, f_1 f_2)$. For practical purposes, one often considers subcategories: Fix a group G and let the category G-$\mathfrak{C}$ have objects (G, ϱ, X) and morphisms (id, f).

The category G-$\mathfrak{C}$ leads to two aspects, according to the variables G or $\mathfrak{C}$.

Firstly, G-$\mathfrak{C}$ is considered to be an invariant of G: One studies G-$\mathfrak{C}$ via its dependence on G (for a suitable class of groups G, e.g. finite, compact Lie, discrete). This could be called the general viewpoint of representation theory.

Secondly, G-$\mathfrak{C}$ is studied using its dependence on $\mathfrak{C}$. A functor $F\colon \mathfrak{C} \to \mathfrak{D}$ induces in an obvious manner a functor G-$F\colon G$-$\mathfrak{C} \to G$-$\mathfrak{D}$. One can use a suitable G-F to define invariants for G-$\mathfrak{C}$.

If $\mathfrak{C}$ is a category of sets with additional structure, then, instead of looking at homomorphisms $G \to \mathrm{Aut}(X)$, one considers actions of G on the set X such that, at least, each left translation $L_g\colon X \to X$ is a morphism in $\mathfrak{C}$. The action $\varrho\colon G \times X \to X$ is usually restricted by further structural conditions. This leads to subcategories of $\mathfrak{C}$. Typical examples are:

(i) G a topological group, X a topological space, and ϱ continuous.
(ii) G a Lie group, X a differentiable manifold, and ϱ differentiable.
(iii) G a discrete group, X a cellular complex, each L_g a cellular map.
(iv) X a vector bundle, each L_g a bundle automorphism.
(v) X a Riemannian manifold, each L_g an isometry.

(2.2) Representation theory. Let R be a commutative ring and R-Mod the category of R-modules. Study of G-R-Mod is representation theory of G over R. Representation theory is a well-developed and deep theory and absolutely indispensible for the theory of transformation groups. Each functor $\mathfrak{C} \to R$-Mod leads to a functor G-$\mathfrak{C} \to G$-R-Mod and thus to representation theoretical invariants for G-$\mathfrak{C}$. As an example: Apply functors of algebraic topology (homotopy, homology) to G-spaces.

(2.3) Natural symmetries. One of the origins of transformation group theory is the consideration of symmetries for naturally occuring mathematical objects, e.g. Galois theory or classical geometry. Later, we shall list a few of the geometrical examples. It turns out that mathematicians like (geometrical) objects which have a large symmetry group. It is generally believed that the most symmetrical individuals contain the deepest mathematical information.

In topology, the emphasis is nowadays more on construction of exotic symmetries and classification of symmetries. But still, the natural objects are ideal models which serve as a starting point for the construction of variations or as guiding examples for the theory. (Nature as justification principle).

(2.4) Lie groups. Some of the most important groups are defined as transformation groups, as automorphism groups of geometric objects. These are the classical matrix groups, which we now review.

The **general linear group** $\mathrm{Gl}(n, \mathbb{R})$, resp. $\mathrm{GL}(n, \mathbb{C})$, over the field of real numbers $\mathbb{R}$, resp. complex numbers $\mathbb{C}$, is the group of all invertible (n, n)-matrices with entries in the field $\mathbb{R}$, resp. $\mathbb{C}$. The **special linear groups** $\mathrm{SL}(n, \mathbb{R}) \subset \mathrm{GL}(n, \mathbb{C})$ and $\mathrm{SL}(n, \mathbb{C}) \subset \mathrm{GL}(n, \mathbb{C})$ are the respective subgroups of matrices with determinant one. Let A^t be the transpose and $\bar{A}$ the complex conjugate of A.

The **orthogonal group** $\mathrm{O}(n)$ is the group $\{A \in \mathrm{GL}(n, \mathbb{R}) \mid AA^t = E\}$ of orthogonal matrices, the **unitary group** $\mathrm{U}(n)$ is the group $\{A \in \mathrm{GL}(n, \mathbb{C}) \mid \bar{A}A^t = E\}$ of unitary matrices. We have the **special orthogonal group** $\mathrm{SO}(n) = \mathrm{O}(n) \cap \mathrm{SL}(n, \mathbb{R})$ and the **special unitary group** $\mathrm{SU}(n) = \mathrm{U}(n) \cap \mathrm{SL}(n, \mathbb{C})$.

The matrices $A \in \mathrm{U}(2n)$ satisfying $A^t J A = E$ with

$$J = \begin{pmatrix} 0 & -E \\ E & 0 \end{pmatrix}, \quad E \text{ identity matrix},$$

are called **symplectic**. They form the **symplectic group** $\mathrm{Sp}(n) \subset \mathrm{U}(2n)$. Thus $\mathrm{Sp}(1) = \mathrm{SU}(2)$ and this group is isomorphic to S^3, the group of quaternions of norm 1; see Bröcker-tom Dieck [1985], I(6.18).

The groups $\mathrm{U}(1)$ and $\mathrm{SO}(2)$ are isomorphic to the circle group $S^1 = \{z \in \mathbb{C} \mid |z| = 1\}$, the multiplicative group of complex numbers with absolut value one. A group isomorphic to the direct product of n copies of S^1 is called an n-dimensional **torus**. The additive group $\mathbb{Z}/m$ of integers modulo m is sometimes identified with the subgroup of S^1 consisting of all m-th roots of unity, $k \bmod m \mapsto \exp(2\pi ik/m)$. An abelian compact Lie group is isomorphic to the direct product of a torus and a finite abelian group.

(2.5) Representations. A real (complex) **representation** of the topological group G on the finite dimensional real (complex) vector space V is a continuous action $\varrho: G \times V \to V$ such that, for each $g \in G$, the left translation $L_g: V \to V$ is a linear map. We also call V or (V, ϱ) a **representation space**.

Thus, representations are symmetries of some of the most basic objects in geometry and algebra, namely vector spaces.

If $V = \mathbb{R}^n$, then a representation is specified by a continuous homomorphism $g \mapsto L_g$ of G into $\mathrm{GL}(n, \mathbb{R})$. In particular, we have the **standard representation**

$$\mathrm{GL}(n, \mathbb{R}) \times \mathbb{R}^n \to \mathbb{R}^n, \quad (A, x) \mapsto Ax$$

(matrix multiplication), and similarly for $\mathrm{GL}(n, \mathbb{C})$ and subgroups of $\mathrm{GL}(n, \mathbb{R})$ and $\mathrm{GL}(n, \mathbb{C})$.

The fixed point sets V^H of a representation V are linear subspaces because

$$V^H = \bigcap_{h \in H} \operatorname{kernel}(L_h - \mathrm{id}).$$

For the representation theory of finite groups, see Serre [1971] or Curtis-Reiner [1962]; for compact Lie groups, see Bröcker-tom Dieck [1985].

(2.6) Representation spheres. If a group G acts on a representation space V by orthogonal transformations (or unitary transformations) L_g, then the unit sphere $S(V)$ of V is invariant and one has an induced action of G on $S(V)$. The G-spaces of this type are called **representation spheres**. If V is a real representation of the compact group G, then there always exists an inner product $\langle -,- \rangle$ on V such that $\langle gv, gw \rangle = \langle v, w \rangle$ for $g \in G$ and $v, w \in V$; see Bröcker-tom Dieck [1985], Ch. II. The unit sphere $S(V) = \{v \in V \mid \langle v, v \rangle = 1\}$ is then a representation sphere.

Representation spheres have the special property that fixed point sets $S(V)^H$ are again spheres. For general group actions on spheres, the fixed point sets can be quite complicated, even in the simplest case that $G = \mathbb{Z}/2$ is the group with two elements.

The fundamental importance of representations for the theory of transformation groups is demonstrated by the following embedding theorem of Mostow [1957a] and Palais [1957] (see also Palais [1960], [1960a] and Bredon [1972], p. 111).

Theorem. *Let G be a compact Lie group and X a G-space. Then X is G-homeomorphic to a G-subspace of an orthogonal G-representation space if and* *only if X is a finite-dimensional, separable metric space with a finite number of conjugacy classes of isotropy groups.* □

For refined embedding and immersion theorems in special situations, see Allen [1979], [1980], Illman [1980], Kister-Mann [1962], Wasserman [1969], Bierstone [1974], Copeland-de Groot [1961].

(2.7) Abelian groups. We recall the representation theory of compact abelian Lie groups G in order to describe some explicit examples; see Bröcker-tom Dieck [1985], Ch. II.

A real representation V is the direct sum $V = V_1 \oplus \ldots \oplus V_r$ of irreducible representations V_i. An irreducible representation W is one- or two-dimensional. If $\dim_{\mathbb{R}} W = 1$, then the representation is given by a homomorphism $G \to \mathbb{Z}/2$, and if $\dim_{\mathbb{R}} W = 2$, then the representation is given by a homomorphism $G \to \mathrm{SO}(2) \cong \mathrm{U}(1)$. In the latter case, W is the restriction of a complex representation U. The complex structure on a representation V is often useful to specify orientations for V and $S(V)$.

We look at some special cases. A complex representation of $\mathbb{Z}/m$ on $V = \mathbb{C}^n$ is given by

$$\mathbb{Z}/m \times \mathbb{C}^n \to \mathbb{C}^n,\ (k, (z_1, \ldots, z_n)) \mapsto (\xi_1^k z_1, \ldots, \xi_n^k z_n),$$

where $\xi_1, \ldots, \xi_n$ are m-th roots of unity. The unit sphere $S(V) = \{(z_1, \ldots, z_n) \mid \sum |z_i|^2 = 1\}$ is invariant. If the ξ_j are primitive m-th roots of unity, i.e. $\xi_j = \exp(2\pi i a_j/m)$ with a_j prime to m, then the action on the unit sphere is free. In this case, the orbit space is denoted by

$$L^{2n-1}(m; a_1, \ldots, a_n)$$

and is called a **lens space**. It is a differentiable manifold of dimension $2n-1$. Lens spaces are important and interesting objects in topology and transformation group theory.

If $\mathbb{Z}/2$ acts on $S^n = \{(x_0, \ldots, x_n) \mid \sum x_i^2 = 1\}$ by $(k, (x_0, \ldots, x_n)) \mapsto (-1)^k (x_0, \ldots, x_n)$ ("the antipodal action"), then the orbit space of this free action is called the **real projective space** $\mathbb{R}P^n$ of dimension n. Conceptually, $\mathbb{R}P^n$ is the manifold of one-dimensional subspaces of $\mathbb{R}^{n+1}$, also denoted by $P(\mathbb{R}^{n+1})$.

The circle group S^1 acts on

$$S^{2n+1} = \{(z_0, \ldots, z_n) \mid \sum |z_i|^2 = 1\} \subset \mathbb{C}^{n+1}$$

by $(z, (z_0, \ldots, z_n)) \mapsto (zz_0, \ldots, zz_n)$. This action is free and the orbit space is a $2n$-dimensional differentiable manifold, the **complex projective space** $\mathbb{C}P^n$. There is another description of $\mathbb{C}P^n$ as the orbit space of

$$\begin{aligned} \mathbb{C}^* \times (\mathbb{C}^{n+1} \setminus \{0\}) &\to \mathbb{C}^{n+1} \setminus \{0\} \\ (z, (z_0, \ldots, z_n)) &\mapsto (zz_0, \ldots, zz_n) \end{aligned}$$

with $\mathbb{C}^* = \mathbb{C}\setminus\{0\}$ as multiplicative group. Conceptually, $\mathbb{C}P^n$ is the manifold of one-dimensional subspaces of $\mathbb{C}^{n+1}$, also denoted by $P(\mathbb{C}^{n+1})$.

(2.8) Projective spaces. Let V be a complex representation of G and $P(V)$ the complex projective space of one-dimensional subspaces of V. Since $L_g: V \to V$ maps lines into lines, we have an induced action $G \times P(V) \to P(V)$. The fixed point set $P(V)^G$ is the set of lines which are invariant under G, i.e. a fixed point corresponds to a one-dimensional subrepresentation. If G is a non-abelian compact Lie group, then G has an irreducible representation V of dimension greater than one. In this case, the fixed point set $P(V)^G$ is empty. If we add a trivial representation, $V \oplus \mathbb{C} = W$, then $P(W)$ has a single fixed point, corresponding to the subspace $\{0\} \times \mathbb{C}$. In general, the fixed point set $P(V)^G$ is a disjoint union of projective spaces $P(U)$, where those $U \subset V$ are G-subspaces in which each line is G-invariant.

(2.9) Covering spaces. The fundamental group $G = \pi_1(X, x)$ of a topological space X with basepoint $x \in X$ describes a hidden symmetry of the space, namely a symmetry of its universal covering space (if it exists). If $\tilde{X} \to X$ is a universal covering of X, then $G = \pi_1(X, x)$ acts freely on $\tilde{X}$ as the group of covering transformations, and the orbit space $\tilde{X}/G$ is canonically homeomorphic to X. Any other covering of X is given as $\tilde{X}/H \to X$ for a subgroup H of G. (This is a situation which is dual to Galois theory). The category of coverings of X is equivalent to the category of $\pi_1(X, x)$-sets. For the theory of covering spaces, see e.g. Massey [1967]. If the discrete group G acts freely on X such that each $x \in X$ has an open neighbourhood U with $U \cap L_g(U) = \emptyset$ for $g \neq e$, then $X \to X/G$ is a covering.

It is often a good strategy to interpret the fundamental group as a symmetry group. The viewpoint of transformation groups thus introduced leads to new invariants of the space. For instance, the chain complex and the homology groups of the universal covering are modules over the fundamental group G. This leads to integral representation theory ($\mathbb{Z}G$-modules), algebraic K-theory, and subtle invariants like finiteness obstructions or Whitehead torsion.

(2.10) Space forms. It is a classical problem to determine the complete connected Riemannian manifolds M of constant curvature K. It is known that the universal cover $\tilde{M}$ of M is one of three standard models: The standard sphere S^n for $K = 1$, the flat Euclidean space $\mathbb{R}^n$ for $K = 0$, and the hyperbolic space H^n for $K = -1$. Therefore, M has the form $\tilde{M}/\Gamma$ where Γ is a discrete group of isometries of M; see Wolf [1967], 2.4.10. In dimension two, this is related to looking at Riemann surfaces from the viewpoint of transformation groups (automorphic forms etc.).

The spherical space forms thus lead to the investigation of orthogonal representations $G \to O(n+1)$ of finite groups G such that the induced action on S^n is free. This advanced exercise in representation theory is carried out in Wolf [1967], Chapters 6 and 7. The concept of spherical space forms has also been applied to the more general situation of a free action of G on a sphere; later, we shall have occasion to mention the fascinating piece of research concerned with such objects.

The topological investigation of free actions of discrete groups on Euclidian spaces is still in its infancy.

(2.11) Lie groups as automorphism groups. The consideration of Lie group actions on manifolds is justified by the fact that many naturally occurring automorphism groups are Lie groups. We mention some examples.

Firstly, we have the fundamental result of Montgomery-Zippin [1955], p. 108 and p. 212.

Theorem 1. *Let G be a locally compact Hausdorff transformation group acting effectively on a connected manifold M of class C^k, $1 \leq k \leq \infty$ or $k = \omega$ (meaning analytic). Suppose that each left translation is C^1. Then G is an (analytic) Lie group and the action is of class C^k.* □

In light of this theorem, it is desirable to find conditions under which a transformation group becomes locally compact. For the next theorem, see van Dantzig-van der Waerden [1928] and Kobayashi-Nomizu [1963], p. 46.

Theorem 2. *Let X be a locally compact metric space with a finite number of components. Then the group of isometries, equipped with the CO-topology, is a locally compact Hausdorff group.* □

For related results, in particular concerning automorphisms of complex manifolds, see Bochner-Montgomery [1946], [1947]. For a general setting of Lie transformation groups, see Palais [1957b].

(2.12) Isometries of Riemannian manifolds. References are Myers-Steenrod [1939], Palais [1957a]. Let M be a connected Riemannian manifold. A surjective mapping of sets $M \to M$ preserving the distance is a diffeomorphism. The group $I(M)$ of isometries $M \to M$ is a Lie group and the canonical action $I(M) \times M \to M$ is of class C^k (by (2.11) Theorem 1). If M is compact, then $I(M)$ is compact. If a compact Lie group G acts smoothly on a differentiable manifold M, then M carries a Riemannian metric with respect to which G is a group of isometries.

The isometry group of an n-dimensional Riemannian manifold cannot be too large. For the next result, see Kobayashi-Nomizu [1963], p. 308.

Theorem. *The isometry group $I(M)$ of a connected n-dimensional Riemannian* *manifold is of dimension at most* $\frac{1}{2}n(n+1)$. *If* $\dim I(M) = \frac{1}{2}n(n+1)$, *then* M *is isometric to one of the following spaces of constant curvature:*

(i) *Euclidian space* $\mathbb{R}^n$; (ii) *sphere* S^n; (iii) *real projective n-space* $\mathbb{R}P^n$; (iv) *an n-dimensional, simply connected hyperbolic space.* □

(2.13) Homogeneous spaces. The viewpoint of transformation groups reveals many of the classical geometrical objects as homogeneous spaces of Lie groups. See also (5.3).

Spheres. The orthogonal group $O(n+1)$ acts transitively on the sphere S^n. This yields an isomorphism of $O(n+1)$-spaces $S^n \cong O(n+1)/O(n)$ where $O(n)$ is an isotropy group of a unit vector. Considering the unit sphere S^{2n-1} in complex n-space, one obtains an isomorphism $S^{2n-1} \cong U(n+1)/U(n)$. There are a few more possibilities for spheres to arise as homogeneous spaces of compact connected Lie groups; for a classification, see Montgomery-Samelson [1943], Borel [1949], [1950].

Stiefel manifolds. Let $V_k(\mathbb{R}^n)$ be the set of orthonormal k-frames $(x_1, \ldots, x_k)$, $x_i \in \mathbb{R}^n$.

The orthogonal group $O(n)$ acts on $V_k(\mathbb{R}^n)$ by $(A, (x_1, \ldots, x_k)) \mapsto (Ax_1, \ldots, Ax_k)$. By linear algebra, this action is transitive. The isotropy group of $(e_1, \ldots, e_k)$, e_i the i-th standard unit vector, is canonically isomorphic to $O(n-k)$. Thus $V_k(\mathbb{R}^n) \cong O(n)/O(n-k)$. Since coset spaces G/H of Lie groups G by closed subgroups H are canonically differentiable G-manifolds, we can use the isomorphism above to turn $V_k(\mathbb{R}^n)$ into a differentiable $O(n)$-manifold. A similar remark applies to the following examples.

Grassmann manifolds. Let $G_k(\mathbb{R}^n)$ be the set of k-dimensional linear subspaces of $\mathbb{R}^n$. The standard action of $O(n)$ on $\mathbb{R}^n$ maps k-spaces to k-spaces and thus induces an action of $O(n)$ on $G_k(\mathbb{R}^n)$. This action is transitive. Let $[e_1, \ldots, e_k]$ be the subspace generated by $e_1, \ldots, e_k$. Its isotropy group is the subgroup $O(k) \times O(n-k)$ of $O(n)$ consisting of all matrices

$$\begin{pmatrix} A & 0 \\ 0 & B \end{pmatrix}, \quad A \in O(k), \ B \in O(n-k).$$

Thus $G_k(\mathbb{R}^n) \cong O(n)/(O(k) \times O(n-k))$.

Similarly, $G_k(\mathbb{C}^n) \cong U(n)/(U(k) \times U(n-k))$ for the Grassmann manifold of complex k-dimensional subspaces of $\mathbb{C}^n$.

Flag manifolds. Let V be an n-dimensional complex vector space. A **flag** in V is a sequence $0 \subset V_1 \subset V_2 \subset \ldots \subset V_{n-1} \subset V$ of subspaces, $\dim V_i = i$. Suppose V carries a Hermitian inner product $\langle -,- \rangle$. Given a flag (V_i), one can choose an orthonormal basis $(v_1, \ldots, v_n)$ of V such that $V_i = [v_1, \ldots, v_i]$, the space generated by $v_1, \ldots, v_i$. One sees that the standard action of the unitary group $U(V)$ of $(V, \langle -,- \rangle)$ is transitive on the set $F(V)$ of all flags. If $V = \mathbb{C}^n$, the isotropy group of $[e_1], [e_1, e_2], \ldots, [e_1, \ldots, e_n]$ is the subgroup $T(n) \subset U(n)$ consisting of diagonal matrices. The group $T(n)$ is a maximal torus of $U(n)$; see Bröcker-tom Dieck [1985], Ch. IV.

Thus $F(\mathbb{C}^n) \cong U(n)/T(n)$.

By letting $GL(n, \mathbb{C})$ act on $F(\mathbb{C}^n)$, one obtains $F(\mathbb{C}^n) \cong GL(n, \mathbb{C})/B$ where B is the subgroup of upper triangular matrices; in this way $F(\mathbb{C}^n)$ is written as the coset space of a complex Lie group modulo a complex subgroup and thus becomes a complex manifold.

Hyperbolic space. Let g: $\mathbb{R}^{n+1} \to \mathbb{R}$, $n \geq 1$, be the quadratic form $q(x_1, \ldots, x_{n+1}) = x_1^2 + \ldots + x_n^2 - x_{n+1}^2$.

Let $Q = \{x \in \mathbb{R}^{n+1} \mid q(x) = -1, x_{n+1} \geq 1\}$. The form q restricted to the tangent spaces of Q is positive definite and thus induces a Riemannian metric on Q. This Riemannian manifold Q can serve as a model for hyperbolic n-space. Let $O(n, 1) \subset GL(n+1, \mathbb{R})$ be the subgroup leaving the form q invariant. By Witt's theorem on quadratic forms (see e.g. Cassels [1978], Chapter 2.4), $O(n, 1)$ acts transitively on the manifold $M = \{x \in \mathbb{R}^{n+1} \mid q(x) = -1\}$. The manifold Q is one of the two components of M. The group $O(n, 1)$ has four components. The group $SO(n, 1) = O(n, 1) \cap SL(n+1, \mathbb{R})$ has two components. Let $G(n) \cong SO(n, 1)^+$ be the component which preserves Q. Then $G(n)$ still acts transitively on Q. The isotropy group of $(0, \ldots, 0, 1) \in Q$ is $SO(n)$.

Thus $Q \cong G(n)/SO(n)$. Matrices $(a_{ij}) \in O(n, 1)$ with $a_{n+1,n+1} > 0$ form a subgroup $O(n, 1)^+$. One has

$$SO(n, 1)^+ = SO(n, 1) \cap O(n, 1)^+ .$$

The group $O(n, 1)^+$ is the full isometry group of the Riemannian manifold Q.

For the particular case $n = 2$, other models are used, e.g. the upper half plane $\{z \in \mathbb{C} \mid \operatorname{Im} z > 0\} = H$.

The action $z \mapsto \dfrac{az+b}{cz+d}$ of $\begin{pmatrix} a & b \\ c & d \end{pmatrix} \in SL(2, \mathbb{R})$ yields $H \cong SL(2, \mathbb{R})/SO(2)$ as a model for the hyperbolic plane; $SO(2)$ is the isotropy group of the point $i \in H$.

Symmetric spaces. A Riemannian manifold M is called **globally symmetric** if each point $p \in M$ is an isolated fixed point of an isometry s_p which is an involution ($s_p^2 = \mathrm{id}$). It turns out that the isometry group $I(M)$ acts transitively on M; see Helgason [1962], Ch. IV. Examples are the space forms; they also have the property of being two-point homogeneous.

Two-point homogeneous spaces. A Riemannian manifold M is called two-point homogeneous if for any two pairs (p_1, p_2) and (q_1, q_2) of points in M with the same distance there exists an isometry f such that $f(p_i) = q_i$. It turns out that such spaces are globally symmetric. They can be classified; see Wolf [1967], 8.12.

(2.12) Invariant theory. This is a classical topic (Weyl [1939]). Let V be a representation space of the Lie group G. The general problem is to analyse polynomials on V (or other types of functions) which are invariant under G. Roughly stated, an equivalent problem is the following: Study the orbit space of V from the viewpoint of algebraic geometry, differential topology etc. If G is compact, the invariant polynomials form a finitely generated algebra (see Helgason [1962], ChX, Theorem 5.6).

(2.15) Fundamental domains. In the case of discrete subgroups of $SL(2, \mathbb{R})$ acting on the upper half plane, the fundamental domains have been studied extensively. The geometry of the fundamental domain provides rich information about the group (generators and relations) and the action. See Beardon [1983] for an introduction. An abstract version of the relation between the structure of discrete groups and certain geometrical data is developed in the theory of trees (Serre [1980]).

(2.16) Homeomorphisms. Each homeomorphism $f: X \to X$ of a space X induces an action of the discrete group $\mathbb{Z}$ of the integers on X such that f is the left translation by 1. The orbit of a point x consists of $\{f^n(x) \mid n \in \mathbb{Z}\}$.

This point of view emphasizes the dynamics of the homeomorphism. However, dynamical systems are not our concern. For some aspects of the differential topology of diffeomorphisms in connection with cobordism theory, see Kreck [1984].

(2.17) Dynamical systems. Let M be a smooth manifold and $\mathbb{R}$ the additive group of real numbers. A dynamical system is a smooth action $\mathbb{R} \times M \to M$. Such actions arise as global flows of certain differential equations (vector fields) on M. Typical for this theory are stability questions for dynamical systems.

(2.18) The Burnside ring. In dealing with transformation groups one often encounters combinatorial problems connected with the lattice of subgroups of a given group together with the conjugation action on the set of subgroups. A useful way of formalizing some of this structure is the Burnside ring $A(G)$ of a finite group G. We are going to define this ring. Later, we shall define and investigate this ring also for compact Lie groups.

Let G be a finite group and $A^+(G)$ the set of isomorphism classes of finite G-sets. Disjoint union (cartesian product with diagonal action) of finite G-sets induces addition (multiplication) on $A^+(G)$, and the resulting structure may be called a commutative semi-ring with identity. Since a finite G-set is a disjoint union of its orbits, each element of $A^+(G)$ can be written in a unique way as

$\sum n(G/H)[G/H]$ where $n(G/H)$ is a non-negative integer and $[G/H]$ is the isomorphism class of G/H, which, by (1.14), depends only on the conjugacy class of H. The non-triviality of the multiplication results from considering the decomposition of $G/H \times G/K$ into orbits.

Proposition 1. *There are canonical bijections between the following sets:*
(i) *G-orbits of $G/H \times G/K$.*
(ii) *H-orbits of G/K with left H-action.*
(iii) *Double cosets HgK, $g \in G$.*

Proof. The orbit through (eH, gK) of $G/H \times G/K$ corresponds to the orbit through gK of G/K and to the double coset HgK. Bijectivity is easily checked. □

The Grothendieck ring constructed from the semi-ring $A^+(G)$ is denoted by $A(G)$ and will be called the **Burnside ring** of G. Additively, it is the free abelian group on the isomorphism classes $[G/H]$ of homogeneous spaces G/H. It is a commutative ring with unit $[G/G]$. If H is a subgroup of G and S, T are finite G-sets, then for the cardinality of the H-fixed point sets we have $|S^H + T^H| = |S^H| + |T^H|$ and $|(S \times T)^H| = |S^H||T^H|$. Hence $S \mapsto |S^H|$ extends to a ring homomorphism $\varphi_H \colon A(G) \to \mathbb{Z}$ depending only on the conjugacy class of H. Let $\Phi(G)$ be the set of conjugacy classes of subgroups of G and let

$$\varphi = (\varphi_H) : A(G) \to \prod_{(H) \in \Phi(G)} \mathbb{Z}$$

be the map with components φ_H.

Proposition 2. *φ is an injective ring homomorphism.*

Proof. By definition, φ is a ring homomorphism. Suppose $x \neq 0$ is in the kernel of φ. Write x in terms of the basis, $x = \sum a_H[G/H]$. The set of all $[G/H]$ is partially ordered, namely $[G/H] \leq [G/K]$ if and only if H is conjugate to a subgroup of K. Let $[G/H]$ be maximal among the basis elements with $a_H \neq 0$. Then $(G/K)^H \neq \emptyset$ implies $[G/H] \leq [G/K]$. Hence $0 = \varphi_H x = a_H|(G/H)^H| = a_H|NH/H| \neq 0$, a contradiction. (Observe: $gK \in (G/K)^H \Leftrightarrow g^{-1}Hg \subset K$.) □

(2.19) Exercises.

1. The orbit space of the action $\mathbb{Z}/2 \times \mathbb{R} \to \mathbb{R}$, $(k, x) \mapsto (-1)^k x$ is $[0, \infty[$. Also, $[0, \infty[$ is a fundamental domain.
2. The orbit space of the action $S^1 \times \mathbb{C} \to \mathbb{C}$, $(x, z) \mapsto xz$ is $[0, \infty[$. A fundamental domain?
3. The orbit space of the action $\mathbb{Z}/m \times \mathbb{C} \to \mathbb{C}$, $(k, z) \mapsto \zeta^k z$ with

$\zeta = \exp(2\pi i/m)$ is $\mathbb{C}$. Describe explicitly the orbit map $\mathbb{C} \to \mathbb{C}$. ($m$-fold ramification in complex function theory.)

4. Let G be a finite group. Show that $(G/H)^H = NH/H$.
5. In the Burnside ring of an abelian group G one has $[G/H][G/K] = c[G/(H \cap K)]$. Determine the integer c.
6. Let G and H be finite groups whose orders are relatively prime. Show that $A(G \times H) \cong A(G) \otimes_{\mathbb{Z}} A(H)$.
7. Show that the cokernel of the map φ in (2.18), Proposition 2, has order $\prod_{(H) \in \Phi(G)} |WH|$. (Hint: Triangular matrix.)
8. Let G be the discrete group of homeomorphisms of $\mathbb{R}$ generated by the reflections in 0 and 1. Show that G is isomorphic to the free product $\mathbb{Z}/2 * \mathbb{Z}/2$. Show that G contains a normal subgroup of index 2 isomorphic to $\mathbb{Z}$, consisting of translations. Show that $[0, 1]$ is a fundamental domain.
9. Let V be a representation space for the compact Lie group G. We have an induced action of G on the Grassmann manifold $G_k(V)$. Describe the fixed point set. (Compare (2.8).)
10. Let $\mathrm{U}(n)$ act on itself by conjugation $(A, B) \mapsto ABA^{-1}$. Let $\mathrm{Conj}(\mathrm{U}(n))$ be the orbit space. Let $T(n) \subset \mathrm{U}(n)$ be the n-dimensional torus of diagonal matrices. The symmetric group S_n acts on $T(n)$ by permuting the diagonal elements. Establish, by linear algebra, a homeomorphism

$$T(n)/S_n \cong \mathrm{Conj}(\mathrm{U}(n)).$$

(For the space of conjugacy classes of a simply connected compact Lie group, see Bourbaki [1968], V. 3.9.)

11. The symmetric group S_n acts on the n-fold cartesian product X^n of a space X by permuting the factors. The orbit space is called the n-fold **symmetric product** $SP^n(X)$ of X. The following is a determination of $SP^n(S^1)$ and $SP^n(S^2)$, $n \geq 2$. (i) Put $T(n) = (S^1)^n$. Let $\det\colon T(n) \to S^1$, $(z_1, \ldots, z_n) \mapsto z_1 z_2 \ldots z_n$. This induces a map of orbit spaces $d\colon SP^n(S^1) \to S^1$. This map is a fibre bundle. Let F be the fibre $d^{-1}(1)$. Use the universal covering $\mathbb{R}^n \to T(n)$, $(x_1, \ldots, x_n) \mapsto (\exp 2\pi i x_1, \ldots, \exp 2\pi i x_n)$ to show that $F = N/\Gamma$ where $N = \{(x_1, \ldots, x_n) \in \mathbb{R}^n \mid \sum x_i = 0\}$ and where Γ, being a semidirect product of the translation group $I = \{(x_1, \ldots, x_n) \in \mathbb{Z}^n \mid \sum x_i = 0\}$ and S_n, is a discrete group acting on N. Show that the $(n-1)$-simplex $D = \{x \in N \mid x_1 \geq x_2 \geq \ldots \geq x_{n-1} \geq 0\}$ is a fundamental domain for the action of Γ on N. Thus $N/\Gamma \cong D$ since D is closed. Show that the fibre bundle $d\colon SP^n(S^1) \to S^1$ is trivial if and only if n is odd. In particular, $SP^2(S^1)$ is the Möbius band.

The space $SP^n(S^1)$ has the following interpretation. Consider real projective forms, i.e. homogeneous polynomials of degree n which are of the form $\prod_{i=1}^{n} (a_i x + b_i y) = c_0 x^n + c_1 x^{n-1} y + \ldots + c_n y^n$ where the a_i, b_i, are real and

not both zero. The set of points $[c_0, \ldots, c_n]$ in projective n-space $\mathbb{R}P^n$ obtainable in this way is homeomorphic to $SP^n(S^1)$. Employing $S^1 \cong \mathbb{R}P^1$, the homeomorphism is induced by $([a_0, b_0], \ldots, [a_n, b_n]) \mapsto [c_0, \ldots, c_n]$.
(ii) Show $SP^n(S^2) \cong \mathbb{C}P^n$ by using $S^2 \cong \mathbb{C}P^1$ and considering homogeneous complex polynomials of degree n as product of n linear homogeneous factors.

12. Show by linear algebra that the space of conjugacy classes of SO(3) is a closed interval.
13. Let the group $\mathrm{GL}(n, \mathbb{R})$ act on the space $\mathrm{Sym}(n)$ of symmetric real (n, n)-matrices by $(A, B) \mapsto ABA^t$, A^t transpose of A. Use the classification theory of quadratic forms to show that the orbit space is finite.
14. An orthogonal complex structure on $\mathbb{R}^{2n}$ is any $J \in \mathrm{O}(2n)$ such that $J^2 = -E$. Let $\mathrm{O}(2n)$ act on the space of complex structures by $(A, J) \mapsto AJA^{-1}$. Show that this is a transitive $\mathrm{O}(2n)$-space. Show that an isotropy group is isomorphic to $\mathrm{U}(n)$.
15. Let $\mathrm{O}(n)$ act on $\mathbb{R}^n \times \mathbb{R}^n$ by $(A, (x, y)) \mapsto (Ax, Ay)$. Show that the isotropy groups are conjugate to $\mathrm{O}(n-1)$ if x and y are linearly dependent, or to $\mathrm{O}(n-2)$ if x and y are linearly independent. Show that (x, y) and (x', y') belong to the same orbit if and only if $|x| = |x'|$, $|y| = |y'|$, $\langle x, y\rangle = \langle x', y'\rangle$. Use this fact and the Cauchy-Schwarz inequality to describe the orbit space as a subspace of $\mathbb{R}^3$.
16. The isometry group $\mathrm{O}(n, 1)^+$ of hyperbolic n-space also acts on the sphere S^{n-1}. One way of seeing this is to take the Poincaré model $B^n = \{x \in \mathbb{R}^n \mid |x| < 1\}$ for hyperbolic space. One observes that the isometries are extendable to homeomorphisms of the boundary S^{n-1}. They are no longer isometries of S^{n-1}. It turns out that $\mathrm{O}(n, 1)^+$ is the group of conformal automorphisms of S^{n-1}. (By definition, conformal maps change the Riemannian metric by a scalar function.)
Here is another way how to obtain conformal automorphisms of S^n. Embed S^n into $\mathbb{R}P^{n+1}$ by mapping $(y_1, \ldots, y_{n+1})$ to $[x_0, \ldots, x_{n+1}]$, where $x_0 = \frac{1}{\sqrt{2}}(1 + y_{n+1})$, $x_i = y_i$ for $1 \leqq i \leqq n$, $x_{n+1} = \frac{1}{\sqrt{2}}(1 - y_{n+1})$. The image consists of the quadric Q defined by $q(x) = x_1^2 + \ldots + x_n^2 - 2x_0 x_{n+1} = 0$. Hence the group G of linear transformations of $\mathbb{R}^{n+2}$ leaving the quadratic form q invariant acts on Q. The embedding of the sphere is isometric if the tangent space of Q at x carries the Riemannian metric

$$u \mapsto 2\frac{\langle x, x\rangle \cdot \langle u, u\rangle - \langle x, u\rangle^2}{\langle x, x\rangle^2}.$$

One verifies that G acts conformally on Q with respect to this metric.

3. Elementary properties.

We collect some elementary results from point set topology of transformation groups. The reader should be aware that, occasionally, point set topology causes problems. Moreover: Compact groups are much easier to handle than topological groups in general. We point out that, in our terminology, compactness does not include the Hausdorff property.

Let G be a topological group and X a G-space. For subsets $A \subset G$ and $B \subset X$ we use the notation

$$AB = \{gx \mid g \in A, x \in B\}$$
$$A^{-1} = \{g^{-1} \mid g \in G\}.$$

(3.1) Proposition.
(i) *If $A \subset G$ and $B \subset X$ are open, then $AB \subset X$ is open.*
(ii) *If $A \subset G$ and $B \subset X$ are compact, then AB is compact.*
(iii) *If $A \subset G$ is compact and $B \subset X$ is closed, then AB is closed.*
(iv) *The projection $p: X \to X/G$ is open.*
(v) *If G is compact and X a Hausdorff space, then X/G is a Hausdorff space.*

Proof.
(i) $AB = \bigcup_{a \in A} aB$. The set $aB = L_a(B)$ is open because $L_a: X \to X, x \mapsto ax$ is a homeomorphism.
(ii) The product $A \times B$ is compact. The set AB is the continuous image under $G \times X \to X$ of $A \times B$.
(iii) Suppose x is not contained in AB. Take $a \in A$. Since the action $r: G \times X \to X$ is continuous and $X \backslash aB$ is open, we find open neighbourhoods V_a of $e \in G$ and W_a of $x \in X$ such that $r(V_a \times W_a) = V_a W_a \subset X \backslash aB$. We conclude that $W_a \cap V_a^{-1} aB = \emptyset$. Since A is compact, we can choose a finite number of elements $a_1, \ldots, a_n \in A$ such that

$$A \subset (V_{a_1}^{-1} a_1 \cup \ldots \cup V_{a_n}^{-1} a_n).$$

Set

$$W = \bigcap_{i=1}^{n} W_{a_i}.$$

Then W is a neighbourhood of x. Moreover, $W \cap AB = \emptyset$ because, for each $ab = y \in AB$, there exists i such that $a \in V_{a_i}^{-1} a_i$; and $W \cap V_{a_i}^{-1} a_i B$ is empty.
(iv) Let $U \subset X$ be open. The set pU is open if and only if $p^{-1}pU$ is open (quotient topology). But $p^{-1}pU = \bigcup_{g \in G} gU$ is open.
(v) We use the following elementary result from point set topology.

(3.2) Lemma. *Let $f\colon X \to Y$ be a continuous surjective open map. Then Y is a Hausdorff space if and only if*

$$R = \{(x_1, x_2) \mid f(x_1) = f(x_2)\}$$

is closed in $X \times X$. □

In our case, we have to show that $R = \{(x, gx) \mid g \in G, x \in X\}$ is closed in $X \times X$. We have the action $t\colon (g, (a, b)) \mapsto (a, gb)$ of G on $X \times X$. The diagonal $D \subset X \times X$ is closed. By part (iii), $R = t(G \times D)$ is closed. □

In (1.4) we have introduced the G-spaces G/H for each subgroup H of G. Let $p\colon G \to G/H$ be the quotient map defining the topology of G/H.

(3.3) Proposition. *G/H is a Hausdorff space if and only if H is closed in G.*

Proof. Since $p\colon G \to G/H$ is surjective and open, the space G/H is separated if and only if $R = \{(x, y) \mid px = py\}$ is closed in $G \times G$. We have $R = \{(g, gh) \mid g \in G, h \in H\}$. The map $f\colon G \times G \to G$, $(x, y) \mapsto x^{-1}y$ is continuous and $f^{-1}H = R$. Hence R is closed if H is closed. Conversely, if G/H is separated, then the point $eH \in G/H$ is closed; hence $H = p^{-1}(eH)$ is closed. □

Proposition 3.3 is one reason for restricting attention to closed subgroups only.

If H is a normal subgroup of G, then G/H, equipped with the quotient structures, is a topological group. The proof is left as an exercise. More generally, the reader may show the following Proposition. Suppose X is a G-space and H a normal subgroup. Restricting the group action to H, we obtain an H-space X. The orbit space $H\backslash X$ carries a G/H-action s such that the diagram

$$\begin{array}{ccc} G \times X & \xrightarrow{\ r\ } & X \\ \downarrow & & \downarrow \\ G/H \times H\backslash X & \xrightarrow{\ s\ } & H\backslash X \end{array}$$

is commutative.

(3.4) Proposition. *The action s is continuous.* □

The Proposition enables us to reduce any action to an effective action.

(3.5) Proposition. *Let X be a Hausdorff space. Then the isotropy groups are closed subgroups.*

Proof. G_x is the pre-image of the diagonal under $G \to X \times X$, $g \mapsto (x, gx)$. □

(3.6) Proposition. *Let G be compact and let X be a G-space with orbit map $p: X \to X/G$. Let G and X be Hausdorff spaces. Then:*
(i) *p is a closed mapping.*
(ii) *p is a proper mapping.*
(iii) *X is compact if and only if X/G is compact.*
(iv) *X is locally compact if and only if X/G is locally compact.*

Proof.
(i) Let $A \subset X$ be closed. Then $p^{-1}pA = GA$ is closed by Proposition (3.1). Hence pA is closed by definition of the quotient topology.
(ii) We recall that a map is proper if it is closed and if the pre-image of any compact set is compact. Let $C \subset X/G$ be compact and let $(U_\alpha | \alpha \in A)$ be an open covering of $p^{-1}C$. For each $c \in C$, we can find a finite subset $A_c \subset A$ such that
$$U_c = \cup (U_\alpha | \alpha \in A_c) \supset p^{-1}(c).$$
The set
$$V_c = X/G \backslash p(X \backslash U_c)$$
is open since p is closed. Moreover, $c \in V_c$ and $p^{-1} V_c \subset U_c$. If C is covered by $V_{c(1)}, \dots, V_{c(n)}$, we have
$$p^{-1}(C) \subset p^{-1} V_{c(1)} \cup \dots \cup p^{-1} V_{c(n)} \subset U_{c(1)} \cup \dots \cup U_{c(n)}.$$
(iii) follows immediately from (ii).
(iv) Let $U \subset X$ be an open neighbourhood of x with compact closure $\bar{U}$. Then $p(x) \in p(\bar{U})$ and $p(\bar{U})$ is a compact neighbourhood of $p(x)$. If, conversely, C is a compact neighbourhood of $p(x)$, then (by (ii) above) $p^{-1}(C)$ is a compact neighbourhood of x. □

An equivariant map $G/H \to G/K$ of sets is always continuous.

(3.7) Proposition. *If G is compact Hausdorff and $H \subset G$ is a closed subgroup, then $gHg^{-1} = H$ if and only if $gHg^{-1} \subset H$. Hence each G-map $G/H \to G/H$ is a homeomorphism.*

Proof. Suppose $gHg^{-1} \subset H$. Consider $c: G \times G \to G$, $(x, k) \mapsto xkx^{-1}$ and let $A = \{g^n | n = 0, 1, 2, \dots\}$. Then $c(A \times H) \subset H$ and by continuity, as H is closed, $c(\bar{A} \times H) \subset H$. If we can show that $g^{-1} \in \bar{A}$, then $g^{-1}Hg \subset H$. Thus it suffices to verify that $\bar{A}$ is a subgroup of G.

Let B be the subgroup $\{g^n | n \in \mathbb{Z}\}$. Then $\bar{B}$, the closure of B, is also a subgroup of G. If e is isolated in $\bar{B}$, then $\bar{B}$ is compact and discrete and hence finite and we must then have $g^n = e$ for some $n > 0$. Thus suppose that e is not isolated in $\bar{B}$. Let U be a neighbourhood of e in G. Then $V = U \cap U^{-1}$ is again a neighbourhood and there exists $n \neq 0$ with $g^n \in V$; so we may assume that $n > 0$. Then $g^{n-1} \in (g^{-1}V) \cap A$. The $g^{-1}V$ form a neighbourhood basis of g^{-1}. Therefore, $g^{-1} \in \bar{A}$ and this implies $\bar{A} = \bar{B}$. □

If X and Y are G-spaces, we let $C_G(X, Y)$ denote the space of G-maps $X \to Y$ with compact-open-topology (see section 1).

(3.8) Proposition. *Let G be compact and $H \subset G$ a closed subgroup. There exists a canonical homeomorphism*

$$X^H \cong C_G(G/H, X).$$

Proof. Let a: $X^H \to C_G(G/H, X)$ be given by $a(x)(gH) = gx$. Then $a(x)$ is a well-defined element in $C_G(G/H, X)$. Let b: $C_G(G/H, X) \to X^H$ be given by $b(f) = f(H)$; then $f(H) \in X^H$. The maps a and b are inverse to each other. We show that they are continuous. The map a is adjoint to the continuous map

$$G/H \times X^H \to X, \quad (gH, x) \mapsto gx$$

and therefore continuous. The map b is a restriction of the evaluation map

$$ev\colon C_G(G/H, X) \times G/H \to X, \ (f, gH) \mapsto f(gH),$$

and ev is continuous since G/H is compact. □

(3.9) Proposition. *Let X be a Hausdorff space. Then X^H is closed in X.*

Proof. One has $X^H = \bigcap_{g \in H} X^g$ with $X^g = \{x \mid gx = x\}$. The space X^g is the pre-image of the diagonal under $X \to X \times X$, $x \mapsto (x, gx)$. □

The fixed point sets of closed subgroups have the convenient property that the intersection of two of them is again a fixed point set. Let H and K be closed subgroups and let $\langle H, K \rangle$ denote the closed subgroup generated by H and K (i.e. the intersection of all closed subgroups containing $H \cup K$).

(3.10) Proposition. $X^H \cap X^K = X^{\langle H, K \rangle}$ *for Hausdorff spaces X.*

Proof. $\langle H, K \rangle \supset H \cup K$ implies $X^{\langle H, K \rangle} \subset X^H \cap X^K$. Conversely, $x \in X^H \cap X^K$ implies $H \cup K \subset G_x$, hence $G_x \supset \langle H, K \rangle$; thus $x \in X^{\langle H, K \rangle}$. □

The reader should note that for $H \subset K$ but $H \neq K$ it may happen that $X^H = X^K$. In general, $X^H \cap X^K$ is the fixed point set of the subgroup generated by H and K.
If $H = G_x$, then $x \in X^H$, so each $x \in X$ is contained in a fixed point set of an

isotropy group. But, in general, there are other fixed point sets. Fixed point sets which do not belong to isotropy groups usually produce difficulties. But there are some general situations where this cannot happen.

If the set $\mathrm{Iso}(X)$ is closed under taking intersections, then for each subgroup K with $X^K \neq \emptyset$ there exists a unique minimal isotropy subgroup $m(K)$ with $K \subset m(K)$. In this case, $X^K = X^{m(K)}$ because X^K is the union $\cup X^H$ over $H \supset K$, $H \in \mathrm{Iso}(X)$.

One of the useful properties of compact groups G is the existence of an invariant integration (Haar integral) for continuous functions on G (see e.g. Pontrjagin [1957]).

(3.11) Proposition. *Let the compact Hausdorff group G act on the normal space X. Let $A \subset X$ be a closed invariant subspace and $f\colon A \to V$ a G-map into a representation space V. Then f has an equivariant extension $F\colon X \to V$.*

Proof. By the Tietze extension theorem of general topology, f has a continuous extension $f'\colon X \to V$. This extension need not be equivariant. Therefore, we use the normalized Haar integral on G to define a new function

$$F(x) = \int_G g^{-1} f'(gx)\, dg.$$

Invariance of the integral yields $F(gx) = gF(x)$, and for $a \in A$ we have $F(a) = \int g^{-1} f(ga)\, dg = \int f(a)\, dg = f(a) \int dg = f(a)$ by normalization of the integral. □

Another application of integration is to partitions of unity. Recall that a **numeration** of a covering $(U_j \mid j \in J)$ of a space X is a locally finite partition of unity subordinate to the covering.

(3.12) Proposition. *Let the compact Hausdorff group G act on the paracompact space X. Each open covering by G-invariant sets has a numeration by G-functions.*

Proof. Let $(U_j \mid j \in J)$ be an open covering by invariant subsets and let (u_j) be a subordinate partition of unity. Consider the functions $v_j(x) = \int u_j(gx)\, dg$. These are G-functions and the support of v_j is again contained in U_j. The family (v_j) is locally finite: Fix $x \in X$. Since (u_j) is locally finite, each point gx has a neighbourhood on which only finitely many u_j are non-zero. Hence there exist neighbourhoods $V_g \subset G$ of g and $W_g \subset X$ of x such that $u_j(V_g W_g) = \{0\}$ for $j \in J_g$, $J \setminus J_g$ finite. Since G is compact, it is covered by finitely many of these V_g. Let W be the intersection of the corresponding W_g. Let J_x be the intersection of the corresponding J_g. Then we have $u_j(GW) = \{0\}$ for $j \in J_x$ and therefore $v_j(W) = \{0\}$ for $j \in J_x$. The v_j again constitute a partition of unity because of normalization of the integral. □

A useful class of group actions, which generalizes actions of compact groups on Hausdorff spaces, are the so-called proper actions. They are particularly useful for locally compact (and for discrete) groups.

First we recall the point set topology of proper maps (see Bourbaki [1961a], § 10). Let $f: X \to Y$ be a continuous map. Then f is called **proper** if one of the following equivalent properties holds:

(3.13) For each topological space Z, the map $f \times \mathrm{id}: X \times Z \to Y \times Z$ is closed.

(3.14) f is closed and, for each $y \in Y$, the pre-image $f^{-1}(y)$ is compact.

If X and Y are Hausdorff spaces and Y is locally compact, then f is proper if and only if for each compact subset $K \subset Y$, the pre-image $f^{-1}(K)$ is compact. In this case, X is also locally compact.

(3.15) Let $f: X \to Y$ be a continuous injective map. Then f is proper if and only if f is closed if and only if f is a homeomorphism onto a closed subspace.

(3.16) Let $f: X \to Y$ and $g: Y \to Z$ be continuous maps.
(i) If f and g are proper, then gf is proper.
(ii) If gf is proper and f surjective, then g is proper.
(iii) If gf is proper and g injective, then f is proper.

(3.17) Let $f: X \to Y$ be continuous and $B \subset Y$. If f is proper, then the induced map $f^{-1}(B) \to B$ is proper.

An **action** $\varrho: G \times X \to X, (g, x) \mapsto gx$ of the topological group G on the space X is called **proper** if the associated map

$$\theta = \theta_\varrho: \; G \times X \to X \times X, \; (g, x) \mapsto (x, gx)$$

is proper in the sense defined above. We are going to collect some properties of proper actions, thereby generalizing some of the previous results.

(3.18) Proposition. *Given a proper action of G on X. Then:*
(i) *X/G is a Hausdorff space.*
(ii) *If G is Hausdorff, then X is Hausdorff.*

Proof.
(i) Let C be the image of θ. Since θ is proper, θ is also closed. Hence $C \subset X \times X$ is closed. The assertion now follows from (3.1, iv) and (3.2).
(ii) $X \to G \times X$, $x \mapsto (e, x)$ is a homeomorphism onto a closed subset and hence is proper by (3.15). The composition with θ is proper (3.16). This composition is the diagonal $X \to X \times X$, which therefore has a closed image. □

The next proposition shows that proper actions have special properties.

(3.19) Proposition. *Let G act properly on X. For each $x \in X$ the following holds:*
(i) *$\omega: G \to X$, $g \mapsto gx$ is proper.*
(ii) *The isotropy group G_x is compact.*
(iii) *The map $\omega': G/G_x \to Gx$ induced by ω is a homeomorphism.*
(iv) *The orbit Gx is closed in X.*

Proof. We have $\theta^{-1}(\{x\} \times X) = G \times \{x\}$. Therefore, by (3.17), the map $G \times \{x\} \to \{x\} \times X$, $(g, x) \mapsto (x, gx)$ is proper. This shows (i). The pre-image $\omega^{-1}(x) = G_x$ is compact by (3.14). Since ω is proper, its image Gx is closed in X. The map ω' is proper by (3.16) and therefore a homeomorphism (3.15). □

(3.20) Proposition. *Let G act freely on X. The following are equivalent:*
(i) *G acts properly.*
(ii) *$C = \operatorname{image} \theta \subset X \times X$ is closed and $\varphi: C \to G$, $(x, gx) \mapsto g$ is continuous.*

Proof. Since G acts freely, the map θ is injective. By (3.15), θ is proper if and only if C is closed and $\theta': G \times X \to C$, $(g, x) \mapsto (x, gx)$ is a homeomorphism. The map $C \to G \times X$, $(x, y) \mapsto (\varphi(x, y), x)$ is inverse to θ'. It is continuous if and only if φ is continuous. Thus θ' is a homeomorphism if and only if φ is continuous. □

The next result is an important characterization of proper actions for locally compact groups.

(3.21) Proposition. *Let the locally compact Hausdorff group G act on the Hausdorff space X. Then G acts properly if and only if the following holds: For each pair x, y of points in X, there exist neighbourhoods V_x of x and V_y of y in X such that $\{g \in G \mid gV_x \cap V_y \neq \emptyset\}$ is relatively compact in G.*

Proof. Suppose the latter condition is satisfied. We show that $\theta: G \times X \to X \times X$ is closed. Let $A \subset G \times X$ be closed. Let $((x_j, y_j) \mid j \in J)$ be a net of points in $\theta(A)$ which converges to $(x, y) \in X \times X$. We have to show that $(x, y) \in \theta(A)$.

Write $y_j = g_j x_j$ with $(g_j, x_j) \in A$. Choose V_x and V_y such that $\{g \mid gV_x \cap V_y \neq \emptyset\}$ is contained in a compact set K. We may assume that $x_j \in V_x$, $y_j \in V_y$ for all j. Then $g_j \in K$ and, by compactness, there exists a subnet (g_α) of (g_j) which converges to $g \in K$. Since A is closed, we have $(g, x) \in A$ and since θ is continuous, $\theta(g, x) = (x, y)$. It is also easy to show that θ has compact pre-images of points. Thus θ is proper.

Conversely, assume that θ is proper. Then $G \times X \to G \times X \times X$, $(g, x) \mapsto (g, x, gx)$ is a homeomorphism onto its image D and it transforms the proper map θ to the proper map $p: D \to X \times X$, $(g, x, gx) \mapsto (x, gx)$. Let F

$= G \cup \{\infty\}$ be the one-point-compactification of G. We show that D is closed in $F \times X \times X$. The set $E = \{(g, g) | g \in G\} \subset F \times G$ is closed, being the graph of the inclusion $G \to F$. Therefore, $(E \times X \times X) \cap (F \times D) =: U$ is closed in $F \times D$. Since p is proper, $u: F \times D \to F \times X \times X$, $(h, g, x, y) \mapsto (h, x, y)$ is closed. Since $u(D) = D$, we conclude that D is closed in $F \times X \times X$, as claimed. We have $(\{\infty\} \times X \times X) \cap D = \emptyset$. Therefore, there exist neighbourhoods V of $\{\infty\}$ in F and W of (x, y) in $X \times X$ such that $(V \times W) \cap D = \emptyset$. By definition of F, we can take V to be of the form $(G \backslash K) \cup \{\infty\}$, $K \subset G$ compact. If $V_x \times V_y \subset W$, then $((G \backslash K) \times (V_x \times V_y)) \cap D = \emptyset$ which is equivalent to $(g \notin K \Rightarrow gV_x \cap V_y = \emptyset)$. □

(3.22) Corollary. *A discrete group G acts properly on the Hausdorff space X if and only if for each pair of points (x, y) in X there exist neighbourhoods V_x of x and V_y of y such that $\{g | gV_x \cap V_y \neq \emptyset\}$ is finite.* □

In the literature, proper actions of discrete groups are often called **properly discontinuous**.

A proper action of a discrete group has locally the orbit space of a finite group action. The next proposition makes this precise.

(3.23) Proposition. *Let the discrete group G act properly on the Hausdorff space X. Then the isotropy group G_x of $x \in X$ is finite. Moreover:*

(i) *There exists an open neighbourhood U of x which is a G_x-subspace and satisfies $U \cap gU = \emptyset$ for $g \notin G_x$.*

(ii) *U can be chosen in such a way that the canonical map $U/G_x \to X/G$ is a homeomorphism onto an open set.*

Proof. G_x is finite by (3.19). By (3.21), there exists an open neighbourhood U_0 of x such that $K = \{g \in G | gU_0 \cap U_0 \neq \emptyset\}$ is finite. We have $G_x \subset K$. Let $g_1, \ldots, g_n$ be the elements of $K \backslash G_x$. The points $x_i = g_i x$ are different from x. Since X is a Hausdorff space, there exist open neighbourhoods V_i of x and V_i' of $g_i x$ such that $V_i \cap V_i' = \emptyset$. Let $U_i = V_i \cap g_i^{-1} V_i'$. This is an open neighbourhood of x satisfying $U_i \cap g_i U_i \subset V_i \cap V_i' = \emptyset$. Let $U' = U_0 \cap U_1 \cap \ldots \cap U_n$. This open neighbourhood of x satisfies $U' \cap gU' = \emptyset$ for $g \notin G_x$. The neighbourhood $U = \bigcap_{g \in G_x} gU'$ is a G_x-subspace and satisfies $U \cap gU = \emptyset$ for $g \notin G_x$.

The canonical map $U/G_x \to X/G$ is injective by construction. Moreover, it is continuous and open, thus a homeomorphism onto its image. □

(3.24) Corollary. *Let the discrete group G act freely and properly on the Hausdorff space X. Then the orbit map $X \to X/G$ is a covering, i.e. a locally trivial map with typical fibre G.* □

(3.25) Exercises. In these exercises, compact groups are assumed to be Hausdorff.

1. Let H be a closed subgroup of the compact group G. The group of G-homeomorphisms of G/H endowed with the compact-open-topology is a topological group isomorphic to NH/H.
2. Explain why doubly periodic continuous functions $\mathbb{C} \to \mathbb{D}$ correspond to continuous functions $S^1 \times S^1 \to \mathbb{C}$.
3. Consider the action $\mathbb{R} \times \mathbb{R}^2 \to \mathbb{R}^2$, $(x, y, z) \mapsto (x + y, z)$. Find closed sets $A \subset \mathbb{R}$, $B \subset \mathbb{R}^2$ such that AB is not closed.
4. Let $\lambda \in \mathbb{R}$ be irrational. Consider the action $\mathbb{Z} \times S^1 \to S^1$, $(n, z) \mapsto \exp(2\pi i n \lambda) z$.
   Show: The action is free. The canonical map $G/G_x \to Gx$ is not a homeomorphism.
5. Let $\lambda \in \mathbb{R}$ be irrational. Show that the subgroup $\{(x, \lambda x) | x \in \mathbb{R}\}$ defines a subgroup $H \subset \mathbb{R}^2/\mathbb{Z}^2 = G$ which is dense in G. (First show that $\{\lambda m + n | m, n \in \mathbb{Z}\}$ is dense in $\mathbb{R}$.) The orbit space G/H is not Hausdorff. The isotropy groups of G/H are not closed in G.
6. Let G be a topological group, H a subgroup. Show that G is connected if H and G/H are connected.
7. Let G be a topological group and $H = \{e\}^-$ the closure of $\{e\}$. Show that G/H is Hausdorff and that each continuous homomorphism $f\colon G \to K$ into a Hausdorff group K is factorizable over $G \to G/H$.
8. Give an example of a G-space X such that $X^H = X^K$, $H \subset K$, $H \neq K$, $K \in \mathrm{Iso}(X)$, $H \notin \mathrm{Iso}(X)$.
9. Let G be compact. If $G \times [0, 1[\to [0, 1[$ is an effective action, then $G = \{e\}$.
10. Let G be compact. Let $G \times [0, 1] \to [0, 1]$ be effective. Show that either $G = \{e\}$ or $G \cong \mathbb{Z}/2$ and the action is G-homeomorphic to $[0, 1] \to [0, 1]$, $x \mapsto 1 - x$.
11. The group of homeomorphism of $[0, 1]$ with compact-open-topology has two components. The component of the identity is contractible.
12. Let $A \subset X$ be a G-subset. Show that A/G, as a subset of X/G, carries the subspace topology. In particular, $X^G \subset X \to X/G$ is an embedding.
13. The category of G-spaces has pull-backs.
14. Let X be a G-space and let $f\colon B \to X/G$ be a continuous map. Let

$$\begin{array}{ccc} C & \longrightarrow & X \\ \Big\downarrow{\scriptstyle r} & & \Big\downarrow{\scriptstyle p} \\ B & \xrightarrow[f]{} & X/G \end{array}$$

    be a pull-back. Then r is canonically homeomorphic to the orbit map $C \to C/G$.
15. Let G be compact and let $A \subset X$ be a closed G-subset. Show that each neighbourhood of A contains a G-invariant neighbourhood.

16. Let $r\colon G \times X \to X$ be an action of a compact group. Show that r is a closed mapping.
17. If X carries a left G-action and a right H-action which commute, then there is an induced G-action $G \times X/H \to X/H$. Why is this a special case of (3.4)?
18. Let the compact group G act on X. Show that if X is regular (resp. completely regular, normal, completely normal, compactly generated, separable metric, paracompact), then X/G has the corresponding property. (Engelking [1977].)
19. Let the compact group G act on X. Let $C \subset X$ be a closed fundamental domain. Then the map $X/G \to X, p(x) \mapsto Gx \cap C$ is a continuous section of $p\colon X \to X/G$ (Bredon [1972], I.3.2).
20. (i) Let G be a finite group and M a connected differentiable G-manifold. Show that a connected set $M^H \neq \emptyset$, $H \subset G$ a subgroup, is the fixed point set of an isotropy group.
(Hint: Use that M^K is a differentiable submanifold; a connected differentiable manifold cannot be the union of a finite number of proper submanifolds.)
(ii) Let $\mathrm{SO}(3) \times S^2 \to S^2$ be the standard action. The fixed point set of $\{1\}$ is not the fixed point set of an isotropy group.
21. Let G act on itself by left translation. Show that this action is proper (3.15).
22. Let G act properly on X and let $H \subset G$ be a closed subgroup. The induced action $H \times X \to X$ is proper.
23. Let $f\colon X \to Y$ be a proper surjective G-map. Let G act properly on X. Then G acts properly on Y. (Use that products of proper maps are proper, (3.16).)
24. Let G be a Lie group and K a compact subgroup. Show that the left translation action of G on G/K is proper (21. and 23.).
If G is a connected Lie group, there is a maximal compact subgroup K, unique up to conjugation, and G/K is diffeomorphic to a Euclidean space (see Hochschild [1965], XV). Thus any discrete subgroup of G acts properly on the Euclidean space G/K. An example of such a situation is the hyperbolic space $\mathrm{SO}(n, 1)^+/\mathrm{SO}(n-1)$, see (2.13). Especially, the action of $\mathrm{PSL}(2, \mathbb{R})$ on the upper half plane is proper.
25. (Bourbaki [1960], Ch. III.4.5, Théorème 1.)
Let the Hausdorff topological group G act on the locally compact Hausdorff space X. Suppose that for compact subsets K and L of X the set $\{g \in G \mid gK \cap L \neq \emptyset\}$ is relatively compact in G. Then G acts properly and is locally compact.
26. Let X be a locally compact metric space with a finite number of components. The action of the isometry group $I(X)$ of X on X (2.11) is proper.
27. The action of the conformal automorphisms of S^n on S^n is not proper. In particular, the action of $\mathrm{SL}(2, \mathbb{C})$ on the Riemannian sphere $S^2 = \mathbb{C} \cup \{\infty\}$,

$$\left(\begin{pmatrix} a & b \\ c & d \end{pmatrix}, z\right) \mapsto \frac{az+b}{cz+d},$$

is not proper. There exist discrete subgroups G of $SL(2, \mathbb{C})$ such that the restricted G-action on S^2 is not proper. A standard example is the Picard group $G = SL(2, \mathbb{Z}[i])$; see Beardon [1983], 5.3.4.

4. Functorial properties.

It is often useful or even necessary to consider actions of various groups simultaneously. We develop some of the relevant formal properties.

Let G-Top denote the category of G-spaces and continuous G-maps. If H is a subgroup of G, then restricting the group action from G to H induces a functor

(4.1) $\quad \mathrm{res}_H^G\colon G\text{-Top} \to H\text{-Top}$

which is called **restriction**. This functor has a left adjoint which is called **induction** (or **extension**) and will be denoted by

(4.2) $\quad \mathrm{ind}_H^G\colon H\text{-Top} \to G\text{-Top}.$

We are going to describe this functor. Let X be an H-space. The product $G \times X$ carries an H-action $(h, (g, x)) \mapsto (gh^{-1}, hx)$. The orbit space is denoted by

$$G \times_H X = \mathrm{ind}_H^G X$$

(analogy with tensor product!). The G-action $(g', (g, x)) \mapsto (g'g, x)$ on $G \times X$ induces a G-action on $G \times_H X$ (see (3.4) and 3. Exercise 17). For any H-map $f\colon X \to Y$, we have an induced G-map

$$G \times_H f = \mathrm{ind}_H^G f\colon G \times_H X \to G \times_H Y, \; (g, x) \mapsto (g, fx).$$

Obviously, ind_H^G can be considered as a functor. If G and H are Hausdorff and compact and if X is separated, then $G \times_H X$ is again separated (3.6).

Let $\mathrm{Top}_G(X, Y)$ denote the set of continuous G-maps $X \to Y$. The adjointness property is

(4.3) Proposition. *There is a canonical bijection*

$$\mathrm{Top}_G(G \times_H X, Y) \cong \mathrm{Top}_H(X, \mathrm{res}_H^G Y)$$

for G-spaces Y and H-spaces X.

Proof. Given an H-map $f\colon X \to Y$, the map $G \times X \to Y$, $(g, x) \mapsto gf(x)$ admits a factorization over $G \times_H X$; this yields a G-map $G \times_H X \to Y$. Given a G-map $F\colon G \times_H X \to Y$, we define an H-map f by $f(x) = F(e, x)$. One verifies that these two constructions are inverses of each other (and „natural" in X and Y). □

Our next aim is to show that G-spaces often have the form $G \times_H A$. Let X be a

G-space and $f\colon X \to G/H$ be a G-map. Then $A = f^{-1}(eH)$ is an H-subspace of X and we obtain a G-map $F\colon G \times_H A \to X$, $(g, a) \mapsto ga$, which satisfies $fF = p$ where p is the canonical projection $G \times_H A \to G/H$. One checks that F is always bijective. Our next proposition gives conditions under which F is a homeomorphism.

We say $q\colon G \to G/H$, $g \mapsto gH$ has a **local section** if there exists an open neighbourhood U of eH in G/H and a section of q over U.

(4.4) Proposition. *The map F defined above is a homeomorphism if one of the following conditions is satisfied:*
(i) *G is compact Hausdorff and H is closed in G.*
(ii) *$q\colon G \to G/H$ has a local section.*

Proof. Let $s\colon U \to G$ be a local section. By applying left translations, it suffices to show that F is a homeomorphism over U. If $x \in f^{-1}(U)$, then $(sfx)^{-1}x \in A$. Define $F^*\colon f^{-1}U \to p^{-1}U$ by $F^*(x) = (sfx, (sfx)^{-1}x)$. Then F^* is continuous and inverse to F over U. If G is compact, then $G \times_H A \to X$ is a closed mapping (use 3. Exercise 16). □

Let B a G-space. Then we have the category $G\text{-Top}_B$ of G-spaces over B: the objects are G-maps $p\colon X \to B$ and the morphisms from $p\colon X \to B$ to $q\colon Y \to B$ are G-maps $f\colon X \to Y$ such that $qf = p$.

If the conditions of (4.4) are satisfied, then we have an equivalence of the categories H-Top and $G\text{-Top}_{G/H}$.

On objects, the equivalences are given by $X \mapsto (G \times_H X \to G/H)$ and $(p\colon G \times_H X \to G/H) \mapsto p^{-1}(eH)$.

Let $f\colon B \to C$ be a G-map. Composition with f yields a covariant functor $f_*\colon G\text{-Top}_B \to G\text{-Top}_C$. Taking pull-backs along f yields a functor $f^*\colon G\text{-Top}_C \to G\text{-Top}_B$. In view of the equivalence of categories noted above, given $f\colon G/H \to G/G$, the functor f_* corresponds to ind_H^G and the functor f^* corresponds to res_H^G.

Let

$$\begin{array}{ccc} C & \xrightarrow{F} & D \\ {\scriptstyle P}\downarrow & & \downarrow{\scriptstyle p} \\ A & \xrightarrow[f]{} & B \end{array}$$

be a pull-back diagram of G-spaces.

(4.5) Proposition. $p^*f_* = F_*P^*$.

Proof. Unravel the assertion and apply transitivity of pull-backs. □

If (4.5) is applied to $A = G/H$, $D = G/K$, and $B = G/G$, it yields the double-coset-formula for the induction-restriction formalism. We make this more explicit for finite G. In this case, p^*f_* corresponds to $\mathrm{res}^G_K \mathrm{ind}^G_H$. The G-space $C = G/H \times G/K$ is the disjoint union of orbits $C = \coprod A(i)$. Let $P(i) = P|A(i)$, $F(i) = F|F(i)$. Then $F_*P^* = \coprod F(i)_* P(i)^*$, a union of objects over D. Set $Y = p^*f_*X$. Let $KgH \subset G$ be a double coset with its left K-action and right H-action given by translation. Then $KgH \times_H X$ is a K-space and Y is the disjoint union, over the double cosets, of such spaces. The space $KgH \times_H X$ has another description as follows. Let $L = K \cap gHg^{-1}$. Let X_g be the L-space

$$L \times X \to X, \ (u, x) \mapsto g^{-1}ugx.$$

Then there is a K-homeomorphism

$$\text{(4.6)} \qquad KgH \times_H X \cong K \times_L X_g$$

given on representatives by $(kgh, x) \mapsto (k, hx)$.

The adjointness (4.3) has its analogue for sets of homotopy classes. For an H-space X and a G-space Y, this reads

$$\text{(4.7)} \qquad [G \times_H X, Y]_G \cong [X, \mathrm{res}^G_H Y]_H.$$

We leave the verification to the reader and turn our attention to spaces with basepoint. The basepoint of a G-space X is assumed to be a stationary point. If Z is any G-space, then Z^+ denotes the G-space Z with a disjoint G-fixed basepoint $+$ added. If X and Y are G-spaces with basepoint and G is locally compact, then (see Ex. 11)

$$\text{(4.8)} \qquad X \wedge Y = X \times Y/X \vee Y,$$

the **smash-product** of X and Y, is again a pointed G-space. We let $[X, Y]^0_G$ denote the set of pointed G-homotopy classes $X \to Y$. If X is a pointed H-space and H a subgroup of G, then H acts on the pointed space $G^+ \wedge X$ by $(h, (g, x)) \mapsto (gh, h^{-1}x)$. The orbit space is denoted by $G^+ \wedge_H X$ and $(g', (g, x)) \mapsto (g'g, x)$ induces a G-action so that $G^+ \wedge_H X$ becomes a pointed G-space.

Let G be locally compact. Then we have

(4.9) Proposition. *For pointed G-spaces Y and pointed H-spaces X, there is a natural bijection*

$$[G^+ \wedge_H X, Y]^0_G \cong [X, \mathrm{res}^G_H Y]^0_H.$$

Proof. Given a pointed H-map $f\colon X \to Y$, consider the G-map $G \times X \to Y$, $(g, x) \mapsto gf(x)$. The G-subset $G \times \{*\}$, $*$ the basepoint, is mapped into the basepoint. Therefore, the induced G-map $G \times X/G \times \{*\} = G^+ \wedge X \to Y$ factors over $G^+ \wedge X \to G^+ \wedge_H X$ and yields a pointed G-map $G^+ \wedge_H X \to Y$. This construction is compatible with homotopies thus producing a map $[X, \mathrm{res}^G_H Y]^0_H \to [G^+ \wedge_H X, Y]^0_G$. An inverse is given by restricting a map $G^+ \wedge_H X \to Y$ to the H-subspace $H^+ \wedge_H X \cong X$. □

Recall that the set of maps $\mathrm{Map}(X, Y)$ between G-spaces carries a G-action defined by $(g \cdot f)(x) = gf(g^{-1}x)$ for $g \in G$ and $f\colon X \to Y$. The fixed point set $\mathrm{Map}(X, Y)^G$ equals the set of G-maps $\mathrm{Map}_G(X, Y)$. (No continuity so far.) With this G-action, one can consider $\mathrm{Map}(X, \mathrm{Map}(Y, Z))$. The adjunction

$$\mathrm{Map}(X \times Y, Z) \to \mathrm{Map}(X, \mathrm{Map}(Y, Z)),\ f \mapsto \bar{f}$$

with $\bar{f}(x)(y) = f(x, y)$ is then a bijection of G-sets as is easily verified. This induces a bijection

$$\mathrm{Map}_G(X \times Y, Z) \to \mathrm{Map}_G(X, \mathrm{Map}(Y, Z)).$$

Of course, $X \times Y$ carries the diagonal action. Similar statements hold for continuous maps and mapping spaces with compact-open-topology provided one has the exponential law for mapping spaces.

The restriction functor also has a right adjoint, called **multiplicative induction**

$$m^G_H\colon\ H\text{-Top} \to G\text{-Top}.$$

We consider in detail only the case that H is a (closed and open) subgroup of finite index in G. Let X be an H-space and let

(4.10) $m^G_H X = \mathrm{Top}_H(G, X)$

be the set of H-maps where G carries the left translation action. The G-action on $m^G_H(X)$ is induced by right translation on G, i.e. $(g \cdot f)(x) = f(xg)$ for $x \in G$, $g \in G, f \in m^G_H(X)$. We have a G-space

$$\prod_{gH \in G/H} (gH \times_H X)$$

with G-action $g'(g, x) = (g'g, x)$ on the components (in general permuting the factors). The map

$$m^G_H(X) \to \prod(gH \times_H X)$$

sending f to $((g, f(g^{-1})) \mid gH \in G/H)$ is well-defined, bijective, G-equivariant and supplies $m^G_H(X)$ with a topology as a product of $|G/H|$ copies of X (note that $X \cong gH \times_H X$ as spaces).

(4.11) Proposition. *For G-spaces Y and H-spaces X, there is a natural bijection*

$$\mathrm{Top}_G(Y, m^G_H X) \cong \mathrm{Top}_H(\mathrm{res}^G_H Y, X).$$

Proof. Let $f\colon Y \to X$ be a continuous H-map. Then $\bar{f}(y)\colon g \mapsto f(gy)$ is contained in $\mathrm{Top}_H(G, X)$ and $F_f\colon y \mapsto \bar{f}(y)$ in $\mathrm{Top}_G(Y, m_H^G X)$. The assignment $f \mapsto F_f$ is the asserted bijection. An inverse is given as follows. Let $F\colon Y \to m_H^G X$ be a continuous G-map. Evaluation at $e \in H$ is a continuous H-map $E\colon m_H^G X \to X$. Take the composition $f_F = EF$. □

(4.12) Example. If V is an H-representation, then $m_H^G V$ is in a natural way a vector space and a G-representation. This is the so-called **induced representation** of representation theory.

Multiplicative induction has a property which is analogous to (4.5). Let $\varphi\colon S \to T$ be a G-map between finite G-sets. Define a functor $\varphi_{(m)}\colon G\text{-}\mathrm{Top}_S \to G\text{-}\mathrm{Top}_T$ on objects by assigning to a G-space X over S with fibre X_s over $s \in S$ the G-space Y with fibre $Y_t = \prod_{s \in \varphi^{-1}(t)} X_s$ over t. For a pull-back diagram between finite G-sets as before (4.5), one has

(4.13) $p^* f_{(m)} = F_{(m)} P^*$.

The verification is left as an exercise. For more details on such matters, see Dress [1973].

(4.14) Exercises.

1. Let H be a subgroup of G and let A be an H-space. The inclusion $A \to G \times_H A$, $a \mapsto (e, a)$ induces a homeomorphism $A/H \cong (G \times_H A)/G$.
2. Let X be a right H-space, Y a left H-, right K-space (with commuting actions), Z a left K-space. There is a canonical homeomorphism

$$(X \times_H Y) \times_K Z \to X \times_H (Y \times_K Z), \quad ((x, y), z) \mapsto (x, (y, z)).$$

   Of course, $X \times_H Y$ is the orbit space of $H \times X \times Y \to X \times Y$, $(h, x, y) \mapsto (xh, h^{-1} y)$ and carries an induced K-action.
3. Let $K \subset H \subset G$ be subgroups. Let X be a K-space. There is a G-homeomorphism

$$G \times_K X \cong G \times_H (H \times_K X).$$

   This yields $\mathrm{ind}_H^G \, \mathrm{ind}_K^H \cong \mathrm{ind}_K^G$.
4. The isotropy group of $G \times_H X$ at (g, x) is $g H_x g^{-1}$.
5. Let $\alpha\colon H \to G$ be a continuous homomorphism of topological groups. There is a functor $\mathrm{res}_\alpha\colon G\text{-}\mathrm{Top} \to H\text{-}\mathrm{Top}$ which maps a G-space X to the H-space X with action $H \times X \to X$, $(h, x) \mapsto \alpha(h)x$. There is a functor $\mathrm{ind}_\alpha\colon H\text{-}\mathrm{Top} \to G\text{-}\mathrm{Top}$ mapping the H-space Y to the G-space $G \times_\alpha Y$ defined as $G \times Y/\sim$, $(g, y) \sim (g\alpha(h), h^{-1} y)$, with G-action $(g', (g, x)) = (g'g, x)$. Show adjointness $\mathrm{Top}_G(\mathrm{ind}_\alpha Y, X) \cong \mathrm{Top}_H(Y, \mathrm{res}_\alpha X)$.
6. Verify (4.5), (4.6), (4.7) and (4.13).
7. Suppose X, Y and Z are pointed G-spaces. Let X and Y be Hausdorff spaces,

Y locally compact. Suppose G is locally compact. Show that there is a natural bijection

$$[X \wedge Y, Z]_G^0 \cong [X, C^0(Y, Z)]_G^0$$

where $C^0(Y, Z)$ is the space of pointed maps $Y \to Z$ with compact-open-topology and G-action induced from $C^0(Y, Z) \subset \mathrm{Map}(Y, Z)$. What is the corresponding statement for unpointed spaces?

8. Let G be locally compact.
Let X be a pointed H-space and let H be a subgroup of finite index in G. In analogy to the multiplicative induction, define a G-action on the iterated smash-product $\bigwedge_{gH \in G/H} (gH \times_H X)$. If V^c denotes the one-point compactification of a representation space V, show that

$$(m_H^G V)^c \cong \bigwedge (gH \times_H V^c)$$

as pointed G-spaces. Taking homotopies yields a map

$$[X, Y]_H^0 \to [\bigwedge (gH \times_H X), \bigwedge (gH \times_H Y)]_G^0,$$

for pointed H-spaces X and Y. (Remark: In general, the smash-product is not associative. In order to obtain associativity, one has either to restrict the class of spaces or to use devices such as compactly generated spaces.)

9. Let H and K be subgroups of G. Let X be an H-space. Verify that the fixed point set $(G \times_H X)^K$ is the set $\{(g, x) \mid g^{-1}Kg \subset H, x \in X^{g^{-1}Kg}\}$. Compare also with (4.6).

10. Let X be a G-space. Show that $G \times_H X$ is naturally G-homeomorphic to $G/H \times X$.

11. Let X be a G-space, A a G-subspace and X/A the space obtained by identifying A with a point. Show that for locally compact G there is an induced G-action on X/A such that the quotient map is equivariant.

12. Investigate under which conditions the bijections (4.3) and (4.11) give homeomorphisms between mapping spaces with CO-topology

$$C_G(G \times_H X, Y) \cong C_H(X, \mathrm{res}_H^G Y)$$
$$C_G(Y, m_H^G X) \cong C_H(\mathrm{res}_H^G Y, X).$$

Similarly, obtain conditions under which there are bijections

$$\mathrm{Top}_G(X \times Y, Z) \cong \mathrm{Top}_G(X, C(Y, Z))$$

or homeomorphisms

$$C_G(X \times Y, Z) \cong C_G(X, C(Y, Z)).$$

5. Differentiable manifolds. Tubes and slices.

Differentiable G-manifolds are important objects in the theory of transformation groups. We recall the basic definition. Let G be a Lie group and M a differentiable manifold (always of class C^∞). A **differentiable action** of G on M is an action $G \times M \to M$ which is a C^∞-differentiable map. A manifold M together with a differentiable action of G on it is called a **differentiable G-manifold**. In this section, the term G-manifold is meant to imply that the action is differentiable. If not specified otherwise, subgroups are assumed to be closed.

This section is not intended to begin a systematic study of G-manifolds. But, occasionally, we have to use some basic results and examples, and so it seems appropriate to introduce them at an early stage in this book. Also, the basic notions of tubes and slices are introduced and the important general slice theorem is proved.

The left translations in a G-manifold are diffeomorphisms. In general, the orbit space M/G of a G-manifold is not a manifold (see (2.19), Ex. 1–3). (M/G carries a structure which may be called a manifold with singularities, at least if G is compact.)

The tangent bundle of M is denoted by $TM \to M$ with $T_m M$ being the tangent space at $m \in M$. The differential of $f\colon M \to N$ is $Tf\colon TM \to TN$. For basic material concerning differentiable manifolds, we refer to Bröcker-Jänich [1973] and Hirsch [1976].

There are a few situations where M/G carries the structure of a differentiable manifold such that $M \to M/G$ is a submersion. In order to describe these, we use the following result from differential topology.

Let M be a differentiable manifold, $R \subset M \times M$ an equivalence relation on M, and $p\colon M \to M/R$ the quotient map.

(5.1) Proposition. *M/R carries the structure of a differentiable manifold such that $M \to M/R$ is a submersion if the following holds: R is a closed submanifold of $M \times M$ and $\mathrm{pr}_1\colon R \to M$ is a submersion.* □

The proof is not difficult. The formulation is taken from Bourbaki [1967], 5.9.5.

Recall from section 3 that a free action $G \times X \to X$ is called **proper** if $C = \{(x, gx) \mid x \in X, g \in G\}$ is closed in $X \times X$ and $t\colon C \to G$, $(x, gx) \mapsto g$ is continuous.

(5.2) Proposition. *Let $r\colon G \times M \to M$ be a proper, free, differentiable action. Then M/G carries a (unique) differentiable structure such that $p\colon M \to M/G$ is a submersion.*

Proof. Using (5.1), we have to show that $C \subset M \times M$ is a submanifold and

$pr_1: C \to M$ is a submersion. The image of $\alpha: G \times M \to M \times M$, $(g, x) \mapsto (x, gx)$ is C. The map α is differentiable. We proceed to show that α is an embedding. The map

$$\beta: C \to G \times M, \ (x, y) \mapsto (t(x, y), x)$$

is continuous since the action is proper. We have $\beta\alpha = \mathrm{id}$ so that $\alpha: G \times M \to C$ is a homeomorphism. If we show that α is an immersion, then C is known to be a submanifold. The differential of $pr_1 \circ \alpha$ at (g, x) has kernel $T_g(G) \times \{0\} \subset T_{(g,x)}(G \times M)$. Therefore, it suffices to prove that the differential of pr_2 is injective on this kernel, i.e. that $f: G \to X$, $g \mapsto gx$ has an injective differential. The action being free, the map f is injective. If we show that Tf has constant rank, then, by the rank theorem of differential calculus (Bröcker-Jänich [1973], (5.4)), Tf must be injective. We have $fr = lf$ with $r(h) = gh$, $l(x) = gx$, and r and l are diffeomorphisms. Passing to differentials, we conclude that rank $T_g f =$ rank $T_1 f$. The map $pr_1: C \to M$ must be a submersion because its composition with α is a projection and hence a submersion. □

As an immediate corollary of (5.2) we obtain

(5.3) Proposition. *Let G be a Lie group and K a subgroup which is a submanifold. Then $K \times G \to G$, $(k, g) \mapsto kg$ is a proper free action. Therefore, $K \backslash G$ carries a canonical differential structure such that $p: G \to K \backslash G$ is a submersion. The canonical action $K \backslash G \times G \to K \backslash G$ is differentiable.* □

In the sequel, the space $K \backslash G$ shall always carry this differential structure. A basic theorem of Lie group theory states that a closed subgroup K of a Lie group G is a submanifold; hence $K \backslash G$ has always the structure of a G-manifold.

(5.4) Proposition. *Let G be a compact Lie group and M a G-manifold. For each $x \in M$, the map $\alpha_x: G/G_x \to M, gG_x \mapsto gx$ is an embedding. Hence the orbit Gx is a submanifold which is G-diffeomorphic to G/G_x.*

Proof. $G \to M$, $g \mapsto gx$ has constant rank and factors over the submersion $G \to G/G_x$; thus it induces an injective differentiable map α_x of constant rank and hence an injective immersion. Since G is compact, α_x is a topological embedding. Both facts taken together imply the assertion. □

Basic examples of differentiable G-manifolds are representations. A continuous homomorphism between Lie groups is always differentiable. As a consequence of the Peter-Weyl theorem in representation theory, one has the following basic result for the theory of transformation groups (see e.g. Bröcker-tom Dieck [1985], III.5; see also Ex. 3).

(5.5) Theorem. *Let G be a compact Lie group and let H be a closed subgroup of G. Then there exists a finite-dimensional representation V of G such that H is the isotropy group of some point $x \in V$.* □

Another way of expressing (5.5) is: The differentiable manifold G/H has an equivariant differentiable embedding into a representation space. Submanifolds of differentiable manifolds have nice neighbourhoods, the so called tubular neighbourhoods, which are diffeomorphic to vector bundles. Of particular importance are such neighbourhoods of orbits.

Let G be a compact Lie group, H a closed subgroup, and V an H-representation space. By (5.2), $G \times_H V$ is a differentiable manifold and $G \times_H V \to G/H$ is a differentiable G-map with fibres isomorphic to V. A basic fact about G-manifolds is that there always exist neighbourhoods of orbits $Gx \cong G/G_x$ which are G-diffeomorphic to $G \times_H V$, $H = G_x$, V suitably chosen. Despite of its importance, we do not prove this result here because it would need too many prerequisites from differential topology. It is more important to understand the phenomenon. In this section, we need the following theorem only for the case that M is an orthogonal representation space. A standard proof for this particular situation is outlined in exercise 5.

Thus we only state

(5.6) Theorem. *Let G be a compact Lie group and M a differentiable G-manifold. For $m \in M$ and $H = G_m$, there exists a unique H-representation V and a G-diffeomorphism $\varphi\colon G \times_H V \to M$ onto an open neighbourhood of Gm such that $\varphi(g, 0) = gm$.* □

The representation V in (5.6) is called the **slice representation** of M at m. If $m \in M^G$, then the slice representation $V \cong T_m M$ is also called the **tangential representation** at m (in this case (5.6) is due to Bochner [1945]).

We now draw some important conclusions from (5.6). Let the compact group G act on the Hausdorff space X. An open G-subset U of X is called a **tube** about $x \in U$ if there exists a G-map $f\colon U \to G/G_x$. By (4.4), a tube is G-homeomorphic to $G \times_H A$ where $A = f^{-1}(eG_x)$ is an $H = G_x$-space. The subset A is called a slice of the G-space X at x. More formally, let $x \in S \subset X$ and $G_x(S) = S$. Then S is called a **slice** at x if the map $G \times_{G_x} S \to X$, $(g, s) \mapsto gs$ is an embedding onto a tube about x. Note that, by (5.6), differentiable G-manifolds have tubes and slices of a very special kind.

The next theorem, the slice theorem, is of fundamental importance.

(5.7) Theorem. *Let G be a compact Lie group and X a G-space which is completely regular. Then there is a tube about each of its points.*

(5.7) was proved for free actions in Gleason [1950]. General actions were treated by Montgomery-Yang [1957], Mostow [1957], and Palais [1960].

Proof of (5.7). A completely regular space X is a subspace of a normal (actually a compact) space Y (Schubert [1964], I.9.2). The orbit Gx through $x \in X$ is compact and hence closed in Y. Choose an embedding $Gx \subset V$ as in (5.5). By the Tietze extension theorem, f has an extension to Y and hence to X. By the proof of (3.11), there exists an equivariant extension $F\colon X \to V$. Let $r\colon U \to G/G_x$ be a G-map from a G-neighbourhood U of $f(Gx)$ in V; its existence follows from (5.6). Then $W = F^{-1}(U)$ is a G-neighbourhood of Gx in X which is mapped under rF to G/G_x. This completes the proof of (5.7). □

(5.8) Remark. We have mentioned in the previous proof that a completely regular space has an embedding into a compact space. For a universal G-embedding of this type, the Čech-G-compactification, see Palais [1960].

Theorem (5.7) has many applications some of which we present now.

(5.9) Theorem. *Let G be a compact Lie group and let H be a closed subgroup of G. Given a neighbourhood U of e in G, there exists a neighbourhood $V \subset U$ of e such that any subgroup $K \subset VH$ is conjugate by an element $u \in U$ to a subgroup of H.*

Proof. The space X of all closed non-empty subsets of G/H carries the Hausdorff metric $d(A, B) = \max d(a, B) + \max d(A, b)$ where d denotes a metric on G/H. The left translation action of G on G/H induces a continuous action on X (exercise 6.) and the isotropy group of the point $x = \{eH\}$ is H. By (5.7), there exists a G-neighbourhood W of x in X and a G-map $f\colon W \to G/H$. Set $W' = f^{-1}(UH)$; this is a neighbourhood of x in X. Choose $V \subset U$ such that any closed set contained in $VH/H \subset G/H$ is a point in W. For $K \subset VH$, the set $y = KH/H$ is an element of W and, moreover, $K \subset G_y$. For any $y \in W'$, there exists $u \in U$ such that $u^{-1} G_y u \subset G_x$, essentially by the definition of W'. Thus $u^{-1} K u \subset u^{-1} G_y u \subset G_x = H$ for some $u \in U$.

If G is an arbitrary Lie group and H a compact subgroup, the statement of (5.9) is due to Montgomery-Zippin [1955]. The present proof is due to Mostow [1957]. The more general case is also proved in Palais [1961].

(5.10) Proposition (Bredon [1972], II.5.7). *Let $K \subset H \subset G$ be compact Lie groups. The NK/K-action on $(G/H)^K$ has only a finite number of orbits.*

Proof. Let $gH \in (G/H)^K$, that is $g^{-1}Kg \subset H$. We apply (5.9) with $g^{-1}Kg \subset H$ in place of $H \subset G$. Thus there exists a neighbourhood W of e in G such that if

$$L \subset (W \cap H) g^{-1} K g = W(g^{-1}Kg) \cap H,$$

then $h^{-1} L h \subset g^{-1} K g$ for some $h \in H$. We can find a neighbourhood V of e such that $V^{-1} g^{-1} K g V \subset W g^{-1} K g$. Now suppose that for $gv \in gV$ we have

$gvH \in (G/H)^K$, i.e. $v^{-1}g^{-1}Kgv \subset H$. Then, for $L = v^{-1}g^{-1}Kgv$, we can find $h \in H$ such that $h^{-1}Lh = h^{-1}v^{-1}g^{-1}Kgvh \subset g^{-1}Kg$, i.e. $gvhg^{-1} \in NK$ or $gvH \subset NKgH$. Consequently, gvH is in the NK-orbit of $gH \in (G/H)^K$. Since gVH/H is a neighbourhood of gH in G/H, we see that the NK-orbits are open in $(G/H)^K$. By compactness, there are only finitely many of them. □

(5.11) Theorem. *Let G be a compact Lie group. A compact differentiable G-manifold has finite orbit type, i.e. it has only finitely many conjugacy classes of isotropy groups.*

Proof. (Induction on the dimension of M.) If $\dim M = 0$, then M consists of a finite number of points and therefore has only finitely many orbits. Suppose $\dim M = n$. By compactness and (5.6), there exists a covering of M by a finite number of open sets which have the form $G \times_H V$. Consider an H-invariant inner product on V (see Bröcker-tom Dieck [1985], II.1). The unit sphere $S(V)$ in V is then a compact H-manifold which has dimension less than n and therefore finite H-orbit type by induction. Then $G \times_H SV$ has also finite G-orbit type. But $G \times_H (V \setminus \{0\})$ and $G \times_H SV$ have the same set of isotropy subgroups. □

(5.12) Remark. For more general theorems implying finite orbit type, see Bredon [1972], IV.10.

(5.13) Proposition. *Let G be a compact Lie group and M a differentiable G-manifold. Let H be any isotropy group of M. Then* $M_{(H)} = \{x \in M \mid (G_x) = (H)\}$ *is a submanifold of M (which may have components of different dimension).*

Proof. $A \subset M$ is a submanifold if and only if each point $x \in M$ has an open neighbourhood U such that $A \cap U$ is a submanifold of U. Therefore, by (5.6), we need only consider the case $M = G \times_H V$. The isotropy group at $(g, v) \in G \times_H V$ is $G_{(g,v)} = gH_v g^{-1}$, $H_v \subset H$. Therefore, $(G_{(g,v)}) = (H)$ if and only if $H_v = H$, i.e. $v \in V^H$. But V^H is a subspace of V and hence a submanifold. The inclusion $G \times_H V^H \subset G \times_H V$ is a differentiable sub-vector-bundle and therefore a differentiable submanifold. □

(5.13) implies in particular that M^G is always a closed submanifold. In general, even if M is compact, the $M_{(H)}$ are neither compact nor closed. For the rest of this section, we assume that G is a compact Lie group and M a differentiable G-manifold.

(5.14) Theorem. *Suppose M/G is connected. Then there exists a unique isotropy type (H) of M such that $M_{(H)}$ is open and dense in M. The space $M_{(H)}/G$ is* *connected. Each other isotropy type (K) satisfies $(H) \leqq (K)$, i.e. H is subconjugate to K. The set M^H intersects each orbit.*

The space G/H, with H as in (5.14), is called the **principal orbit** type of M and $M_{(H)}$ is called the **principal orbit bundle** of M.

Proof. (Induction on the dimension of M.) If $\dim M = 0$, the space M/G is a point and M consists of a single orbit. Thus suppose $\dim M = n \geqq 1$. By induction, the theorem is true for the sphere bundle $G \times_H SV$, $H = G_x$, $G \times_H V$ a tube as in (5.6), as long as the orbit space is connected. The only case where the orbit space could be disconnected occurs for $\dim V = 1$ and the trivial representation $H \to \mathrm{GL}(V) = \mathrm{Aut}(V)$. In this case, $G \times_H V \cong G/H \times \mathbb{R}$, and the theorem is obviously true. In the remaining case, let H/K be the principal orbit of SV. Then either $V_{(K)} \cong SV_{(K)} \times]0, \infty[$; or $0 \in V_{(K)}$, i.e. $K = H$, and a dense subspace of SV is fixed by H whence $H \to GL(V)$ is trivial. In both instances, the theorem is true for V and hence for $G \times_H V$. Now cover M by open G-sets diffeomorphic to $G \times_H V$. If two of them have non-empty intersection, then the same holds for their open and dense principal orbit bundles; therefore, the corresponding orbit types are equal. Since M/G is connected, it follows that any two of these orbit types are equal. It also follows that the union of these principal orbit bundles is open and dense in M and has connected orbit space.

Now let K be any isotropy group of M, say $K = G_x$. Choose a neighbourhood U of Gx isomorphic to $G \times_K V$. Then, by denseness, there exists a principal orbit contained in U. Using $G \times_K V \to G/K$, we see that $(H) \leqq (K)$. It follows that $G/K^H \neq \emptyset$ and hence $M^H \cap Gx \neq \emptyset$. □

Suppose again that M/G is connected. If an orbit has dimension less than the dimension of the principal orbit, it is called a **singular orbit**. If it has the same dimension as the principal orbit but is different from a principal orbit, then it is called an **exceptional orbit**.

Example. Let $G = S^1$ be the circle group acting on the real projective plane P^2 as follows. Let S^1 act on (x, y, z)-space by rotations about the z-axis. Let P^2 be the unit sphere S^2 with antipodal points identified. Then P^2 has one fixed point, a singular orbit, and one exceptional orbit corresponding to $x^2 + y^2 = 1$ with isotropy group $\mathbb{Z}/2$. All other orbits are free.

(5.15) Proposition. *Suppose M/G is connected. If G/U is a singular orbit, then $M_{(U)}$ has at least codimension two in M.*

Proof. (Induction on the dimension of M.) In the zero-dimensional case there are no singular orbits. Suppose $\dim M = n$. It suffices to consider spaces of the type $G \times_U V$. Then

$$(G \times_U V)_{(U)} \cong G \times_U V^U \cong G/U \times V^U.$$

We thus have to show that V^U has at least codimension 2 in V. Consider the U-manifold SV. The set SV cannot be empty, for otherwise $\dim G/U = n$ which is impossible for a singular orbit. Suppose $\dim SV = 0$ so that $\dim G/U = n-1$. In this case, the principal orbits are of dimension n and the manifold consists of a single orbit. So we may assume that $\dim V \geqq 2$, and SV and SV/U are connected. We apply the induction hypothesis to SV. The fixed points SV^U are singular orbits: otherwise, all orbits would be zero-dimensional and all orbits of $G \times_U V$ would have the dimension of G/U; this contradicts the denseness of the principal orbit bundle and the fact that G/U is singular. Therefore, SV^U consists of singular orbits and $\dim V^U \leqq \dim V - 2$ by induction hypothesis. □

(5.16) Remark. For the principal orbit type theorem for non-smooth actions, see Montgomery, Samelson, Zippin [1956], Montgomery, Samelson, Yang [1956], Montgomery, Yang [1958], Montgomery [1960]. For locally smooth actions, see Bredon [1972], IV., in particular IV.3.

(5.17) Proposition. *Let the compact Lie group G act differentiably and effectively on the connected n-dimensional manifold M. Then* $\dim G \leqq \frac{1}{2}n(n+1)$.

Proof. (Induction on n.) The case $n = 0$ is trivial. We may assume that G is connected because the component of e acts effectively and has the same dimension. Let G/H be the principal orbit type of M. Then $\dim G \leqq n + \dim H$. If we knew that H acts effectively on a manifold of dimension at most $(n-1)$, then we could use the induction hypothesis to prove the theorem for n because $n + \frac{1}{2}(n-1)n = \frac{1}{2}n(n+1)$. We have $\dim G/H = k \leqq n$ and G acts effectively on G/H; in fact, if $g \in G$ acts as identity on G/H, then it acts as identity on any principal orbit and hence on M. Thus H_0, the component of e in H, acts effectively on G/H and on the principal orbit of the H_0-manifold G/H which is connected. If the dimension of this principal orbit is less than k, we are done. Otherwise, $H_0 eH = G/H$ and G/H is a point; then $G = \{e\}$ since it acts effectively. □

Manifolds for which equality holds in (5.17) are to be considered the most symmetric ones. The group $O(n+1)$ of dimension $\frac{1}{2}n(n+1)$ acts effectively on the n-sphere and on the n-dimensional real projective space. No wonder that mathematicians like these spaces (see (2.12)).

Let X be a connected differentiable manifold. The maximal dimension $N(X)$ of a compact Lie group G which can act effectively on X is called the **degree of symmetry** of X. Certain values cannot occur as degree of symmetry; see Mann [1966]. For the degree of symmetry of spheres with exotic differentiable structure, see Wu-yi Hsiang [1967] and Wu-chung Hsiang and Wu-yi Hsiang

[1967]. For many manifolds, $N(X) = 0$; see Atiyah-Hirzebruch [1970]. There even exist compact manifolds without effective action of non-trivial finite groups; see Conner-Raymond-Weinberger [1972], Bloomberg [1975].

These manifolds are not simply connected. It is an open problem whether simply connected manifolds always have non-trivial actions. For a large class of manifolds, Löffler and Raußen [1985] have shown the existence of free $\mathbb{Z}/p$-actions for almost all p.

(5.18) Exercises.

1. Let $h\colon H \times [0,1] \to K$ be a homotopy of continuous homomorphisms between compact Lie groups (i.e. for each $t \in [0,1]$, the map $h_t\colon x \mapsto h(x,t)$ is a homomorphism). Then there exists $k \in K$ such that $h(x,0) = kh(x,1)k^{-1}$. (Homotopic homomorphisms are conjugate). Hint: Look at the subgroups $\mathrm{Graph}(h_t) \subset H \times K$ and apply (5.9).
2. Show by an example that for non-compact Lie groups a result as in exercise 1 cannot be true in general.
3. Let G be a finite group, $H \subset G$ a subgroup and $V = \mathbb{R}(G/H)$ the $\mathbb{R}$-vector space (permutation representation) with basis G/H and induced G-action. Show that there exist $v \in V$ with $G_v = H$. For compact G, find an infinite-dimensional representation V of G with v such that $G_v = H$ (Hint: Continuous functions on G/H.)
4. Supply a proof of (5.1).
5. Let G be a compact Lie group, V an orthogonal representation of G and $M \subset V$ a G-submanifold with tangent space $T_m M$ canonically identified with a subspace of V. Let $N_m(M)$ be the orthogonal complement of $T_m M$ in V. Let $E = \{(x,v) \mid x \in M, v \in N_x(M)\} \subset M \times V$. Show that E is a G-submanifold of $M \times V$ and $E \to M$, $(x,v) \mapsto x$ a G-vector bundle. Show that $E \to V$, $(x,v) \mapsto x+v$ maps a neighbourhood of the zero section diffeomorphically. Use this fact to find an equivariant tubular neighbourhood. (Compare Hirsch [1976], 4.5).
6. Verify that G acts continuously on the space of closed subsets of G/H (by the action defined in the proof of (5.9)).
7. Let the compact connected Lie group G act on itself by conjugation. Determine the principal orbit type. (Use Bröcker-tom Dieck [1985], Chapter IV.)
8. Show that a compact Lie group has only a countable number of conjugacy classes of closed subgroups. (Hint: There is a countable number of isomorphism classes of representations; (5.5), (5.11)).
9. Let the compact Lie group G act on the space X. The action is called **locally smooth** if for each $x \in X$ there exists a representation space V of $H = G_x$ and a G-homeomorphism $\varphi\colon G \times_H V \to X$ with $\varphi(g,0) = gx$ onto an open G-neighbourhood of x. If X is a locally smooth G-space and K a closed

subgroup of G, then X is locally smooth as a K-space. (Remark: Since there exist non-isomorphic but homeomorphic representations, the representation V is not uniquely determined by x; see Cappell-Shaneson [1979].)

6. Families of subgroups.

In this section we consider only compact Hausdorff groups acting on Hausdorff spaces. Subgroups are assumed to be closed.

The **orbit type** of a G-space X is the set of isomorphism classes of homogeneous spaces which are isomorphic to orbits. If this set is finite, we say X has **finite orbit type**. The **isotropy type** of X is the set of conjugacy classes of isotropy groups.

An **isotropy family** (or family for short) for the group G is a set $\mathfrak{F}$ of closed subgroups of G such that $H \in \mathfrak{F}$, $H \sim K$ implies $K \in \mathfrak{F}$.

The family $\mathfrak{F}$ is called **closed** if $H \in \mathfrak{F}$, $H \subset K$ implies $K \in \mathfrak{F}$ and **open** if $H \in \mathfrak{F}$, $K \subset H$ implies $K \in \mathfrak{F}$. If X is a G-space and $\mathfrak{F}$ a family, we let

(6.1) $X(\mathfrak{F}) = \{x \in X \mid G_x \in \mathfrak{F}\}.$

This is a G-subspace and a union of orbit bundles. A G-space X is called **$\mathfrak{F}$-trivial** if it admits a G-map $X \to G/H$, $H \in \mathfrak{F}$ and **$\mathfrak{F}$-locally trivial** if it has an open covering by $\mathfrak{F}$-trivial G-subspaces. Recall that an open G-subset U of X is called a tube about $x \in U$ if there exists a G-map $U \to G/G_x$. A G-space is called **strongly locally trivial** if it has a tube about each of its points.

(6.2) Proposition. *Let X be strongly locally trivial. If $\mathfrak{F}$ is a closed (resp. open) family, then $X(\mathfrak{F})$ is closed (resp. open) in X.*

Proof. Let $\mathfrak{F}$ be open, $x \in X(\mathfrak{F})$, and $f\colon U \to G/G_x$ a G-map from a G-neighbourhood U of x. Let $f(x) = gG_x$. If $y \in U$, then, by equivariance, $G_y \subset gG_xg^{-1}$. Hence all isotropy groups of points in U are subconjugate to $G_x \in \mathfrak{F}$ so that $U \subset X(\mathfrak{F})$. Let $\mathfrak{F}$ be closed and $\mathfrak{F}'$ the complement of $\mathfrak{F}$ in the set of all closed subgroups. Then $\mathfrak{F}'$ is open and $X(\mathfrak{F}) = X \backslash X(\mathfrak{F}')$ is closed in X. □

We point out that a G-map $f\colon X \to Y$ induces a G-map

(6.3) $f(\mathfrak{F})\colon X(\mathfrak{F}) \to Y(\mathfrak{F})$

for each closed $\mathfrak{F}$. Therefore, it is useful to look at the filtration of X defined by the subspaces $X(\mathfrak{F})$, $\mathfrak{F}$ closed. Of course, the term filtration in this context does not imply a total ordering of the subsets $X(\mathfrak{F})$ by inclusion. Proofs and constructions in transformation group theory often proceed via induction on the filtration $X(\mathfrak{F})$ by starting with the fixed point set and adding one orbit bundle at a time. Since the important paper of Conner and Floyd [1966], this has become a standard method.

A G-space X is called **$\mathfrak{F}$-numerable** if the following holds: There exists an open covering $U = (U_j \mid j \in J)$ of X by G-subspaces with the following properties:

(6.4) For each $j \in J$, there exists a G-map

$$f_j\colon U_j \to G/G_j, \quad G_j \in \mathfrak{F}.$$

(6.5) There exists a locally finite partition of unity $(t_j \mid j \in J)$ subordinate to U by G-functions $t_j\colon X \to [0,1]$.

If $f\colon X \to Y$ is a G-map and Y is $\mathfrak{F}$-numerable, then X is $\mathfrak{F}$-numerable.

Let G be a compact Lie group. We show that the homotopy category of $\mathfrak{F}$-numerable G-spaces has a terminal object $E(\mathfrak{F})$ (see tom Dieck [1972]).

(6.6) Theorem. *Let $\mathfrak{F}$ be a family of closed subgroups of G which contains the intersection of any two of its members. There exists an $\mathfrak{F}$-numerable G-space $E(\mathfrak{F})$ such that each $\mathfrak{F}$-numerable G-space X admits, up to G-homotopy, a unique G-map $X \to E(\mathfrak{F})$.*

The proof of (6.6) is contained in (6.11), (6.13), and (6.14).

The space $E(\mathfrak{F})$, which, by universality, is unique up to G-homotopy equivalence, is called the **classifying space for the family** $\mathfrak{F}$. It generalizes the classifying space for principal bundles as we shall explain in section 8. For simplicity, we use Milnor's construction of classifying spaces (Milnor [1956]).

Let $(X_j \mid j \in J)$ be a family of topological spaces X_j. The **join**

$$X = *_{j \in J} X_j$$

is defined to be the following space: Elements of X are represented by J-tuples

$$(t_j x_j \mid j \in J), \quad t_j \in [0,1], \quad x_j \in X_j, \quad \textstyle\sum t_j = 1,$$

with only finitely many t_j non-zero. The elements $(t_j x_j)$ and $(u_j y_j)$ define the same element of X if and only if:

(i) for all $j \in J$: $t_j = u_j$

(ii) for all $j \in J$: $t_j \neq 0$ implies $x_j = y_j$.

One has coordinate maps

(6.7) $\quad t_j\colon X \to [0,1], \ (t_i x_i) \mapsto t_j$

and

(6.8) $\quad p_j\colon t_j^{-1}]0,1] \to X_j, \ (t_i x_i) \mapsto x_j.$

The topology on X is the coarsest topology which makes the maps t_j and

p_j continuous. This topology is characterized by the following property: A map $f\colon Y \to X$ is continuous if and only if the maps $t_j f\colon Y \to [0,1]$ and $p_j f\colon f^{-1} t_j^{-1}]0,1] \to X_j$ are continuous. Thus, if the X_j are G-spaces, then $(g, \sum t_j x_j) \mapsto (t_j g x_j)$ defines a continuous G-action on the join.

The sets $V_j = t_j^{-1}]0,1]$, $j \in J$, form an open covering of X. The functions t_j are a point-finite partition of unity, i.e. for each $x \in X$, the set $\{j \in J \mid t_j(x) \neq 0\}$ is finite and $\sum t_j(x) = 1$. We quote the following Lemma from Dold [1972], Appendix A2, Proposition 2.8.

(6.9) Lemma. *If $\pi = (\pi_j \mid j \in J)$ is a partition of unity (not necessarily point-finite), then there exists a locally finite partition of unity $\varrho = (\varrho_j \mid j \in J)$ such that $\varrho_j^{-1}]0,1] \subset \pi_j^{-1}]0,1]$ for all $j \in J$. If the functions π_j are G-invariant, then the ϱ_j can be chosen to be G-invariant.* □

Now let $\mathfrak{F}$ be a family as in (6.6) and let $(G/H_a \mid a \in A)$ be a family of homogeneous spaces such that each G/H, $H \in \mathfrak{F}$, is isomorphic to a space G/H_a. Let $X = \coprod_{a \in A} G/H_a$ be the topological sum of the G/H_a. We form

(6.10) $E(\mathfrak{F}) = X * X * X * \ldots,$

the join of a countable number of copies X. We have coordinate functions $t_j\colon E(\mathfrak{F}) \to [0,1]$ for $j = 1,2,3,\ldots$ and $p_j\colon V_j \to X$ with $V_j = t_j^{-1}]0,1]$. Let $V_{j,a}$ be $p_j^{-1}(G/H_a)$ so that $V_j = \coprod_{a \in A} V_{j,a}$. The $V_{j,a}$ are open G-sets in $E(\mathfrak{F})$ and, by construction, we have G-maps $V_{j,a} \to G/H_a$. Using these data and (6.9), the reader should verify (exercise 3):

(6.11) Lemma. *$E(\mathfrak{F})$ is an $\mathfrak{F}$-numerable G-space.* □

We claim that $E(\mathfrak{F})$ is a classifying space for the family $\mathfrak{F}$. The proof needs some preparation. First we develop some facts about partitions of unity.

Let $(U_b \mid b \in B)$ be an open covering of Z with subordinate partition of unity $(t_b \mid b \in B)$. For each finite subset $S \subset B$ let

$$U(S) = \{z \in Z \mid i \in S, j \in B \setminus S \Rightarrow t_i(z) > t_j(z)\}.$$

If $|S| = |T| = n$ and $U(S) \cap U(T) \neq \emptyset$, then $S = T$. We form $U_n = \coprod_{|S|=n} U(S)$.

(6.12) Lemma. *The U_n, $n = 1,2,3,\ldots$, are a numerable covering of Z.*

Proof. On $U(S)$, take the function

$$q_S(z) = \max(0, \min_{i \in S} t_i(z) - \max_{j \in J \setminus S} t_j(z))$$

and define $q_n\colon U_n \to [0, \infty[$ by $q_n | U(S) = q_S$. Then $v_n = q_n / \sum_1^\infty q_i$ is a numeration of (U_n). □

If for each U_b there is a G-map into some G/H_a, then there exists a G-map $\varphi_n\colon U_n \to X$. Using this, we can prove half of (6.6).

(6.13) Lemma. *Let Z be an $\mathfrak{F}$-numerable G-space. Then Z admits a G-map $Z \to E(\mathfrak{F})$.*

Proof. Using (6.12), we find a covering (U_n) with numeration v_n and G-maps $\varphi_n\colon U_n \to X$. The required G-map is given by

$$z \mapsto (v_1(z)\varphi_1(z), v_2(z)\varphi_2(z), \ldots).$$

If $\varphi_n(z)$ is not defined, then $v_n(z)\varphi_n(z)$ is to be interpreted as $0 \cdot x$, for any $x \in X$. □

The proof of (6.6) will be completed with the following proposition.

(6.14) Proposition. *Let X be a G-space and $E = X * X * \ldots$ the join of a countable number of copies X. Then any two G-maps $Z \to E$ are G-homotopic.*

Proof. Given two G-maps $f, g\colon Z \to E$, we look at coordinates

$$(t_1(x)f_1(x), t_2(x)f_2(x), \ldots) \text{ and } (u_1(x)g_1(x), u_2(x)g_2(x), \ldots)$$

of $f(x)$ and $g(x)$. We show that f and g are G-homotopic to maps with coordinates

(6.15) $$(t_1(x)f_1(x), 0, t_2(x)f_2(x), 0, \ldots)$$
$$(0, u_1(x)g_1(x), 0, u_2(x)g_2(x), \ldots)$$

where 0 denotes any element of the form $0 \cdot y$. If this is the case, then two maps of this form are homotopic by a homotopy

$$((1-t)t_1f_1, tu_1g_1, (1-t)t_2f_2, tu_2g_2, \ldots).$$

In order to achieve the form (6.15), one constructs a homotopy in an infinite number of steps. The first step is

$$(t_1f_1, tt_2f_2, (1-t)t_2f_2, tt_3f_3, (1-t)t_3f_3, \ldots)$$

which removes the first 0 in (6.15). Now iterate this procedure. One obtains a continuous homotopy because at each place only a finite number of homotopies are effective. □

(6.16) Remark. There are, of course, many other ways of constructing spaces $E(\mathfrak{F})$. The systematic way of constructing classifying spaces is to use the classifying space of a (topological) category as explained in Segal [1968a]. In the situation of (6.6), one takes the category with objects G/H, $H \in \mathfrak{F}$, and morphisms the G-maps $G/H \to G/K$. For a comparison of different constructions of classifying spaces, see tom Dieck [1974]. Also, Elmendorf [1983] can be used.

(6.17) Proposition. *Let $\mathfrak{F}$ be as in* (6.6) *and let X be an $\mathfrak{F}$-numerable G-space. Then $X * E(\mathfrak{F})$ is G-homotopy equivalent to $E(\mathfrak{F})$.*

Proof. The proof of (6.15) shows that any two G-maps $Z \to E(\mathfrak{F}) * X * X * \ldots = Y$ are homotopic. Therefore, Y can be used as a model for $E(\mathfrak{F})$. This yields the homotopy equivalences

$$E(\mathfrak{F}) * X \simeq Y * X \simeq Y \simeq E(\mathfrak{F}). \quad \square$$

Classifying spaces for families can be useful in different contexts. For proper actions of discrete groups, see Serre [1971a], Connolly-Koźniewski [1986].

Remark. The notation $E(\mathfrak{F})$ is slightly ambiguous because, given any G-space E, we could consider the subspace $E(\mathfrak{F})$ with isotropy in $\mathfrak{F}$. But, of course, $(E(\mathfrak{F}))(\mathfrak{F}) = E(\mathfrak{F})$.

(6.18) Exercises.

1. Prove or disprove: The join with Milnor's topology is associative.
2. Let X and Y be compact Hausdorff spaces. Let CX denote the cone over X, i.e. $CX = X \times [0,1]/X \times 0$. Define the join $X * Y$ as a subspace of $CX \times CY$.
3. Show that the covering $(V_{j,a})$ described before (6.11) is numerable.
4. Differentiable G-manifolds M with $\operatorname{Iso}(M) \subset \mathfrak{F}$ are $\mathfrak{F}$-numerable.
5. For $H \in \mathfrak{F}$, the space $E(\mathfrak{F})^H$ is contractible.
6. Let $\mathfrak{F}$ be a closed family for G. For a subgroup $L \subset G$, put $\mathfrak{F} \cap L = \{L \cap H \mid H \in \mathfrak{F}\}$. Then $\mathfrak{F} \cap L$ is a closed family for L. If X is an $\mathfrak{F}$-numerable G-space, then $\operatorname{res}^G_L X$ is an $(\mathfrak{F} \cap L)$-numerable L-space. Moreover, $\operatorname{res}^G_L E(\mathfrak{F}) = E(\mathfrak{F} \cap L)$.
7. Let E be a G-complex (see Chapter II) with $\operatorname{Iso}(E) \subset \mathfrak{F}$, $\mathfrak{F}$ closed. Suppose E^H is contractible for all $H \in \mathfrak{F}$. Then E is G-homotopy equivalent to $E(\mathfrak{F})$.

7. Equivariant maps.

Let X and Y be G-spaces. For a G-map $f: X \to Y$, we have $G_x \subset G_{fx}$ for all $x \in X$. Therefore, we consider the subspace

$$I(X, Y) = \{(x, y) \mid G_x \subset G_y\} \subset X \times Y.$$

This is a G-subspace of $X \times Y$ with diagonal action. The orbit space is denoted by $M(X, Y) = I(X, Y)/G$. The projection $X \times Y \to X$ induces

$$q\colon\ M(X, Y) \to X/G$$

by passing to orbit spaces. A G-map $f\colon X \to Y$ induces $X \to I(X, Y)$, $x \mapsto (x, fx)$ and passage to orbit spaces yields

$$s_f\colon\ X/G \to M(X, Y).$$

This map s_f is a section of q, i.e. $qs_f = \mathrm{id}$. We want to show that, under certain conditions, sections of q correspond to equivariant maps. The consideration of the map q in this context is due to G. Segal. For our present purposes, we use the following definition.

A pair (X, Y) of G-spaces is called **admissible** if the diagram

(7.1)
$$\begin{array}{ccc} I(X, Y) & \xrightarrow{\ Q\ } & X \\ {\scriptstyle P}\downarrow & & \downarrow{\scriptstyle p} \\ M(X, Y) & \xrightarrow[\ q\ ]{} & X/G \end{array}$$

(which defines q) is a pull-back.

The relevance of this property is seen from

(7.2) Proposition. *Let (X, Y) be admissible. Then the assignment $f \mapsto s_f$ is a bijective correspondence between G-maps $X \to Y$ and sections of q. Two G-maps $f_0, f_1\colon X \to Y$ are G-homotopic if and only if the corresponding sections are homotopic as sections.*

Proof. Let $s\colon X/G \to M(X, Y)$ be a section. The pull-back diagram of (7.1) provides us with an induced section $t\colon X \to I(X, Y)$. We compose t with the projection $I(X, Y) \to Y$ and obtain a map $f_s\colon X \to Y$. This map is equivariant: $Qt(gx) = gx = gQt(x) = Q(gt(x))$ because t is a section and Q a G-map. Furthermore, $Pt(gx) = sp(gx) = sp(x)$ because t is an induced section and p an orbit map, and $P(gtx) = Pt(x) = sp(x)$ because P is an orbit map. Thus $gt(x)$ and $t(gx)$ have the same image under P and Q, hence they are equal. The assignment $s \mapsto f_s$ is easily checked to be inverse to $f \mapsto s_f$. A G-homotopy $H\colon X \times I \to Y$ induces a section $s_H\colon (X \times I)/G \to M(X \times I, Y)$. The map $M(X \times I, Y) \to (X \times I)/G$ is canonically homeomorphic to $q \times \mathrm{id}_I\colon M(X, Y) \times I \to X/G \times I$ (exercise) and s_H is transformed by this homeomorphism into a homotopy of sections of q. The pair $(X \times I, Y)$ is again admissible (exercise). □

We now look for conditions which imply admissibility of (X, Y).

(7.3) Proposition. *Let G be compact and suppose X and Y are Hausdorff spaces. Then the pair (X, Y) is admissible.*

Proof. We have to show that (7.1) is a pull-back. Let $Z \to M(X, Y)$ be the pull-back of p along q. The universal property of the pull-back yields a G-map $I(X, Y) \to Z$ over $M(X, Y)$. This map is bijective on fibres and hence bijective. In each pull-back diagram

$$\begin{array}{ccc} C & \longrightarrow & X \\ {\scriptstyle r}\downarrow & & \downarrow{\scriptstyle p} \\ B & \longrightarrow & X/G \end{array}$$

the map r is canonically homeomorphic to the orbit map $C \to C/G$ (3.25, Ex. 14). Since X and Y are Hausdorff spaces, the spaces $I(X, Y)$ and Z are Hausdorff spaces and so are their orbit spaces (3.1, v). Moreover, the orbit maps are proper (3.6). Now one uses the following result from general topology (Bourbaki [1961a], I.10.5): If gf is proper and the range of f is separated, then f is proper. Consequently, $I(X, Y) \to Z$ is a proper bijective map and therefore a homeomorphism. □

Another useful special case arises when the action on X is free. Here, $I(X, Y) = X \times Y$ and the action is again free. In the next section we deal with the case that $X \to X/G$ is a G-principal bundle. Then any pair (X, Y) is admissible (exercise 3).

Proposition (7.2) can be used to study equivariant maps using methods from algebraic topology, in particular obstruction theory.

In the previous section we have mentioned the important principle of induction over orbit types. This reduces many problems to (relatively) free group actions. We describe another instance of this principle.

Let $\mathfrak{F}$ be a closed family and let X be a strongly locally trivial G-space (see section 6). Let H be maximal among the isotropy groups of X not in $\mathfrak{F}$ and let $\mathfrak{F}'$ be the smallest closed family containing $\mathfrak{F}$ and (H).

Suppose we are given a G-map $f\colon X(\mathfrak{F}) \to Y(\mathfrak{F})$. We ask for extensions $F\colon X(\mathfrak{F}') \to Y(\mathfrak{F}')$ of f.

(7.4) Proposition. *The extensions F of f correspond bijectively to the NH/H-extensions $h\colon X(\mathfrak{F}')^H \to Y(\mathfrak{F}')^H$ of f^H.*

Proof. Suppose F_1 and F_2 are extensions which agree on $X(\mathfrak{F}')^H$. Then they agree on $X(\mathfrak{F})$ and $(X(\mathfrak{F}')\backslash X(\mathfrak{F}))^H = X^H_{(H)} = X_H$. But $X_{(H)} = GX_H$. By equivariance, they agree on $X_{(H)}$. Our assumption on H implies that $X(\mathfrak{F}') = X_{(H)} \cup X(\mathfrak{F})$.

Now let an NH/H-map h be given which extends f^H. We define

$$E\colon X(\mathfrak{F}') \to Y(\mathfrak{F}')$$

as

$$E(x) = f(x), \quad x \in X(\mathfrak{F})$$
$$E(x) = gh(y), \quad x = gy, \quad y \in X(\mathfrak{F}')^H.$$

We have to show that E is well-defined, continuous, and equivariant. Suppose $x = g_1 y_1$ with $y_1 \in X(\mathfrak{F})^H$. Then $g_1 h(y_1) = g_1 f(y_1) = f(g_1 y_1) = f(x)$. If $x = g_1 y_1 = g_2 y_2$ and $y_1, y_2 \in X_H$, then $g_1 = g_2 n$ with suitable $n \in NH$. We conclude that

$$g_1 h(y_1) = g_2 nh(y_1) = g_2 h(ny_1) = g_2 h(y_2)$$

because h is NH-equivariant. Therefore, E is well-defined. It is continuous because it is continuous on the closed subsets $X(\mathfrak{F})$ and $GX(\mathfrak{F}')^H$ (see (3.1), (3.9), (6.2)). Equivariance is easy. □

Let $p\colon E \to B$ be a G-map. It is called a **G-fibration** if the following holds: Given a commutative diagram

$$\begin{array}{ccccc} x & & X & \xrightarrow{\ f\ } & E \\ \downarrow & & \downarrow{\scriptstyle i_0} & & \downarrow{\scriptstyle p} \\ (x, 0) & & X \times I & \xrightarrow{\ \varphi\ } & B \end{array}$$

of G-maps, there exists a G-map $\phi\colon X \times I \to E$ such that $p\phi = \varphi$ and $\phi i_0 = f$. The elementary theory of fibrations (see e.g. tom Dieck – Kamps – Puppe [1970]) easily extends to G-fibrations. The exercises provide a sample collection of relevant facts.

(7.5) Exercises.

1. Verify the statements in the proof of (7.2) that were left as an exercise.
2. Let G be a compact Lie group. Show that the fibre of $M(X, Y) \to X/G$ over the orbit Gx is homeomorphic to Y^H, $H = G_x$.
3. Let X be a principal G-bundle. Show that any pair (X, Y) is admissible. One has $M(X, Y) = (X \times Y)/G$ and $q\colon M(X, Y) \to X/G$ is a fibre bundle with fibre Y and structure group G. (See section 8).
4. Let $p\colon E \to B$ be a G-fibration and $H \subset G$ a subgroup.
   (i) $p^H\colon E^H \to B^H$ is an ordinary fibration.
   (ii) $p\colon E \to B$ is an H-fibration.
   (iii) The restriction of p, $p^{-1}(B^H) \to B^H$, is an NH-fibration.
5. Let $p\colon E \to B$ be a G-map and G a compact group. Let $(V_j \mid j \in J)$ be a

numerable covering of B by open G-sets. Assume that the restrictions of p, $p^{-1}(V_j) \to V_j$, are G-fibrations. Then p is a G-fibration.

6. Let G be a compact Lie group and E a completely regular G-space with orbits only of type G/H. Suppose E/G is paracompact. Then $E \to E/G$ is a G-fibration.
7. Let H be a closed subgroup of the compact Lie group. Any G-map $p: E \to G/H$ is a G-fibration.
8. Let G be a compact Lie group and H a closed subgroup. Let $p: E \to B$ be a G-map and suppose B has only orbits of type G/H. Suppose the orbit map $B^H \to B^H/NH$ is a numerable bundle. Then p is a G-fibration if and only if $p: p^{-1}B^H \to B^H$ is an NH-fibration.
9. Let G be a compact Lie group. Let $p: E \to B$ and $p': E' \to B$ be G-maps and let $f: p \to p'$ be a G-map over B.
   Let $(V_j | j \in J)$ be a numerable covering of B by open G-sets. Let $f_j: p^{-1}V_j \to (p')^{-1}V_j$ be the map induced by f. Suppose f_j is a G-fibre-homotopy equivalence for each j. Then f is a G-fibre-homotopy equivalence.

8. Bundles.

We recall that an action of the topological group G on a space X is called free if $gx = x$ always implies $g = e$ or, in other words, if $\mathrm{Iso}(X)$ consists of the trivial group $\{e\}$.

The principle of induction on the orbit type leads to a situation (X, A) with free action on $X \backslash A$. Therefore, free actions play a basic role. There is also a well established theory of free group actions in algebraic topology, in the guise of bundle theory. Since every now and then one has to work with bundles, some basic material is collected in this section. Suppose U is a G-space and $f: U \to G$ a G-map into the left translation space G. Then, by (4.4), U is G-homeomorphic to $G \times V$, $V = f^{-1}(e)$. One concludes that $V \cong U/G$ and $U \cong G \times U/G$, the latter being a G-homeomorphism. This has the following consequence.

Let E be a strongly locally trivial free G-space (in the sense of section 6) and let $p: E \to E/G = B$ be the canonical projection. Then each point $b \in B$ has an open neighbourhood V such that $p^{-1}(V)$ is G-homeomorphic over V to $G \times V$. This situation is axiomatized in the notion of a G-principal bundle. It is customary to consider right actions in this context.

A ***G*-principal bundle** consists of a free right G-action $E \times G \to E$ and a surjective continuous map $p: E \to B$ such that

(8.1) $p(eg) = p(e)$ for all $e \in E$, $g \in G$.

(8.2) For each $b \in B$, there exists an open neighbourhood V of b in B and a G-homeomorphism $\varphi: p^{-1}(V) \to V \times G$ such that

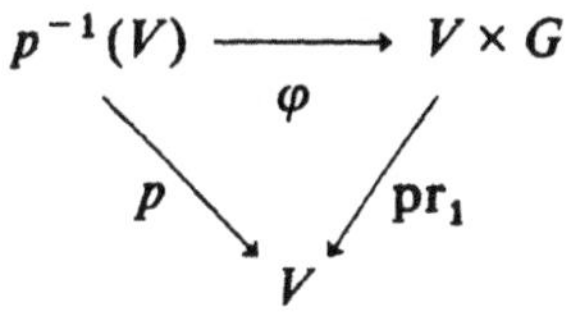

is commutative.

The space E is called the **total space** of the bundle, B is called the **base space** and φ a **trivialisation over** V.

The map p induces a homeomorphism $E/G \to B$ because p is an open mapping. A bundle is called **trivial** if there exists a trivialisation over B.

Let $E \times G \to E$ be a free action. Put $C(E) = \{(x, xg) | x \in E, g \in G\}$ and define τ_E: $C(E) \to G, (x, xg) \mapsto g$. We call τ_E the **translation function** of the G-space E (compare with (3.20)).

(8.3) Proposition. *Suppose E is locally trivial. Then τ_E is continuous.*

Proof. Let U be a G-invariant neighbourhood of $x \in E$ which admits a trivialisation ψ: $V \times G \to U$. The pre-image of $(U \times U) \cap C(E)$ under $\psi \times \psi$ is $\{(v, g, v, h) | v \in V, g, h \in G\}$, and $\tau_E(\psi \times \psi)$ is the mapping $(v, g, v, h) \mapsto g^{-1}h$ which is continuous. □

Now let us consider the following situation: Given a commutative diagram over a trivial G-space B

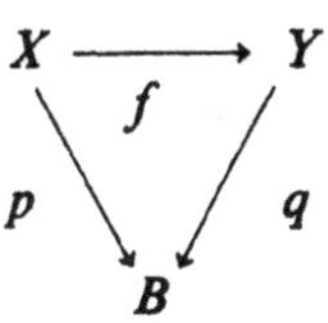

with free G-spaces X and Y, a G-map f, and maps p and q which are assumed to induce homeomorphisms $X/G \cong B \cong Y/G$. Then we have

(8.4) Proposition. *Suppose τ_Y is continuous. Then f is a G-homeomorphism.*

Proof. Since f is bijective on orbits and fibres, f is bijective. It remains to be shown that f^{-1} is continuous. Let $f(x) = y$. Let U be an open neighbourhood of x and choose open neighbourhoods V of x and N of e such that $VN \subset U$. Since τ_Y is continuous, there exists an open neighbourhood W of y such that $\tau_Y((W \times W) \cap C(Y)) \subset N$. Since f is continuous, $V_1 = V \cap f^{-1}(W)$ is an open neighbourhood of x and $f(V_1) \subset W$. The set $p(V_1)$ is an open neighbourhood of $p(x)$ because p is an open map. Let $W_1 = W \cap q^{-1}pV_1$. Then W_1 is an open neighbourhood of y and $q(W_1) = p(V_1), f(V_1) \subset W_1$. For $z \in W_1$, choose $u \in V_1$ such that $q(z) = p(u)$. Then $f(u) \in W$ and $z \in W$. Any $g \in G$ with $f(u)g = z$ is contained in N and therefore $z = f(u)g = f(ug)$ with $f^{-1}(z) = ug \in V_1 N \subset U$. Hence $f^{-1}(W_1) \subset U$ and we have shown continuity of f^{-1} in z. □

Let p: $X \to B$ be a G-principal bundle and f: $C \to B$ a continuous map. Let

(8.5)
$$\begin{array}{ccc} Y & \xrightarrow{F} & X \\ {\scriptstyle q}\downarrow & & \downarrow{\scriptstyle p} \\ C & \xrightarrow[f]{} & B \end{array}$$

be a pull-back. We can consider Y as the subspace $\{(c, x) \mid fc = px\}$ of $C \times X$. It carries the induced G-action. If p is trivial over $V \subset B$, then q is trivial over $f^{-1}(V)$. Hence q is a G-principal bundle. It is called the bundle **induced** by f from p, for short: the **induced bundle**.

Conversely, assume a commutative diagram (8.5) is given with G-principal bundles q and p and G-map F. Then F is called a **bundle map**. Let q^*: $X^* \to C$ be the bundle induced from p by f. The universal property of the pull-back yields a map $Y \to X^*$ over C which, by (8.3) and (8.4), is a G-homeomorphism. Hence we have

(8.6) Proposition. *If F: $Y \to X$ is a bundle map, then Y is canonically isomorphic to the induced bundle X^*.* □

We now consider G-principal bundles together with an automorphism group Γ of bundle maps.

Let Γ be a compact Lie group, G a topological group, and α: $\Gamma \to \mathrm{Aut}(G)$ a homomorphism from Γ into the automorphism group of G. We denote $\alpha(\gamma)$ also by α_γ and require the map $\Gamma \times G \to G$, $(\gamma, g) \mapsto \alpha_\gamma(g)$ to be continuous.

(8.7) A (Γ, α, G)-bundle consists of a locally trivial G-principal bundle p: $E \to B$ together with left Γ-actions on E and B such that the following holds:
(i) p is Γ-equivariant.
(ii) For $\gamma \in \Gamma$, $g \in G$, and $e \in E$, the relation $\gamma(eg) = (\gamma e)\alpha_\gamma(g)$ holds.

If α is trivial, then Γ acts as a group of bundle automorphisms and, in this case, we talk about (Γ, G)-bundles or **Γ-equivariant G-principal bundles**. A **bundle map** between (Γ, α, G)-bundles is a bundle map of the underlying G-bundles which is also Γ-equivariant. The data Γ, α, and G give rise to a semi-direct product $\Gamma \times_\alpha G$: The topological product $\Gamma \times G$ carries the multiplication

$$(\gamma, g)(\gamma', g') = (\gamma\gamma', \alpha_\gamma g' \cdot g).$$

The topological group $\Gamma \times_\alpha G$ acts from the left on the total space E of a (Γ, α, G)-bundle

$$((\gamma, g), e) \mapsto (\gamma e)g.$$

(8.8) Example. Let $\alpha: \mathbb{Z}/2 \to \operatorname{Aut} \mathrm{U}(n)$ be the homomorphism which assigns to the non-trivial element of $\mathbb{Z}/2$ the passage to the complex-conjugate matrix. The $(\mathbb{Z}/2, \alpha, \mathrm{U}(n))$-bundles are the principal bundles which correspond to real vector bundles in the sense of Atiyah [1966].

The (Γ, α, G)-bundles are by definition locally trivial as G-bundles. We also want to define locally trivial (Γ, α, G)-bundles. As preparation, we analyse (Γ, α, G)-bundles with Hausdorff total space over orbits Γ/Λ: For the purpose of this section, such bundles are called **local objects**.

Let $p: E \to \Gamma/\Lambda$ be a local object and $x \in E$ such that $px = \Lambda$. Let $H_x \subset \Gamma \times_\alpha G$ be the isotropy group in x. This is a closed subgroup since E is a Hausdorff space. We consider H_x as a subspace of $\Gamma \times G$. If $(\gamma, g) \in H_x$, then $\gamma \in \Lambda$ since $(\gamma x)g = x$ implies $\Lambda = px = p((\gamma x)g) = \gamma\Lambda$. If (γ, g) and (γ, h) are in H_x, then $g = h$ because G acts freely on E. Therefore, H_x is the graph of a map $t: \Lambda \to G$; it satisfies $t(\lambda\lambda') = \alpha_\lambda(t\lambda')t\lambda$. If α is trivial, t is an antihomomorphism. The map t is continuous because, by assumption, p is a locally trivial G-principal bundle. The map $f: (\Gamma \times_\alpha G)/H_x \to E$, $u \mapsto ux$ is $\Gamma \times_\alpha G$-equivariant. In virtue of the next lemma it is seen that f is a homeomorphism.

(8.9) Lemma. *Suppose the closed subgroup H of $\Gamma \times_\alpha G$ is graph of a continuous map $t: \Lambda \to G$. Then the canonical quotient map $r: (\Gamma \times_\alpha G)/H \to \Gamma/\Lambda$, considered as G-principal bundle, is locally trivial.*

Proof. We use the basic fact that $q: \Gamma \to \Gamma/\Lambda$ has a local section. Let $s: U \to \Gamma$ be a section of q over the open subset $U \subset \Gamma/\Lambda$. We have the following diagram

$$\begin{array}{ccc}
q^{-1}U \times G & \underset{\psi}{\overset{\phi}{\rightleftarrows}} & U \times \Lambda \times G \\
\Big\downarrow c & & \Big\downarrow b \\
r^{-1}U & \underset{\varphi}{\longleftarrow} & U \times G
\end{array}$$

with maps $b(u, \lambda, g) = (u, g)$, $\varphi(u, g) = a(su, g)$, c restriction of $\Gamma \times G \to (\Gamma \times_\alpha G)/H$ and

$$\phi(u, \lambda, g) = (su \cdot \lambda, \alpha_{s(u)}(t\lambda) \cdot g)$$
$$\psi(x, y) = (qx, (sqx)^{-1}x, \alpha_{sqx}^{-1} t((sqx)^{-1}x) \cdot y).$$

One has $c\phi = \varphi b$. Moreover, $\phi\psi = \mathrm{id}$ and $\psi\phi = \mathrm{id}$. Therefore, φ is a G-equivariant homeomorphism over U. □

A (Γ, α, G)-bundle $p\colon E \to B$ is called **locally trivial** if B admits an open covering $\mathfrak{U} = (U_j | j \in J)$ by Γ-sets U_j such that each restriction $p^{-1}U_j \to U_j$ admits a (Γ, α, G)-bundle map into a local object. (This local object is, of course, not uniquely determined.) If, moreover, $\mathfrak{U}$ is numerable, then we call the (Γ, α, G)-bundle **numerable**.

(8.10) Proposition. *Suppose G is also a compact Lie group. Let $p\colon E \to B$ be a (Γ, α, G)-bundle with completely regular total space E. Then p is locally trivial as (Γ, α, G)-bundle.*

Proof. Let $x \in E$ be given. By (5.7), there exists an open $(\Gamma \times_\alpha G)$-neighbourhood U of x and an equivariant map $f\colon U \to (\Gamma \times_\alpha G)/H_x$. The map f yields the required bundle map into a local object. □

Our next task is the construction of universal (Γ, α, G)-bundles. Let $(p_j\colon X_j \to Y_j | j \in J)$ be a family of local objects such that every local object is isomorphic to one in this family. We set $L = \coprod_{j \in J} X_j$ and form the countable join

$$E(\Gamma, \alpha, G) = L * L * L * \ldots$$

as in section 6. Since L is a $\Gamma \times_\alpha G$-space, so is $E(\Gamma, \alpha, G)$. We set $B(\Gamma, \alpha, G) = E(\Gamma, \alpha, G)/G$ and obtain the quotient map

(8.11) $\quad p\colon E(\Gamma, \alpha, G) \to B(\Gamma, \alpha, G)$

which is a Γ-map.

(8.12) Theorem.
(i) *p is a numerable (Γ, α, G)-bundle.*
(ii) *Each numerable (Γ, α, G)-bundle admits a bundle-map into p. Any two such bundle maps are homotopic as (Γ, α, G)-bundle maps.*

Proof. The join $L * L * L * \ldots$ has coordinate maps $t_n\colon E(\Gamma, \alpha, G) \to [0, 1]$, $n = 1, 2, \ldots$, as in (6.7), (6.8). Let $V_n = t_n^{-1}]0, 1]$. There exists a $\Gamma \times_\alpha G$-map $f_n\colon V_n \to L$ with components $f_{n,j}\colon V_{n,j} \to X_j$, say. The image $V_{n,j}/G$ of $V_{n,j}$ in $B(\Gamma, \alpha, G)$ is open and the existence of $f_{n,j}$ shows p to be locally trivial.
The functions $(t_n | n = 1, 2, \ldots)$ are a partition of unity for the covering $(V_n | n = 1, 2, \ldots)$. The functions $t_{n,j} = t_n | V_{n,j}$ are a partition of unity for the covering $(V_{n,j} | n = 1, 2, \ldots; j \in J)$. Since these functions are G-invariant, they yield a partition of unity for $(V_{n,j}/G)$. This shows that p is numerable (compare also Lemma (6.11)).

Let $q\colon E \to B$ be any numerable (Γ, α, G)-bundle. There exists a numerable covering $(U_n | n = 1, 2, \ldots)$ of B by open Γ-sets and a corresponding numeration (v_n) by Γ-invariant functions v_n such that $q\colon q^{-1}U_n \to U_n$ admits a bundle map

$\varphi_n: q^{-1}U_n \to L$; see (6.12). A bundle map $E \to E(\Gamma, \alpha, G)$ is then given by $f(e) = (v_1(qe)\varphi_1(e), v_2(qe)\varphi_2(e), \dots)$. That any two bundle maps $E \to E(\Gamma, \alpha, G)$ are homotopic is shown as in Lemma (6.14). □

A basic property of induced bundles is given by the homotopy theorem: homotopic maps induce equivalent (i.e. isomorphic) bundles. The proof of this theorem is based on a structure theorem for bundles over $B \times [0,1]$ along the lines of Milnor [1964], Lemma (6.9).

(8.13) Proposition. *Let $p: E \to B \times [0,1]$ be a numerable (Γ, α, G)-bundle. Then there exists a bundle map $E \to E$ over $r: B \times [0,1] \to B \times [0,1]$, $(b,t) \mapsto (b,1)$.*

Proof. Suppose first that p is induced by $f: B \times [0,1] \to \Gamma/\Lambda$ from a local object. Then the space B has the form $\Gamma \times_\Lambda B_0$, by (4.4), where $B_0 = f_0^{-1}(e\Lambda)$, and the whole bundle has the form $\Gamma \times_\Lambda p_0: \Gamma \times_\Lambda E_0 \to \Gamma \times_\Lambda (B_0 \times [0,1])$ where p_0 is the restriction of p to $B_0 \times [0,1] \subset B \times [0,1]$. If $\Gamma = \Lambda$, then the bundle is induced from a bundle over a point and the assertion is obvious. If $\Lambda \neq \Gamma$, then we can assume by induction on Γ that the assertion is true.

In order to handle the general case, one uses the following lemma (see Dold [1972], Appendix Proposition 2.19).

(8.14) Lemma. *If $\mathfrak{W}$ is a numerable covering of $B \times [0,1]$, there exists a numerable covering $\mathfrak{U}$ of B and a function $r: \mathfrak{U} \to \mathbb{Z}$ with values $rU > 1$ such that every set*

$$U \times \left[\frac{i-1}{rU}, \frac{i+1}{rU}\right], \quad U \in \mathfrak{U}, \ i \in \mathbb{Z}, \ 0 < i < rU$$

is contained in some $W \in \mathfrak{W}$. □

Using this, one shows that there exists a numerable covering $\mathfrak{U}$ of B such that for $U \in \mathfrak{U}$ the bundle p over $U \times [0,1]$ is isomorphic to $(E | U \times 0) \times [0,1]$. Thus, over $U \times [0,1]$, the bundle admits bundle maps over any map $U \times [0,1] \to U \times [0,1]$, $(u,t) \mapsto (u, \alpha(u,t))$. Now one expresses r as a „composition" of such maps as in Milnor [1964], p. 69. □

As a corollary of (8.13), one obtains in the usual manner using (8.6).

(8.15) Theorem. *Let $p: E \to B$ be a numerable (Γ, α, G)-bundle and $f_t: B' \to B$ a Γ-homotopy. Then the induced bundles $f_0^* p$ and $f_1^* p$ are isomorphic as (Γ, α, G)-bundles.* □

A numerable (Γ, α, G)-bundle p for which (8.12, ii) is true is called a **universal bundle**. Using (8.15), one sees that for a universal bundle

$E(\Gamma, \alpha, G) \to B(\Gamma, \alpha, G)$ and any G-space X the homotopy set $[X, B(\Gamma, \alpha, G)]_\Gamma$ is canonically isomorphic to the set of isomorphism classes of numerable (Γ, α, G)-bundles over X. The space $B(\Gamma, \alpha, G)$ is called the **classifying space** for (Γ, α, G)-bundles. If α is trivial, we write $B(\Gamma, G)$, if Γ is trivial, we write BG. The latter space is the classifying space for principal G-bundles. If $E \to B$ is a principal G-bundle and F a G-space, then the bundle $E \times_G F \to B$ is called a **fibre bundle** with **typical fibre** F and **structure group** G.

In general, a (Γ, α, G)-bundle $p: E \to B$ is called **universal** if it is numerable and if each numerable (Γ, α, G)-bundle $p': E' \to B'$ has, up to homotopy, a unique (Γ, α, G)-bundle map into $E \to B$. Any map $B' \to B$ inducing p' from p (up to isomorphism over B') is called a **classifying map** for p'.

If $\mathfrak{B}(\Gamma, \alpha, G)(C)$ is the set of isomorphism classes (over C) of numerable (Γ, α, G)-bundles over C, one obtains a bijection

$$\mathfrak{B}(\Gamma, \alpha, G)(C) \to [C, B(\Gamma, \alpha, G)]_\Gamma$$

by mapping a bundle to its classifying map. The classifying spaces $B(\Gamma, \alpha, G)$ therefore represent the functor $C \mapsto \mathfrak{B}(\Gamma, \alpha, G)(C)$, defined on the homotopy category of Γ-spaces.

References for the classification of equivariant bundles are Bierstone [1973], tom Dieck [1969], Lashof [1982]. The classification of bundles has been generalized to the classification of G-spaces in Palais [1960]; see also Bredon [1972]. For the more subtle classification of differentiable G-spaces, see Bierstone [1975] and G. Schwarz [1980].

Consider the bundle $EG \to BG$ as (Γ, G)-bundle with trivial Γ-action. It is numerable and therefore has a classifying map $j: BG \to B(\Gamma, G)$. Let $J = \mathrm{id} \times_\Gamma j: E\Gamma \times_\Gamma BG \to E\Gamma \times_\Gamma B(\Gamma, G)$. Note that $E\Gamma \times_\Gamma BG = B\Gamma \times BG$.

(8.15.1) Proposition. *The maps j and J are homotopy equivalences.*

Proof. Forgetting the Γ-structure, $E(\Gamma, G) \to B(\Gamma, G)$ is a universal G-principal bundle (Ex. 6). Since j induces a universal bundle from another one, it must be a homotopy equivalence. The assertion about J is a special case of the next proposition. □

(8.16) Proposition. *Let $E \to B$ be a numerable principal G-bundle. Let the G-map $f: X \to Y$ be an ordinary homotopy equivalence. Then*

$$\mathrm{id} \times_G f: E \times_G X \to E \times_G Y$$

is a homotopy equivalence over B.

Proof. This follows from general facts of fibre bundle theory: A fibrewise map which is a homotopy equivalence on each fibre is a fibre-homotopy equivalence provided the fibrations involved are locally trivial over a numerable covering. See Dold [1963], Theorem 6.3; tom Dieck-Kamps-Puppe [1970], (9.2). □

Let $h^*(-)$ be a cohomology theory which is defined on all spaces, e.g. singular cohomology. Elements of $h^*(BG)$ are called **universal characteristic classes** (with values in $h^*(-)$) for G-principal bundles. Given $x \in h^*(BG)$ and a classifying map $f\colon B \to BG$ for a principal bundle $p\colon E \to B$, then $f^*(x) \in h^*(B)$ is called a **characteristic class** for p of type x. See Milnor-Stasheff [1974] for the most important cases $G = \mathrm{O}(n)$, $\mathrm{U}(n)$ and ordinary cohomology. A similar device can be used for (Γ, G)-bundles provided one uses a type of equivariant cohomology theory defined for Γ-spaces. But employing Proposition (8.15.1), a more simple minded definition of characteristic classes for (Γ, G)-bundles is possible. Using the homotopy equivalence $E\Gamma \times_\Gamma B(\Gamma, G) \simeq B\Gamma \times BG$, one can define universal characteristic classes in $h^*(B\Gamma \times BG)$.

Suppose, in particular, that $p\colon V \to X$ is a numerable (Γ, G)-bundle over a free Γ-space X such that the quotient map $q\colon X \to X/\Gamma$ is numerable. Then $p/\Gamma\colon V/\Gamma \to X/\Gamma$ is a numerable G-principal bundle (Ex. 9). Let $k\colon X \to B(\Gamma, G)$, $k'\colon X/\Gamma \to BG$, and $(L, l)\colon (X, X/\Gamma) \to (E\Gamma, B\Gamma)$ be classifying maps for p, p/Γ, and q, respectively. The orbit map $r\colon E\Gamma \times_\Gamma X \to X/\Gamma$ of the projection $E\Gamma \times X \to X$ is a homotopy equivalence because r is a numerable fibre bundle with contractible fibre. A homotopy inverse to r is given by $s\colon X/\Gamma \to E\Gamma \times_\Gamma X$, $x \mapsto (Lx, x) \bmod \Gamma$.

(8.17) Proposition. *The following diagram is commutative up to homotopy*

$$\begin{array}{ccc} E\Gamma \times_\Gamma X & \xrightarrow[\mathrm{id} \times_\Gamma k]{} & E\Gamma \times_\Gamma B(\Gamma, G) \\ \downarrow{\scriptstyle r} & & \downarrow{\scriptstyle J} \\ X/\Gamma & \xrightarrow[{[l, k']}]{} & B\Gamma \times BG. \end{array}$$

Proof. The map $jk'q$ is Γ-homotopic to k because both maps induce the same (Γ, G)-bundle over X. Therefore, $\mathrm{id} \times_\Gamma jk'q$ is homotopic to $\mathrm{id} \times_\Gamma k$. If we compose $\mathrm{id} \times_\Gamma k'q$ with s, we obtain $[l, k']$. Since s and r are inverse equivalences, the assertion follows. □

As a corollary we see that the characteristic classes in $h^*(E\Gamma \times_\Gamma X)$ coming from $h^*(B\Gamma \times BG)$ may be identified with characteristic classes in $h^*(X/\Gamma)$ induced by the two classifying maps l and k'.

The functor $X \mapsto EG \times_G X$ from G-spaces to spaces has several important applications. The functor converts a transformation group into the fibre bundle $EG \times_G X \to BG$. One can apply methods from algebraic topology in order to analyse this fibre bundle. The chapter on cohomology and localisation presents some applications of this idea. Right now, we develop the elementary homotopy theory of this construction. We use the following categories:

G-Top : category of G-spaces
G-Top^0: category of pointed G-spaces
G-Bun: full subcategory of G-Top of objects E such that $E \to E/G$ is a numerable principal G-bundle.

We use the functors which are defined on objects as follows:

$$\begin{aligned} Q&: G\text{-Bun} \to \text{Top}, \quad X \mapsto X/G \\ P&: G\text{-Top} \to \text{Top}, \quad Y \mapsto EG \times_G Y \\ P^0&: G\text{-Top}^0 \to \text{Top}^0, \quad Z \mapsto EG^+ \wedge_G Z. \end{aligned}$$

(8.18) Proposition.
(i) *Let $F\colon X \to Y$ be a map in* G-Bun *such that $f = Q(F)$ is a homotopy equivalence. Then F is a G-homotopy equivalence.*
(ii) *Let $f\colon M \to N$ be a G-map which is an ordinary homotopy equivalence. Then*

$$\mathrm{id} \times f\colon EG \times M \to EG \times N$$

is a G-homotopy equivalence.
(iii) *Let Y be in* G-Bun. *Then* pr: $EG \times Y \to Y$ *is a G-homotopy equivalence.*

Proof.
(i) (F, f) is a bundle map between numerable G-principal bundles. Let h be homotopy inverse to f. By the homotopy theorem (8.15), there exists a bundle map (H, h) covering h. By the covering homotopy property (Dold [1963], 7.8 on p. 250), a homotopy $fh \simeq \mathrm{id}$ can be lifted to a G-homotopy $FH \simeq H_1$. Since H_1 lies over the identity, it is a bundle automorphism. Thus F has a G-homotopy right inverse HH_1^{-1}.
(ii) $Q(\mathrm{id} \times f)$ is a map over BG which is a homotopy equivalence on each fibre. Therefore, $Q(\mathrm{id} \times f)$ is a homotopy equivalence (Dold [1963], 6.3). Now apply (i).
(iii) $Q(\mathrm{pr})$ is a numerable fibration with contractible fibre and therefore, in particular, a homotopy equivalence. Now apply (i). □
We now define certain sets of natural transformations which are then shown to be isomorphisms. Suppose $N \in G$-Top and $M \in$ Top are given. Let $A(N, M)$ be the set of natural transformations

$$\alpha\colon [?, N]_G \to [P(?), M]$$

of functors G-Top $\to$ Sets which are bijective for $? \in G$-Bun. Let $B(N, M)$ be the set of natural transformations

$$\beta\colon [?, N]_G \to [Q(?), M]$$

of functors G-Bun $\to$ Sets which are bijective for each object.* Let $C(N, M)$ be the set of homotopy classes γ over BG which make the diagram

$$\begin{array}{ccc} P(N) & \xrightarrow{\ \gamma\ } & BG \times M \\ & {\scriptstyle p_N}\searrow \quad \swarrow{\scriptstyle \mathrm{pr}} & \\ & BG & \end{array}$$

commutative and are homotopy equivalences. Here, p_N is the orbit map of pr: $EG \times N \to EG$.

(8.19) Proposition.
(i) *There are canonical bijections*

$$A(N, M) \cong B(N, M) \cong C(N, M).$$

(ii) *If $A(N, M) \neq \emptyset$, then M is homotopy equivalent to N without group action.*

Proof. We begin by defining maps between the sets A, B, and C. A map a: $A(N, M) \to B(N, M)$ is obtained as follows: Given $\alpha \in A(N, M)$, put $a(\alpha)(X)$: $[X, N]_G \xrightarrow[\alpha]{} [P(X), M] \xrightarrow[s_X^*]{} [X/G, M]$ where s_X is a homotopy inverse to $r_X = Q(\mathrm{pr})$, pr: $EG \times X \to X$. Then, clearly, $a(\alpha) \in B(N, M)$. We define a map b: $B(N, M) \to C(N, M)$ as follows: The object $EG \times N$ is contained in G-Bun. Given $\beta \in B(N, M)$, let f: $P(N) \to M$ represent the image of pr: $EG \times N \to N$ under β. Let γ: $P(N) \to BG \times M$ have components p_N and f. Then γ is a map over BG. If we show that γ is a homotopy equivalence, we can define $b(\beta) = [\gamma]$. We show that

$$\gamma_*\colon [Y, P(N)] \to [Y, BG \times M]$$

is bijective for each space Y; this implies that γ is a homotopy equivalence. It suffices to prove that in the commutative diagram

$$\begin{array}{ccc} [Y, P(N)] & \xrightarrow{\ \gamma_*\ } & [Y, BG \times M] \\ & {\scriptstyle p_{N*}}\searrow \quad \swarrow{\scriptstyle \mathrm{pr}_*} & \\ & [Y, BG] & \end{array}$$

γ_* maps fibres bijectively. Given $[h] \in [Y, BG]$; then h induces a space $Y_h \in G$-Bun over Y. Let ξ be the composition

* The notation $B(N, M)$ which conflicts with the notation for classifying spaces is only used in (8.19) and its proof.

$$[Y_h, N]_G \xrightarrow{\cong} [Y_h, EG \times N]_G \xrightarrow{Q} [Y, P(N)].$$

Given $[g] \in [Y, P(N)]$ in the fibre of p_{N*} over $[h]$, there exists $[g_1] \in [Y_h, EG \times N]_G$ such that $Q[g_1] = [g]$. The relation $\xi[\mathrm{pr}_2 \circ g_1] = [g]$ holds. Therefore, ξ maps onto the fibre over $[h]$. Since β is a natural transformation, the following diagram is commutative

$$\begin{array}{ccc}
[Y_h, N]_G & \xrightarrow[\beta]{\cong} & [Y, M] \\
\downarrow \xi & & \downarrow \mathrm{pr}_* \\
[Y, P(N)] & \xrightarrow[\gamma_*]{} & [Y, BG \times M]\,.
\end{array}$$

The map pr_* sends the fibre over $[h]$ bijectively onto $[Y, M]$. We conclude that γ_* maps the fibre over $[h]$ bijectively. We define a map $c\colon C(N, M) \to A(N, M)$ as follows. Let $[\gamma] \in C(N, M)$ be given. This induces a G-homotopy equivalence $\gamma_1\colon EG \times N \to EG \times M$; see (3.18). This is used in the following composition $\alpha(X)$

$$[X, N]_G \xrightarrow[\mathrm{pr}^*]{} [EG \times X, N]_G = [EG \times X, EG \times N]_G$$
$$\cong [EG \times X, EG \times M]_G \cong [EG \times X, M]_G \cong [P(X), M].$$

It is easy to verify that α is a natural transformation in $A(N, M)$.

In order to show that a, b, and c are bijective, one verifies that cba, acb, and bac are the identity maps. The following observation can be used in this verification:

$$\alpha(\mathrm{id}(N)) = a(\beta)(\mathrm{pr}\colon EG \times N \to N).$$

This follows from naturality of α and

$$P(\mathrm{pr}) \circ s_{EG \times N} \simeq r_{EG \times N} \circ s_{EG \times N} \simeq \mathrm{id}.$$

The map ba sends α to the class with components $[p_N]$ and $\alpha(\mathrm{id}(N))$. □

For (8.19), see tom Dieck [1969a].

Now let G and A be compact Lie groups and let A be abelian. Then the product map $m\colon A \times A \to A$ and the inverse map $i\colon A \to A$ are group homomorphisms. They induce maps

$$B(m)\colon\ B(G, A) \times B(G, A) \cong B(G, A \times A) \to B(G, A) \quad \text{and}$$
$$B(i)\colon\ B(G, A) \to B(G, A)$$

which induce on $\mathfrak{B}(G, A)(X) \cong [X, B(G, A)]_G$ the structure of an abelian group. Let

(8.20) $\alpha\colon \mathfrak{B}(G,A)(X) \to \mathfrak{B}(A)(EG\times_G X)$

be the natural transformation which sends a (G,A)-bundle $f\colon E \to X$ to the A-bundle $EG\times_G f\colon EG\times_G E \to EG\times_G X$. The map α is a homomorphism. By the considerations above, on the classifying space level α is given by the composition

$$[X, B(G,A)]_G \to [EG\times_G X, EG\times_G B(G,A)] \to [EG\times_G X, BA];$$

the first map is $[f] \mapsto [EG\times_G f]$ and the second is induced by the projection $EG\times_G B(G,A) \cong BG\times BA \to BA$. The following is proved in Lashof-May-Segal [1983].

(8.21) Theorem. *If X is a G-complex, then α in* (8.20) *is an isomorphism. (G-complexes are defined in Chapter II.)* □

This theorem implies, or is related to, results of Hattori-Yoshida [1976], Steward [1961], Su [1963], and Liulevicius [1978].

(8.22) Exercises.

1. Show that (5.18), Ex. 1 may be obtained as a special case of (8.15).
2. Let $p\colon E \to B$ be a (Γ,α,G)-bundle such that for each local object $q\colon X \to \Gamma/\Lambda$ the associated bundle $q'\colon (X\times E)/G \to \Gamma/\Lambda$ is Γ-shrinkable, i.e. admits a Γ-section s such that sq' is Γ-homotopic over Γ/Λ to the identity. Show that each numerable (Γ,α,G)-bundle admits a bundle map into p and any two such bundle maps are homotopic. Show that for the bundle (8.11), the maps q' are shrinkable. If p above is also numerable, then it is universal. In particular, a numerable principal G-bundle $E \to B$ is universal if and only if E is contractible. (tom Dieck [1969], Dold [1963])
3. Let X be a completely regular G-space (G compact Lie). Assume that all orbits of X are isomorphic to G/H. Then the canonical map $p\colon X \to X/G$ is a fibre bundle with fibre G/H and structure group NH/H (with canonical right action on the fibre G/H). (Bredon [1972], II.5.8).
4. Let G be a compact Lie group. Let $\mathfrak{F}$ be the family of closed subgroups H of $\Gamma\times_\alpha G$ such that $(\Gamma\times_\alpha G)/H$ has free G-action. Then the classifying space $E(\mathfrak{F})$ of (6.6) is a model for $E(\Gamma,\alpha,G)$.
5. Let Λ be a closed subgroup of Γ. Let $E \to B$ be a numerable (Λ,G)-bundle. Show that $\Gamma\times_\Lambda E \to \Gamma\times_\Lambda B$ is a numerable (Γ,G)-bundle.
6. Let $E \to B$ be a universal (Γ,G)-bundle and $\Lambda\subset\Gamma$ a closed subgroup. Show that $\mathrm{res}_\Lambda E \to \mathrm{res}_\Lambda B$ is a universal (Λ,G)-bundle.
7. $G \to G/H$ is an H-principal bundle if and only if it has a local section.
8. Let $G \to G/H$ have a local section. Let $E \to B$ be a G-principal bundle. Show that:

   (i) $E \to E/H$ is an H-principal bundle.

(ii) If $G \to G/H$ is numerable, then $E \to E/H$ is numerable.

(iii) If $E \to B$ is universal and $G \to G/H$ numerable, then $E \to E/H$ is universal. Thus E/H is a model for BH. In particular, one obtains a map $E/H = BH \to BG = B$; this is a fibre bundle with fibre G/H and structure group G.

9. (i) Let B be a free Γ-space such that $B \to B/\Gamma$ is a numerable bundle. Show that the assignment $(E \to B) \mapsto (E/\Gamma \to B/\Gamma)$ is an equivalence between the category of numerable (Γ, G)-bundles over B and numerable G-bundles over B/Γ.

(ii) The assignment $(E \to B) \mapsto (E\Gamma \times_\Gamma E \to E\Gamma \times_\Gamma B)$ is a natural transformation from numerable (Γ, G)-bundles to numerable G-bundles.

(iii) On the classifying space level, the natural transformation in (ii) defines a natural transformation $[B, B(\Gamma, G)] \to [E\Gamma \times_\Gamma B, BG]$. By (8.19), there is a corresponding homotopy equivalence
$\gamma\colon E\Gamma \times_\Gamma B(\Gamma, G) \to B\Gamma \times BG$ over $B\Gamma$. Show that the inverse of γ is the map J of (8.15.1).

10. Show that $B(\Gamma, G_1) \times B(\Gamma, G_2)$ is a model for $B(\Gamma, G_1 \times G_2)$.

11. Let $\varrho\colon G_1 \to G_2$ be a continuous homomorphism. Define a left G_1-action on G_2 by $(g_1, g_2) \mapsto \varrho(g_1)g_2$. Let $E \to B$ be a numerable (Γ, G_1)-bundle. Show that $E \times_{G_1} G_2 \to B$ is a numerable (Γ, G_2)-bundle. Apply this to universal bundles to obtain a Γ-map $B(\varrho)\colon B(\Gamma, G_1) \to B(\Gamma, G_2)$, unique up to Γ-homotopy. If $\sigma\colon G_2 \to G_3$ is another homomorphism, then $B(\sigma)B(\varrho) \simeq_\Gamma B(\sigma\varrho)$.

12. Let H be a normal subgroup of G. Then $E(G/H) \times E(G)$ is a numerable free G-space. Therefore, $(E(G/H) \times EG)/G$ is a model for BG. Using this model and the orbit map of the projection $E(G/H) \times EG \to E(G/H)$, one obtains a map $BG \to B(G/H)$ which is a fibre bundle with structure group G/H and fibre BH, arising as $(EG)/H$ with induced G/H-action. Thus one has a fibration $BH \xrightarrow[i]{} BG \xrightarrow[p]{} B(G/H)$. Show that i (resp. p) are maps $B\varrho$ in the sense of exercise 11 where ϱ is the inclusion $H \subset G$ (resp. the projection $G \to G/H$). Similarly for (Γ, G)-bundles.

13. Let Γ be a Lie group. Show that the path components of $B(G, \Gamma)^G$ correspond to classes of continuous homomorphisms $G \to \Gamma$, up to conjugation in Γ.

14. Show that (8.20) for $X = G/H$ may be identified with the homomorphism B: $\mathrm{Hom}(H, A) \to [BH, BA]$ which sends a continuous homomorphism ϱ to the induced map $B\varrho$. Use this to show that the map B is an isomorphism for finite or connected G.

15. Let G be a compact Lie group. Show that EG can be obtained as a geometric realization of a simplicial space whose space of n-simplices is the $(n+1)$-fold product G^{n+1} (see Milgram [1967], Steenrod [1968], May [1975]).

9. Vector bundles.

In this section we present some of the special properties of vector bundles with a compact Lie group G of automorphisms. For basic material about vector bundles and topological K-theory, see Atiyah [1967], [1968], Karoubi [1978], Milnor-Stasheff [1974], Segal [1968].

Let $p\colon E \to X$ be a G-map and a (real or complex) vector bundle. If, for each $g \in G$, the left translation by g is a bundle map, then these data define a **G-vector bundle**.

In a G-vector bundle, $G_x = H$ acts linearly on the fibre $E_x = p^{-1}(x)$ over x. Thus E_x becomes an H-representation. The variation of this representation with x is one of the objects of study. The tangent bundle of a G-manifold is a (differentiable) G-vector bundle.

A **bundle map** between G-vector bundles is an ordinary bundle map which is equivariant. Therefore, one has the category of G-vector bundles. The usual constructions with vector bundles like Whitney sum, tensor product, dual bundle, exterior powers etc. applied to G-vector bundles yield G-vector bundles again by functoriality. A G-vector bundle is called **trivial** if it has the form $\mathrm{pr}\colon X \times V \to X$, $(x, v) \mapsto x$ for a G-representation V. Thus there are quite different trivial bundles. A trivial bundle in this sense is induced from a bundle over a point. But since the smallest pieces of G-spaces are orbits, one asks for bundles over orbits.

Let $p\colon E \to G/H$ be a G-vector bundle and $V = p^{-1}(eH)$ the fibre over eH considered as H-representation. The canonical map $G \times_H V \to E$, $(g, v) \mapsto gv$ is a G-homeomorphism by (4.4). Since $G \times_H V \to G/H$ is locally trivial, we see that $G \times_H V \to E$ is an isomorphism of G-vector bundles. Conversely, for each H-representation V, the canonical map $G \times_H V \to G/H$ is a G-vector bundle. Such bundles are called **local objects**.

(9.1) Definition. A G-vector bundle $p\colon E \to X$ is called **locally trivial** if each $x \in X$ has a G-neighbourhood U such that the restriction $p_U\colon E_U \to U$ of p to U is induced from a local object. In this case, p is called **trivial** over U. The bundle is called of **finite type** if it is trivial over the sets of a finite open covering by G-sets. It is called **numerable** if it is trivial over the sets of a G-numerable covering of X.

Let S be an H-space and $p\colon E \to G \times_H S$ a G-vector bundle. Let p_S be the restriction of p to the H-subspace $S = H \times_H S \subset G \times_H S$. This is an H-vector bundle.

(9.2) Proposition. *The assignment $p \mapsto p_S$ is an equivalence between the category of G-vector bundles over $G \times_H S$ and the category of H-vector bundles over S.*

Proof. The inverse equivalence maps $p\colon U \to S$ to $G \times_H p\colon G \times_H U \to G \times_H S$.

This map defines a G-vector bundle once local triviality is shown. If $G \to G/H$ has a local section over W, then $G \times_H p$ is isomorphic over W to $\mathrm{id} \times p$: $W \times E \to W \times S$ and hence locally trivial as an ordinary vector bundle. In order to verify that the two assignments are inverse to each other, use (4.4) again. □

(9.3) Corollary. *A G-vector bundle over a locally trivial G-space (in the sense of section 6) is locally trivial.* □

Let X be a free G-space such that $X \to X/G$ is a locally trivial G-principal bundle. Let p: $E \to X$ be a vector bundle. We have the induced map p/G: $E/G \to X/G$. We show that this is again a vector bundle. Let φ: $U \to G \times U/G$ be a local trivialisation of $X \to X/G$. Let p_0 be the restriction of p to $\varphi^{-1}(\{e\} \times U/G)$. Then $p_U \cong G \times p_0$. Hence, over U/G, p/G is the bundle p_0 and therefore p/G is locally trivial. The quotient map

$$\begin{array}{ccc} E & \longrightarrow & E/G \\ \downarrow & & \downarrow \\ X & \underset{q}{\longrightarrow} & X/G \end{array}$$

is then a bundle map and therefore p is induced from p/G along q. Inducing along q maps bundles to G-bundles. These constructions are used to show

(9.4) Proposition. *Let X be a locally trivial free-G-space. Assigning $E/G \to X/G$ to $E \to X$ yields an equivalence between the category of G-vector bundles over X and vector bundles over X/G.* □

We now consider the other extreme, namely bundles over trivial G-spaces X. In this case, each fibre E_x is a G-representation. The decomposition of E_x according to irreducible representations yields a corresponding decomposition of vector bundles. Let us look at complex vector bundles first. Let $\mathrm{Irr}(G, \mathbb{C})$ be the set of isomorphism classes of irreducible complex G-representations. If W is any complex representation, we consider the map

$$\bigoplus_{V \in \mathrm{Irr}(G,\mathbb{C})} \mathrm{Hom}_G(V, W) \otimes_{\mathbb{C}} V \to W, \; f \otimes v \mapsto fv.$$

This is an isomorphism of representations (compare Bröcker-tom Dieck [1985]). The image of the V-summand consists of those irreducible subrepresentations which are isomorphic to V, the **V-isotypical part** of W. We obtain an analogous decomposition for vector bundles p: $E \to X$ over trivial G-spaces X by applying this construction fibrewise

$$(9.5) \qquad E \cong \bigoplus_{V \in \mathrm{Irr}(G,\mathbb{C})} \mathrm{Hom}_G(V, E) \otimes_{\mathbb{C}} V.$$

Here, $\mathrm{Hom}_G(V, E)$ is an ordinary vector bundle and $\mathrm{Hom}_G(V, E) \otimes_{\mathbb{C}} V$ is the resulting G-vector bundle. Thus E splits into the **V-isotypical subbundles**. For real vector bundles, the situation is slightly more complicated. Let V be a real irreducible G-module. The endomorphism ring $D(V) = \mathrm{Hom}_G(V, V)$ is then a division algebra over $\mathbb{R}$ hence $\mathbb{R}$, $\mathbb{C}$, or $\mathbb{H}$. Let $\mathrm{Irr}(G, \mathbb{R})$ be the set of isomorphism classes of real irreducible G-modules. Then the analogue of (9.5) is an isomorphism (Bröcker-tom Dieck [1985], II (6.9))

$$(9.6) \qquad E \cong \bigoplus_{V \in \mathrm{Irr}(G,\mathbb{R})} \mathrm{Hom}_G(V, E) \otimes_{D(V)} V.$$

An important fact is the existence of inverse bundles. A bundle $p_2\colon E_2 \to X$ is called **inverse** to $p_1\colon E_1 \to X$ if the Whitney sum $p_1 \oplus p_2\colon E_1 \oplus E_2 \to X$ is isomorphic to a trivial bundle $X \times V \to X$, V a G-representation. If a G-manifold M is a submanifold of V, then it can be shown that the normal bundle N of this embedding $M \subset V$ is a G-vector bundle inverse in this sense to the tangent bundle.

(9.7) Proposition. *$p\colon E \to X$ has an inverse bundle if and only if p is numerable of finite type.*

Proof. Let $q\colon F \to X$ be inverse to p. Choose an isomorphism $E \oplus F \cong X \times V$. If p is k-dimensional, each fibre E_x of p is mapped onto a k-dimensional subspace V_x of V. The mapping $x \mapsto V_x$ is a G-map from X into the Grassmann manifold $G_k(V)$ of k-dimensional subspaces of V; and p is induced via this map from the canonical k-dimensional vector bundle over $G_k(V)$. A G-vector bundle over the compact G-manifold $G_k(V)$ is numerable of finite type and so is every induced bundle.

Conversely, assume that $p\colon E \to X$ is numerable of finite type. Locally, we then have a bundle map into $G \times_H V \to G/H$. Choose a G-representation W which contains V as an H-direct summand (Bröcker-tom Dieck [1985], III (4.5)). Then $G \times_H W \cong G/H \times W$, see 4. Ex. 10. Therefore, $G \times_H V$ has a fibrewise-injective G-map into a G-representation W and, similarly, p has such a map locally. Using a partition of unity, such local maps can be glued together to yield a fibrewise-injective G-map $E \to U$ into a G-representation U. Looking at orthogonal complements yields an inverse bundle. (See exercises for some details.) □

The classifying space for n-dimensional complex G-vector bundles is the space $B(G, \mathrm{U}(n))$. Let $f_n(G)\colon E_n(G) \to B_n(G)$ denote the universal n-dimensional complex G-vector bundle. Let $H \lhd G$, let $X = B_n(G)^H$ and denote by $p\colon E \to X$ the restriction of $f_n(G)$ to X. Since H is normal in G, the bundle p inherits a G-action. Since H acts trivially on X, each fibre of p is an H-representation.

The path-components of X correspond bijectively to the isomorphism classes of n-dimensional complex H-modules; $p^{-1}(b)$ and $p^{-1}(c)$ are isomorphic if and only if b and c belong to the same component, and each isomorphism type appears in a suitable fibre of p. All this follows directly from the universal property of $f_n(G)$.

The bundle p is universal for the following type of vector bundles: Numerable n-dimensional complex G-vector bundles with trivial H-action on the base.

If V is a complex H-module and $g \in G$, then V with H-action $(h, v) \mapsto ghg^{-1}v$ is again an H-module, denoted by ${}^gV$. The map $V \mapsto {}^gV$ induces an action of $g \in G$ on the set $\mathrm{Rep}(H)$ of isomorphism classes of complex H-modules. Let $D_n(H)$ be the orbit set of this G-action on n-dimensional H-modules. Let

$$X = \coprod (X_j | j \in D_n(H))$$

be the partition of X into X_j where X_j collects the components belonging to representations in j. The X_j are G-subspaces of X. Each space X_j has, up to G-homotopy type, the form

$$X_j = B\mathrm{U}(j_0) \times Y_j;$$

here, G acts trivially on the classifying space $B\mathrm{U}(j_0)$ of the unitary group $\mathrm{U}(j_0)$ and j_0 denotes the complex dimension of the H-trivial direct summand of j (i.e. of any representative). This can be seen as follows. As in (9.5), the bundle p splits as H-bundle. In particular, the H-trivial part splits off as a G-bundle and hence, by universality, X_j must have the indicated form.

Let H be a closed subgroup of the compact Lie group G. Let V be a real H-module. Then $G \times_H V \to G/H$, $(g, v) \mapsto gH$ is a real G-vector bundle over G/H. Any G-vector bundle over G/H has this form. For, if $p\colon E \to G/H$ is such a bundle, then the fibre $p^{-1}(eH) =: W$ is an H-module and $G \times_H W \to E$, $(g, w) \mapsto gw$ is an isomorphism (see (5.6), (8.9), and (9.2)).

Let M be a differentiable G-manifold. The orbit Gx through $x \in M$ is diffeomorphic to G/G_x. The normal bundle of the submanifold $Gx \subset M$ is a G-vector bundle isomorphic to $G \times_{G_x} S_x$. The G-module

$$S_x = T_x M / T_x Gx$$

is called the **slice representation** at x (see (5.6)). Let F_x be the subspace of S_x on which G_x acts trivially. Then $V_x = S_x/F_x$ is called the **normal representation** at x. Compare Davis [1978].

Suppose we have an isomorphism of bundles $\phi\colon G \times_H V \to G \times_K W$. Suppose $\phi(e, v) = (a, Lv)$. Then $L\colon V \to W$ is a linear map and we have

(9.8) $\quad a^{-1}Ha = K, \quad L(hv) = a^{-1}ha\,L(v)$

for $h \in H$ and $v \in V$. Conversely, any pair a, L satisfying (9.8) defines an isomorphism ϕ by setting $\phi(g, v) = (ga, Lv)$.

Consider pairs (H, V) consisting of a closed subgroup H of G and an

H-module V such that $V^H = \{0\}$. Call two such pairs (H, V) and (K, W) equivalent if $G \times_H V$ and $G \times_K W$ are isomorphic as G-bundles. An equivalence class is called a **normal orbit G-type**. Let $[H, V]$ denote the equivalence class of (H, V).

If M is a differentiable G-manifold, then (G_x, V_x) and (G_{gx}, V_{gx}) define the same normal orbit type. An equivalence is induced by the pair (g, L) where $L: V_x \to V_{gx}$ is induced by the differential of the left translation by g. Thus the orbits (G_x) are further partitioned into normal orbit types.

Suppose α is a normal orbit type. The **α-stratum** M_α is the G-subset of M defined by

$$M_\alpha = \{x \in M \mid [G_x, V_x] = \alpha\}.$$

Let $\pi: M \to M/G =: B$ be the orbit map. Set $B_\alpha = \pi(M_\alpha)$ and let $\pi_\alpha: M_\alpha \to B_\alpha$ be the restriction of π. Then we have

(9.9) Proposition. *M_α and B_α are smooth manifolds and $\pi_\alpha: M_\alpha \to B_\alpha$ is a smooth fibre bundle.*

Proof. Let α be represented by (H, V). By (5.6), $x \in M_\alpha$ has a G-neighbourhood U diffeomorphic to $G \times_K S$ with $K = G_x$ and $S = S_x$. Since $x \in M_\alpha$, we have $[H, V] = [K, V_x]$. One verifies that $(G \times_K S)_{(K)} \cong G/K \times S^K \cong (G \times_K S)_\alpha$.

Thus we see that the smooth submanifold $M_{(H)}$ of M (see (5.13)) is the disjoint union of open and closed manifolds M_α where α has a representative of the form (H, V), for some V. The map $M_{(H)} \to M_{(H)}/G$ is a smooth fibre bundle with typical fibre G/H and structure group NH/H and π_α is the part of this bundle over $B_\alpha \subset M_{(H)}/G$. □

Let us now look at the equivariant normal bundle $\nu(M_{(H)}, M)$ of the embedding $M_{(H)} \subset M$. Locally, this is isomorphic to the normal bundle of $G \times_H S^H \subset G \times_H S$. We see that the fibre of $\nu(M_{(H)}, M)$ over $x \in M_{(H)}$ is the G_x-module V_x. Thus the partition of $M_{(H)}$ into the M_α corresponds to the different isomorphism types of H-module fibres.

Let $\mathrm{Norm}(G)$ be the set of normal G-orbit types. Define a partial ordering of $\mathrm{Norm}(G)$: $[H, V] < [K, W]$ if and only if the G-manifold $G \times_H V$ contains an orbit of type $[K, W]$.

(9.10) Exercises.

1. Let $K_G(X)$ be the Grothendieck group of complex G-vector bundles over X. Use (9.4) to produce a natural isomorphism $K_G(X) \cong K(X/G)$ for a locally trivial free G-space. Use (9.5) to produce a natural isomorphism $K_G(X) \cong R(G) \otimes_{\mathbb{Z}} K(X)$ for trivial G-spaces X; here, $R(G) = K_G(\mathrm{Point})$ is the complex representation ring of G.

2. Use Haar integration over G to show that a numerable G-vector bundle $E \to X$ has a G-invariant inner product, i.e. a continuous inner product $\langle -,- \rangle_x$ on the fibres E_x such that $\langle gv, gw \rangle_{gx} = \langle v, w \rangle_x$ holds for all $g \in G$ and $v, w \in E_x$. Use this to show that short exact sequences of numerable G-vector bundles split (orthogonal complement).
3. Let V be a representation, $G_k(V)$ the Grassmann manifold of k-dimensional subspaces in V with G-action as in 2. Ex. 11. Let $E_k(V) = \{(x, v) \mid x \in G_k(V), v \in x\} \subset G_k(V) \times V$. Describe $E_k(V)$ as k-dimensional G-vector bundle over $G_k(V)$. (The canonical bundle.)
4. Let p: $E \to X$ be a G-vector bundle. Show that for $H \subset G$ the map p^H: $E^H \to X^H$ is an NH/H-vector bundle. Show that p^H is a direct summand in the restriction of p to X^H.
5. Let $\alpha \in \mathrm{Norm}(G)$ and let M be a smooth G-manifold. Show that $\bar{M}_\alpha = \bigcup_{\beta < \alpha} M_\beta$.
6. Let M be a smooth G-manifold and let M/G be connected. Show that there exists a maximal normal orbit type γ for M. The submanifold M_γ is open and dense in M. (See section 5.)

10. Orbit categories, fundamental groups, and coverings.

For a thorough understanding of a transformation group a large amount of data are needed, e.g. the fixed point sets, the orbit bundles, group actions on these subspaces, inclusion relations among them, their components and fundamental groups, etc. Handling these data requires a certain amount of categorical and combinatorial machinery. We introduce some of the basic notions in this context. We begin by describing several important categories. The group G is a compact Lie group; subgroups are assumed to be closed.

(10.1) The orbit category $\mathrm{Or}(G)$. The objects are the homogeneous spaces G/H and the morphisms are the G-maps between them. If $\mathfrak{F}$ is any family of subgroups, one has the full subcategory $\mathrm{Or}(G, \mathfrak{F})$ with objects G/H for H in $\mathfrak{F}$. Objects G/H and G/K are isomorphic if and only if H and K are conjugate. Sometimes it is convenient to have only one object in each isomorphism class. Equivalent to $\mathrm{Or}(G)$ is the category of Hausdorff transitive G-spaces.

(10.2) The orbit category $\mathrm{Or}(G, X)$ **over** X. Let X be a G-space. The objects of $\mathrm{Or}(G, X)$ are the G-maps x: $G/H \to X$ for arbitrary H. A morphism from x: $G/H \to X$ to y: $G/K \to X$ is a G-map σ: $G/H \to G/K$ such that $y\sigma = x$. We also write $\sigma^* y$ for $y\sigma$. The objects of $\mathrm{Or}(G, X)$ thus correspond to points in fixed point sets, see (3.8). As in (10.1), one may consider only those G/H with H in a family $\mathfrak{F}$. A morphism σ: $G/H \to G/K$ from x to y is an isomorphism if and only if σ is an isomorphism. The automorphism group $\mathrm{Aut}(x)$ of $x: G/H \to X$ in $\mathrm{Or}(G, X)$ is the subgroup of NH/H which is the isotropy group of $x(eH) \in X^H$. (Recall that $\mathrm{Aut}_G(G/H) \cong NH/H$.)

(10.3) The component category $\pi_0(G, X)$. Objects are G-homotopy classes $[x]: G/H \to X$. A morphism from $[x]: G/H \to X$ to $[y]: G/K \to X$ is a G-map $\sigma: G/H \to G/K$ such that $y\sigma \simeq x$ (G-homotopic). Therefore, objects may be considered as pairs (H, c) where $c \in \pi_0(X^H)$ is a component of X^H. Each endomorphism is an automorphism. The automorphism group of (H, c) is the isotropy group of the $WH = NH/H$-action on $\pi_0(X^H)$ at c; it will be denoted by $W_c H$.

(10.4) The transport category $\mathrm{Tr}(G, X)$. A G-map $\sigma: G/H \to G/K$ has the form $\sigma(uH) = ugK$ for suitable $g \in G$ such that $g^{-1}Hg \subset K$. It is sometimes useful to have the dependence of g built into the category. Thus objects of $\mathrm{Tr}(G, X)$ are again G-maps $x: G/H \to X$ and a morphism from $x: G/H \to X$ to $y: G/K \to X$ is an element $g \in G$ such that $gHg^{-1} \subset K$ and, with $\sigma(g^{-1}): G/H \to G/K$, $uH \mapsto ug^{-1}K$, the equality $y\sigma(g^{-1}) = x$ holds. (Strictly speaking, since morphism sets have to be disjoint, a morphism from x to y is a triple (x, y, g) such that ...). Composition of morphisms corresponds to multiplication in the group:

$$(y, z, h) \circ (x, y, g) = (x, z, hg).$$

There is a covariant functor

$$\mathrm{Tr}(G, X) \to \mathrm{Or}(G, X)$$

which is the identity on objects and maps (x, y, g) to $\sigma(g^{-1})$. Endomorphisms in $\mathrm{Tr}(G, X)$ are automorphisms. The automorphism group of $x: G/H \to X$ is $\{g \in NH \mid gx(eH) = x(eH)\}$.

(10.5) The component transport category $\mathrm{Tr}_0(G, X)$. The objects are the same as for $\pi_0(G, X)$ in (10.3). A morphism from $[x]: G/H \to X$ to $[y]: G/K \to X$ is an element $g \in G$ such that $gHg^{-1} \subset K$ and $y\sigma(g^{-1}) \simeq x$. If $c \in \pi_0(X^H)$ resp. $d \in \pi_0(X^K)$ is the component corresponding to $[x]$ resp. $[y]$, then $y\sigma(g^{-1}) \simeq x$ is equivalent to $gc \supset d$.

This justifies the name of the category: Left translation by g^{-1} maps one fixed point component into another one. As in (10.4), there is a functor

$$\mathrm{Tr}_0(G, X) \to \pi_0(G, X)$$

which maps $([x], [y], g)$ to $\sigma(g^{-1})$. Endomorphisms are automorphisms. The components c and d above represent isomorphic objects if and only if there exists $g \in G$ such that $gHg^{-1} = K$ and $gc = d$.

(10.6) The component category $Q_0(G, X)$. This is a quotient category of $\mathrm{Tr}_0(G, X)$. Given two G-maps σ_i: $G/H \to G/K$, $uH \mapsto ug_i^{-1}K$, they induce the same homomorphisms $H \to K$, $h \mapsto g_i h g_i^{-1}$ if and only if g_1 and g_2 differ by an element in the centralizer $C(H)$ of H, i.e. $g_1 g_2^{-1} \in C(H)$. Therefore, we have a category $Q_0(G, X)$ with the same objects as $\mathrm{Tr}_0(G, X)$, but a morphism from (H, c), $c \in \pi_0(X^H)$ to (K, d), $d \in \pi_0(X^K)$ is a homomorphism $H \to K$ of the form $h \mapsto ghg^{-1}$ with $g \in G$ such that $gHg^{-1} \subset K$ and $gc \supset d$.

(10.7) The fundamental group category $\pi_1(G, X)$. This category extends the orbit category and all of the fundamental groupoids of the fixed point sets. Objects of $\pi_1(G, X)$ are again G-maps x: $G/H \to X$. A morphism from x: $G/H \to X$ to y: $G/K \to X$ is a pair (σ, ϕ) consisting of a G-map σ: $G/H \to G/K$ and a G-homotopy class relative $G/H \times \partial I$ of G-homotopies ϕ: $G/H \times I \to X$ such that $\phi_0 = x$ and $\phi_1 = y\sigma$. (We thus describe a class ϕ through one of its representatives.) The composition of the morphisms (σ_1, ϕ_1): $x_1 \to x_2$ and (σ_2, ϕ_2): $x_2 \to x_3$ is defined as

$$(\sigma_2\sigma_1, \sigma_1^*\phi_2 * \phi_1) = (\sigma_2, \phi_2) \circ (\sigma_1, \phi_1)$$

where $\phi_2' := \sigma_1^*\phi_2 = \phi_2 \circ (\sigma_1 \times \mathrm{id}(I))$ and $\phi_2' * \phi_1$ is the usual composition of homotopies

$$\phi_2' * \phi_1(g, t) = \begin{cases} \phi_1(g, 2t) & \text{for} \quad 0 \le t \le 1/2 \\ \phi_2'(g, 2t-1) & \text{for} \quad 1/2 \le t \le 1. \end{cases}$$

The composition is well-defined and associative; (id, κ) with constant homotopy κ is the identity (exercise 1).

If G is the trivial group, then $\pi_1(G, X)$ is the fundamental groupoid of X. If we identify x and $\sigma^* y = y\sigma$ with points in X^H, then a morphism $x \to y$ is a homotopy class of paths in X^H from x to $\sigma^* y$.

If (σ, ϕ) is an endomorphism of x, then it is an automorphism with inverse $(\sigma^{-1}, (\sigma^{-1})^*\phi^-)$ where ϕ^- denotes the homotopy inverse to ϕ, i.e. $\phi^-(u, t) = \phi(u, 1-t)$.

Two objects of $\pi_1(G, X)$ are isomorphic if and only if they define isomorphic objects in $\pi_0(G, X)$. There is an obvious functor $\pi_1(G, X) \to \pi_0(G, X)$, on objects $x \mapsto [x]$, forgetting the homotopies ϕ.

Let $\mathrm{Aut}_i(x)$ denote the automorphism group of x: $G/H \to X$ in $\pi_i(G, X)$ $(i = 0, 1)$. Then $\mathrm{Aut}_1(x) \to \mathrm{Aut}_0(x)$, $(\sigma, \phi) \mapsto \sigma$ is surjective. Let $X(x)$ denote the component of X^H containing $x(eH) = x$. The kernel of the homomorphism $\mathrm{Aut}_1(x) \to \mathrm{Aut}_0(x)$ can be identified with the fundamental group $\pi_1(X(x), x)$. Altogether, we obtain an exact sequence

(10.8) $1 \to \pi_1(X(x), x) \to \mathrm{Aut}_1(x) \to \mathrm{Aut}_0(x) \to 1.$

In general, $\mathrm{Aut}_0(x)$ is a compact Lie group. One can topologize $\mathrm{Aut}_1(x)$ in such a way that (10.8) becomes an exact sequence of topological groups, with $\pi_1(X(x), x)$ a discrete group.

(10.9) The discrete fundamental group category $\pi_1^d(G, X)$. Objects are again G-maps $x: G/H \to X$. In contrast to the previously defined categories, we now identify suitably homotopic maps $\sigma: G/H \to G/K$. Let $(\sigma_i, \psi_i): x \to y$ be two morphisms ($i = 0,1$). They are called equivalent if there exists a G-homotopy $\sigma: G/H \times I \to G/K$ between σ_0 and σ_1 and a G-homotopy $\Lambda: G/H \times I \times I \to X$ such that

$$\Lambda(gH, 0, t) = x(gH)$$
$$\Lambda(gH, 1, t) = y\sigma(gH, t)$$
$$\Lambda(gH, s, i) = \Psi_i(gH, s) \quad i = 0,1.$$

Here is a figure indicating Λ

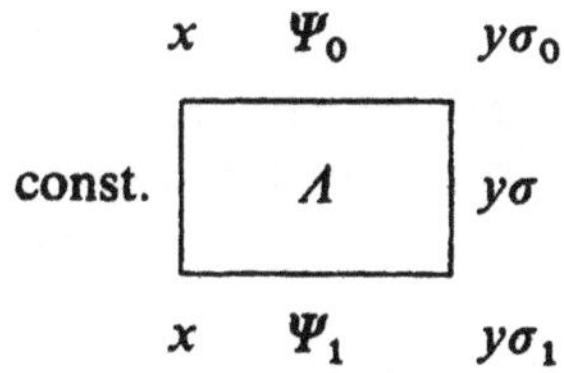

This equivalence relation is compatible with composition in $\pi_1(G, X)$; see exercise 2. The quotient category shall be $\pi_1^d(G, X)$.

Let $\mathrm{Aut}_1^d(x)$ be the automorphism group of x in $\pi_1^d(G, X)$. The projection $(\sigma, \psi) \mapsto \sigma$ induces a surjective homomorphism $\mathrm{Aut}_1^d(x) \to \pi_0 \mathrm{Aut}_0(x)$. The homomorphism $\pi_1(X(x), x) \to \mathrm{Aut}_1(x)$ of (10.8) induces a homomorphism $\pi_1(X(x), x) \to \mathrm{Aut}_1^d(x)$. There is a homomorphism $\pi_1(\mathrm{Aut}_0(x), e) \to \pi_1(X(x), x)$ defined as follows: Let $w: I \to \mathrm{Aut}_0(x)$ be a loop in $\mathrm{Aut}_0(x)$ with endpoint e. Then $t \mapsto w(t)(eH)$ is a loop in $X(x)$ with endpoint $x(eH) = x$. It is not difficult to check that the following sequence is exact (exercise 3).

(10.10) $\pi_1(\mathrm{Aut}_0(x), \mathrm{id}) \to \pi_1(X(x), x) \to \mathrm{Aut}_1^d(x) \to \pi_0 \mathrm{Aut}_0(x) \to 1.$

For the moment, let us abbreviate $W = \mathrm{Aut}_0(x)$. The group W acts from the right on $X(x) \subset X^H$ via $x = x(eH) \mapsto x \circ \sigma = (x \circ \sigma)(eH)$. The sequence (10.10) looks like the beginning of the exact homotopy sequence of the fibration

(10.11) $X(x) \to EW \times_W X(x) \to BW$

associated to the universal principal W-bundle. Indeed we have

(10.12) Proposition. *There exists an isomorphism*

$$\beta: \mathrm{Aut}_1^d(x) \cong \pi_1(EW \times_W X(x))$$

which transforms (10.10) *into the exact homotopy sequence of* (10.11).

Proof. Let (σ, ψ) represent an element in $\mathrm{Aut}_1^d(x)$. We can view ψ as a path from x to $x\sigma$ in $X(x)$. Fix a base point $u \in EW$ and let ω be a path from u to $u\sigma$. Then (ω, ψ) is a path from (u, x) to $(u\sigma, x\sigma)$ in $EW \times X(x)$ and thus gives rise to a loop in $EW \times_W X(x)$ with base point (u, x). The assignment $(\sigma, \psi) \mapsto (\omega, \psi)$ induces a well-defined homomorphism (exercise 4)

$$\beta\colon \mathrm{Aut}_1^d(x) \to \pi_1(EW \times_W X(x)).$$

This homomorphism is the desired isomorphism. In order to see this, consider the following diagram

$$\begin{array}{ccccccccc}
\pi_1(W) & \longrightarrow & \pi_1(X(x)) & \longrightarrow & \mathrm{Aut}_1^d(x) & \longrightarrow & \pi_0(W) & \longrightarrow & 1 \\
\cong \downarrow \beta_1 & & \downarrow \mathrm{id} & & \downarrow \beta & & \cong \downarrow \beta_0 & & \\
\pi_2(BW) & \longrightarrow & \pi_1(X(x)) & \longrightarrow & \pi_1(EW \times_W X(x)) & \longrightarrow & \pi_1(BW) & \longrightarrow & 1.
\end{array}$$

The bottom row is the exact homotopy sequence of the fibration (10.11). The maps β_0 and β_1 are the canonical isomorphisms (inverse to the boundary morphism in the exact sequence of $W \to EW \to BW$; note that EW is contractible). The diagram is commutative (exercise 4). Apply the five-lemma. □

(10.13) Universal coverings. Here is an application of the category $\pi_1(G, X)$. Let X be a G-space and assume that components $X(x)$ have suitable local properties such that universal coverings can be defined in the standard manner using equivalence classes of paths (Massey [1967], p. 175). For each $x\colon G/H \to X$, let $\tilde{X}(x)$ be the space of homotopy classes rel ∂I of paths $I \to X(x)$ which start in $x(eH)$. Then $\tilde{X}(x)$ is a model for the universal covering of $X(x)$; the covering projection $p\colon \tilde{X}(x) \to X(x)$ maps a path to its endpoint.

Let $(\sigma, \phi)\colon x \to y$ be a morphism in $\pi_1(G, X)$. We obtain an induced map

$$\tilde{X}(\sigma, \phi)\colon \tilde{X}(y) \to \tilde{X}(x)$$

by mapping a path v starting in y to the composition $(\sigma * v) * \phi$. The map $\tilde{X}(\sigma, \phi)$ is compatible with the covering projections. The construction $\tilde{X}$ is a contravariant functor

$$\tilde{X}\colon \pi_1(G, X) \to \mathrm{Top}$$

into the category Top of topological spaces and continuous maps. This functor $\tilde{X}$ is called the **universal covering of the G-space X**.

(10.14) Exercises.

1. Verify the axioms of a category for $\pi_1(G, X)$.
2. Show that composition of morphism in $\pi_1^d(G, X)$ is well-defined.
3. Prove exactness of (10.10).
4. Verify the statements that were left as an exercise in the proof of (10.12).

5. Topologize $\mathrm{Aut}_1(x)$ such that (10.8) becomes an exact sequence of topological groups.
6. The categories introduced in this section have the property that endomorphisms are isomorphisms. Verify this. Let $\bar{x}$ be the isomorphism class of the object x. Show that ($\bar{x} \leq \bar{y} \Leftrightarrow$ there exists a morphism $x \to y$) defines a partial ordering on the set of isomorphism classes of objects in these categories. Interpret this partial ordering for $\pi_0(G, X)$ in terms of components of fixed point sets.
7. Determine the automorphism group of the objects in $\mathrm{Tr}(G, X)$, $\mathrm{Tr}_0(G, X)$, and $Q_0(G, X)$.
8. Let $H \subset G$. Show that by mapping $\alpha\colon H/K \to X$ to the adjoint map $G \times_H H/K \cong G/K \to X$ a functor $\pi_1^d(H, \mathrm{res}_H X) \to \pi_1^d(G, X)$ is induced.

11. Elementary algebra of transformation groups.

If G is a discrete group, then representation theory of G over a commutative ring R is the study of modules over the group ring RG. In transformation group theory, one often has to consider the representation theories of various subquotients of G simultaneously. The book-keeping problems which arise in such cases can only be overcome if one uses a suitably abstract language. It is the purpose of this section to indicate that algebra in functor categories is a satisfactory approach. We begin with some generalities which are then applied to the analysis of projective modules and their algebraic K-theory.

A. Algebra in functor categories.

Let R be a commutative ring. The category of (left) R-modules is denoted by R-Mod. Let Γ be a small category.

If $\mathfrak{C}$ is any category, we denote by

$$[\Gamma, \mathfrak{C}]$$

the **functor category** of (covariant) functors $\Gamma \to \mathfrak{C}$. Recall that the objects of $[\Gamma, \mathfrak{C}]$ are (covariant) functors $\phi\colon \Gamma \to \mathfrak{C}$ and a morphism from ϕ_1 to ϕ_2 is a natural transformation $\alpha\colon \phi_1 \to \phi_2$.

We are particularly interested in the category

$$[\Gamma, R\text{-Mod}].$$

The objects of this category are called **left $R\Gamma$-modules** and the category itself will also be denoted by

$$R\Gamma\text{-Mod}$$

and will be called the **category of left $R\Gamma$-modules**. The **dual category** of Γ is named Γ^{op}. We have the category

$$[\Gamma^{op}, R\text{-Mod}] =: \text{Mod-}R\Gamma$$

of contravariant functors $\Gamma \to R$-Mod, alias **right $R\Gamma$-modules**. Another symbol for this category would be $R\Gamma^{op}$-Mod.

See Mitchell [1972] for some algebra in functor categories.

We begin by describing a few examples.

(11.1) Example. Let G be a discrete group. Consider G as a category with a single object and a morphism for each group element. Composition of morphisms is group multiplication. A covariant functor $M\colon G \to R$-Mod is essentially the same thing as a left module over the group ring RG of G over R.

(11.2) Example. Let G/H be a homogeneous G-set for the finite group G. Consider the category $\Gamma(G/H)$ with a single object G/H and G-maps $G/H \to G/H$ as morphisms. The morphisms are automorphisms forming a group $\mathrm{Aut}(G/H)$ under composition. A covariant functor $\Gamma(G/H) \to R$-Mod is essentially the same thing as a left module over the group ring $R\,\mathrm{Aut}(G/H)$. Note that we have an anti-isomorphism $NH/H \to \mathrm{Aut}(G/H)$, mapping $n \in NH$ to $G/H \to G/H$, $gH \mapsto gnH$. This transforms left $R\,\mathrm{Aut}(G/H)$-modules into right $R(NH/H)$-modules.

(11.3) Example. Let $G = \mathbb{Z}/p$ be the group of prime order p. The orbit category $\mathrm{Or}(G)$ has two objects G and G/G, see (10.1). Consider a right $R\,\mathrm{Or}(G)$-module, i.e. a contravariant functor $M\colon \mathrm{Or}(G) \to R$-Mod. We describe M explicitly as follows: There are given two R-modules $M(G)$ and $M(G/G)$. The morphism $r_g\colon G \to G$, $u \mapsto ug$ induces $M(r_g)\colon M(G) \to M(G)$ and we have $M(r_h)M(r_g) = M(r_g r_h) = M(r_{hg})$. Thus

$$G \times M(G) \to M(G), \quad (g, m) \mapsto M(r_g)m$$

is the structure of a left (!) G-module. The morphism $p\colon G \to G/G$ induces an R-linear map $M(G/G) \to M(G)$ which, since $p \circ r_g = p$, has an image contained in the G-fixed point set $M(G)^G$. One can see easily that giving a right $R\,\mathrm{Or}(G)$-module amounts to specifying a left RG-module $M(G)$, a left R-module $M(G/G)$, and an R-linear map $M(G/G) \to M(G)^G$.

(11.4) Example. Let $\pi_1(X)$ be the fundamental groupoid of the space X. A left $R\pi_1(X)$-module is a (covariant) local coefficient system with values in R-modules (see, e.g., G. W. Whitehead [1978], p. 257).

The terminology of $R\Gamma$-modules has the advantage that most notions from the linear algebra of modules have a straight-forward extension to this more general setting. This will be apparent from the following collection of definitions.

The notions **submodule, quotient module, kernel, image**, and **cokernel** for $R\Gamma$-modules are defined object-wise. A sequence of $R\Gamma$-modules is called **exact** if it is exact at each object of Γ. The **direct sum** (= **coproduct**) of $R\Gamma$-modules is given by taking the usual direct sum object-wise. (Compare Schubert [1970], 7.5 and 8.5.)

If S is any set, let RS be the free R-module over S. Let Hom denote Hom-sets in Γ. We have the right $R\Gamma$-module

$$R\Gamma(?, x): \Gamma \to R\text{-Mod}, \ ? \mapsto R\mathrm{Hom}(?, x)$$

and, similarly, the left $R\Gamma$-module $R\Gamma(x, ?)$. These are defined for each object $x \in \mathrm{Ob}(\Gamma)$.

(11.5) Hom_Γ. If M and N are $R\Gamma$-modules, then $\mathrm{Hom}_\Gamma(M, N)$ is the R-module of all natural transformations $M \to N$. The same notation is used for right modules. This notation should not be confused with the Hom-sets in the category Γ which are often denoted by Hom_Γ, too. In this respect, a more appropriate notation for the objects we have in mind would be

$$\mathrm{Hom}_{R\Gamma\text{-Mod}}(M, N).$$

This is obviously too clumsy and rarely necessary.

(11.6) Tensor product. Let M be a right and N a left $R\Gamma$-module. Their tensor product

$$M \otimes_\Gamma N$$

is the following R-module. Form the direct sum

$$F = \bigoplus_{x \in \mathrm{Ob}(\Gamma)} M(x) \otimes_R N(x).$$

Let F' be the R-submodule generated by all elements of the form $mf \otimes n - m \otimes fn$; here, $f\colon x \to y \in \mathrm{Mor}(\Gamma)$, $m \in M(y)$, $n \in N(x)$, $mf = M(f)(m)$, and $fn = N(f)(n)$. Define $M \otimes_\Gamma N = F/F'$.

This tensor product has the familiar adjointness property with respect to the Hom-functor. We recall this in our terminology. Given two small categories Γ and Λ, one has the category of $R\Gamma$-$R\Lambda$-bimodules

$$[\Gamma \times \Lambda^{\mathrm{op}}, R\text{-Mod}] = [\Gamma, [\Lambda^{\mathrm{op}}, R\text{-Mod}]] =: R\Gamma\text{-Mod-}R\Lambda.$$

For a right $R\Gamma$-module A, a $R\Gamma$-$R\Lambda$-bimodule B, and a right $R\Lambda$-module C, one has a natural isomorphism

(11.7) $\mathrm{Hom}_\Lambda(A \otimes_\Gamma B, C) \cong \mathrm{Hom}_\Gamma(A, \mathrm{Hom}_\Lambda(B, C)).$

(11.8) Restriction. Let $F\colon \Gamma \to \Lambda$ be a covariant functor. The induced functor

$$\mathrm{Res}_F\colon [\Lambda^{\mathrm{op}}, R\text{-Mod}] \to [\Gamma^{\mathrm{op}}, R\text{-Mod}]$$

is called **restriction along** F.

(11.9) Induction. Let $F\colon \Gamma \to \Lambda$ be a covariant functor. We define an $R\Gamma$-$R\Lambda$-bimodule

$$R(??, F(?))\colon \Gamma \times \Lambda^{\mathrm{op}} \to R\text{-Mod}$$

on objects $(x, y) \mapsto R\mathrm{Hom}(y, F(x))$. For any right $R\Gamma$-module $M \in \mathrm{Ob}[\Gamma^{\mathrm{op}}, R\text{-Mod}]$, we have the right $R\Lambda$-module

$$M \otimes_\Gamma R(??, F(?)).$$

This construction defines the functor **induction along** F

$$\mathrm{Ind}_F\colon [\Gamma^{\mathrm{op}}, R\text{-Mod}] \to [\Lambda^{\mathrm{op}}, R\text{-Mod}].$$

The names for the functors (11.8) and (11.9) stem from representation theory (= module theory over group rings).

(11.10) Example. Let G and H be groups, considered as categories; see example (11.1). Then a functor $F\colon G \to H$ is nothing else but a homomorphism from G to H. If M is a right RH-module, then $\mathrm{Res}_F M$ is the right RG-module given by $(m, g) \mapsto mF(g)$. If $F\colon G \subset H$, then this is the restriction of the H-action to G in the ordinary sense.

Now, considered as an R-module, the bimodule $R\mathrm{Hom}_H(y, F(x))$ is just RH because $\mathrm{Hom}_H(y, F(x)) = H$ in this case (y and x the unique objects of the categories). This becomes a left RG-module via $(g, h) \mapsto F(g)h$ and

$$\mathrm{Ind}_F M = M \otimes_{RG} RH$$

as a right RH-module in the ordinary sense (induced representation).

Induction and restriction have the expected properties.

(11.11) Proposition.
(i) Ind_F *and* Res_F *are compatible with direct sums.*
(ii) *If F and F' are composable functors, then* $\mathrm{Ind}_{FF'} = \mathrm{Ind}_{F'}\mathrm{Ind}_F$ *and* $\mathrm{Res}_{F'F} = \mathrm{Res}_{F'}\mathrm{Res}_F$ *(up to canonical natural isomorphism of functors).*
(iii) Ind_F *and* Res_F *are adjoint, i.e. there exists a natural isomorphism*

$$\mathrm{Hom}_\Lambda(\mathrm{Ind}_F M, N) \cong \mathrm{Hom}_\Gamma(M, \mathrm{Res}_F N).$$

A *proof* of (iii) uses (11.7) and the isomorphism $\mathrm{Hom}_\Lambda(R(??, F(x)), N(??)) = N(F(x))$ (Yoneda-Lemma). (i) and (ii) are easily verified. □

(11.12) Free $R\Gamma$-modules. A **Γ-set** is a family $(B_x \mid x \in \mathrm{Ob}(\Gamma))$ of sets B_x. A Γ-

map between two Γ-sets (B_x) and (C_x) is a family $(f_x\colon B_x \to C_x \mid x \in \mathrm{Ob}(\Gamma))$ of set maps. We also think of Γ-sets as pairs (B, t), $t\colon B \to \mathrm{Ob}(\Gamma)$, with $B_x = t^{-1}(x)$. An $R\Gamma$-module defines a Γ-set by forgetting structure.

The $R\Gamma$-module M is called **free with** Γ-set $B = (B_x)$ as **basis** if B is a Γ-subset of M, i.e. $B_x \subset M(x)$, and if each Γ-map $f\colon B \to N$ into an $R\Gamma$-module N has a unique extension $F\colon M \to N$ which is an $R\Gamma$-homomorphism.

Given a Γ-set (B, t), we can construct the **free $R\Gamma$-module over** (B, t) as

$$\bigoplus_{b \in B} R\mathrm{Hom}(?, t(b));$$

we identify $b \in B$ with $\mathrm{id}(t(b)) \in R\mathrm{Hom}(t(b), t(b))$. Note that, by the Yoneda lemma, the module $R\mathrm{Hom}(?, y)$ is free with $\mathrm{id}(y)$ as basis element.

Let M be an $R\Gamma$-module and $E \subset M$ be a Γ-subset. The submodule generated by E is defined to be the smallest $R\Gamma$-submodule containing E. We call M **finitely generated** if it is generated by a finite Γ-subset. A module is finitely generated if and only if it is a quotient of a finitely generated free module.

(11.13) Projective $R\Gamma$-modules. Projective $R\Gamma$-modules are defined by the usual lifting property:

$$\begin{array}{ccccc} & & P & & \\ & \swarrow^{w} & \downarrow{\scriptstyle v} & & \\ M & \xrightarrow[u]{} & N & \longrightarrow & 0 \end{array}$$

Given a morphism v and a surjection u, there exists w such that $uw = v$.

In the sequel, we describe projective $R\Gamma$-modules for geometrically relevant categories Γ. Note that free modules are projective.

B. Projective modules.

The various categories defined in 10. have the property that endomorphisms are isomorphisms. Therefore, let Γ be an EI-**category**, which, by definition, is a small category in which each endomorphism is an isomorphism. We define a pre-order on the set $\mathrm{Ob}(\Gamma)$ of objects in the following way: $x \leq y \Leftrightarrow \mathrm{Hom}_\Gamma(x, y) \neq \emptyset$ and $x < y$ if $x \leq y$ and x not isomorphic to y. This induces a partial ordering on the set $\mathrm{Is}(\Gamma)$ of isomorphism classes of objects. The aim of this section is to analyse projective $R\Gamma$-modules. It turns out that they can be constructed from projective modules over group rings. We work with right modules, i.e. contravariant functors $M\colon \Gamma \to R$-Mod. Let Mod-$R\Gamma$ be the category of such modules. If x is an object of Γ with automorphism group $\mathrm{Aut}(x)$, we let $R[x] = R\mathrm{Aut}(x)$ be the group ring of $\mathrm{Aut}(x)$ and write Mod-$R[x]$ for the category of right $R[x]$-modules.

Given $x \in \mathrm{Ob}(\Gamma)$, we define a **splitting functor** S_x: Mod-$R\Gamma \to$ Mod-$R[x]$ as follows. Let M be an $R\Gamma$-module. Let $M(x)_S$ be the R-submodule of $M(x)$ which is generated by the images of all $M(f)$: $M(y) \to M(x)$ where f: $x \to y$ runs through non-isomorphisms in Γ. Each automorphism $g \in \mathrm{Aut}(x)$ induces a map $M(g)$: $M(x) \to M(x)$ which maps $M(x)_S$ into itself. We call $M(x)_S$ the singular part of $M(x)$. We set $S_x(M) = M(x)/M(x)_S$ which, by the action of the $M(g)$, becomes a right $R[x]$-module. It is clear how S_x is defined on morphisms. If we apply S_x to the Hom-functor $R\Gamma(?, y) = R\mathrm{Hom}(?, y)$, we obtain

(11.14) $S_x R\Gamma(?, y) \cong R[x]$ as right $R[x]$-module if $y \cong x$.
$S_x R\Gamma(?, y) = 0$ otherwise.

Associated with each $x \in \mathrm{Ob}(\Gamma)$ is another functor, the **extension functor** E_x: Mod-$R[x] \to$ Mod-$R\Gamma$. Let A be a right $R[x]$-module. For $y \in \mathrm{Ob}(\Gamma)$, we consider $R\mathrm{Hom}(y, x)$ as a left $R[x]$-module; the module structure is induced by composition $\mathrm{Hom}(y, x) \times \mathrm{Aut}(x) \to \mathrm{Hom}(y, x)$ in Γ. Thus we can define the functor

$$E_x(A) = A \otimes_{R[x]} R\mathrm{Hom}(?, x).$$

If we apply E_x to $R[x]$ as right $R[x]$-module, we obtain

(11.15) $E_x R[x] = R[x] \otimes_{R[x]} R\mathrm{Hom}(?, x) \cong R\mathrm{Hom}(?, x).$

The next result follows easily from the definitions.

(11.16) Proposition. *Let $x, y \in \mathrm{Ob}(\Gamma)$. Then*

(i) *$S_y E_x = 0$ if $x \not\cong y$.*
If $x \cong y$, then $S_y E_x$ is the natural equivalence Mod-$R[x] \to$ Mod-$R[y]$, $A \mapsto A \otimes_{R[x]} R\mathrm{Hom}(y, x)$.

(ii) *S_x and E_x are compatible with direct sums. They map finitely generated (free, projective) modules to modules with the same property.*

Proof. (i) To verify $S_y E_x = 0$, consider $E_x A(y) = A \otimes_{R[x]} R\mathrm{Hom}(y, x)$; observe that the element $a \otimes f$ for f: $y \to x$ is the image $E_x A(f)(a \otimes \mathrm{id}(x))$ and, since for $y \not\cong x$ the morphism f is a non-isomorphism, $a \otimes f \in E_x A(y)_S$.

If $y \cong x$ and f: $y \to z$ is a non-isomorphism, then $\mathrm{Hom}_\Gamma(z, x) = \emptyset$ because Γ is an EI-category. Therefore, $E_x A(z) = 0$ and hence $E_x A(y)_S = 0$ and $S_y E_x A \cong E_x A(y) = A \otimes_{R[x]} R\mathrm{Hom}(y, x)$.

(ii) Compatibility with direct sums is easy. The typical free modules $R[x]$ and $R\mathrm{Hom}(?, y)$ are mapped to a typical free module or zero. Hence an arbitrary free module, being a direct sum of typical ones, is mapped to a free module. A surjection is mapped onto a surjection; therefore, a quotient of a finitely generated free module is mapped to a quotient of a finitely generated free module. A direct summand of a free module, i.e. a projective module, is mapped to a direct summand of a free module. □

We let $\mathrm{Is}(\Gamma)$ be the set of isomorphism classes of objects in Γ. A direct sum $\bigoplus_{x \in \mathrm{Is}(\Gamma)}$ means that we sum over a representative set of $x \in \mathrm{Ob}(\Gamma)$ for $\mathrm{Is}(\Gamma)$. We call an $R\Gamma$-module M of **finite type** if it is a quotient of a module of the form

(11.17) $$F = \bigoplus_{j=1}^{n} \bigoplus_{I_j} R\mathrm{Hom}(?, x_j)$$

for non-isomorphic objects $x_1, \dots, x_n$ of Γ and arbitrary index sets $I_1, \dots, I_n$. A finitely generated module is of finite type.

(11.18) Theorem. *Let P be a projective $R\Gamma$-module of finite type. Then P is isomorphic to*

$$\bigoplus_{x \in \mathrm{Is}(\Gamma)} E_x S_x(P).$$

If P is a quotient of (11.17), *then a sum over $x_1, \dots, x_n$ suffices.*

The idea of the proof is to split off inductively the $E_x P_x(P)$ beginning with a maximal x where x, in case of finite $\mathrm{Is}(\Gamma)$, is called maximal if $\mathrm{Hom}_\Gamma(x, y) \neq \emptyset$ implies $x \cong y$. We prepare the proof by an analysis of what essentially is the induction step.

For $x \in \mathrm{Ob}(\Gamma)$, let Γ_x be the full subcategory of Γ with the single object x and let $i\colon \Gamma_x \to \Gamma$ be the inclusion. Then E_x can be interpreted as induction along i (see 11.9), $\mathrm{Ind}_i = E_x$. We write Res_x for Res_i. The adjointness (11.10, iii) leads to a natural transformation $I_x\colon E_x \mathrm{Res}_x \to \mathrm{Id}$ between functors $[\Gamma, R\text{-Mod}]^{\mathrm{op}} \to [\Gamma, R\text{-Mod}]^{\mathrm{op}}$. Explicitly, if M is a right $R\Gamma$-module then

$$I_x M\colon\ E_x \mathrm{Res}_x M = M(x) \otimes_{R[x]} R\mathrm{Hom}(?, x) \to M(?)$$

is given by mapping $m \otimes f$ to $M(f)(m)$. Here, $I_x M$ is a natural transformation between functors $\Gamma^{\mathrm{op}} \to R$-Mod. The cokernel, denoted by $\mathrm{Cok}_x M$, is again such a functor and $M \mapsto \mathrm{Cok}_x M$ defines a functor Cok_x from right $R\Gamma$-modules to right $R\Gamma$-modules. Thus we have an exact sequence

(11.19) $$E_x \mathrm{Res}_x M \xrightarrow[I_x M]{} M \xrightarrow[Pr_x M]{} \mathrm{Cok}_x M \to 0$$

which is natural in M.

In order to prove (11.18), we look for a situation were $I_x M$ is injective and (11.19) splits. We first consider the simplest examples, namely free modules. We call an object $x \in \mathrm{Ob}(\Gamma)$ **M-maximal** if $x < y$ implies $M(y) = 0$. If x is M-maximal, then $M(x)_S = 0$ and $S_x M = \mathrm{Res}_x M = M(x)$; therefore, in (11.19) $\mathrm{Res}_x M$ can be replaced by $S_x M$.

If M is the typical free module $R\mathrm{Hom}(?, y)$, for $x \cong y$, then y is M-maximal. In this case, the map $I_x M$ is the canonical isomorphism

$$E_x S_x M = S_x M \otimes_{R[x]} R\mathrm{Hom}(?, y) = R[x] \otimes_{R[x]} R\mathrm{Hom}(?, y) \rightarrow R\mathrm{Hom}(?, y) = M.$$

If y is not isomorphic to x, then $E_x S_x R\mathrm{Hom}(?, y) = 0$ by (11.14). Applying these facts to a direct sum $M = \bigoplus_{j \in J} R\mathrm{Hom}(?, x_j)$ with x maximal among the x_j, then x is M-maximal and

(11.20) $0 \rightarrow E_x S_x M \rightarrow M \rightarrow \mathrm{Cok}_x M \rightarrow 0$

can be canonically identified with the split exact sequence

(11.21) $0 \rightarrow \bigoplus_{j \in I} R\mathrm{Hom}(?, x_j) \rightarrow M \rightarrow \bigoplus_{j \in J \setminus I} R\mathrm{Hom}(?, x_j) \rightarrow 0$

where $I = \{j \mid x_j \cong x\}$.

(11.22) Proposition. *Let M be an $R\Gamma$-module and $x \in \mathrm{Ob}(\Gamma)$ be M-maximal. Then the following holds:*

(i) *If M is projective, then*

$$0 \rightarrow E_x S_x M \rightarrow M \rightarrow \mathrm{Cok}_x M \rightarrow 0$$

is exact and splits.

(ii) *If M is finitely generated (free, projective), then* $\mathrm{Cok}_x M$ *is finitely generated (free, projective).*

Proof. We have just seen that (11.22) is true for free modules. Let M be projective. There exists a free module F which has x as an F-maximal object and such that M is a direct summand $M \xrightarrow{i} F \xrightarrow{r} M$, $ri = \mathrm{id}$. Let ϱ be a retraction of $I_x F$. Then $E_x S_x(r)\varrho i$ is a retraction for $I_x M$. Thus the sequence in (i) splits. Since $E_x S_x M$ is projective by (11.16, ii), so is $\mathrm{Cok}_x M$. □

Proof of (11.18). We write P as a quotient of (11.17), $q\colon F \rightarrow P$, and use induction on n. We can assume that x_n is maximal among the x_i. Then $x = x_n$ is P-maximal and F-maximal for F as in (11.17). Using the identification of (11.20) and (11.21), we obtain a commutative diagram

$$\begin{array}{ccccc}
\bigoplus_{I_n} R\mathrm{Hom}(?, x_n) & \longrightarrow & F & \longrightarrow & \bigoplus_{j=1}^{n-1} \bigoplus_{I_j} R\mathrm{Hom}(?, x_j) \\
\downarrow = & & \downarrow = & & \downarrow = \\
E_x S_x F & \longrightarrow & F & \longrightarrow & \mathrm{Cok}_x F \\
\downarrow & & \downarrow q & & \downarrow \mathrm{Cok}_x q \\
E_x S_x P & \longrightarrow & P & \longrightarrow & \mathrm{Cok}_x P.
\end{array}$$

By (11.22), the bottom sequence splits. Since $\mathrm{Cok}_x q$ is surjective and $\mathrm{Cok}_x P$ is projective, we can apply the induction hypothesis. □

The splitting (11.18) induces corresponding splittings of algebraic K-groups like projective class groups or Whitehead groups. We define some of these groups and their corresponding direct sum decompositions. We always use *right* $R\Gamma$-modules.

(11.23) $K_0(R\Gamma)$.

This is the Grothendieck group of finitely generated projective $R\Gamma$-modules. It may be constructed as the free abelian group on the isomorphism classes of such modules modulo the subgroup generated by those elements $[P_1]-[P]+[P_2]$ for which there is an exact sequence $0 \to P_1 \to P \to P_2 \to 0$.

(11.24) $K_0(R\Gamma, \text{free})$.

This is the Grothendieck group of finitely generated free $R\Gamma$-modules.

(11.25) $\tilde{K}_0(R\Gamma)$.

Inclusion of categories defines a homomorphism $K_0(R\Gamma, \text{free}) \to K_0(R\Gamma)$. The cokernel is denoted by $\tilde{K}_0(R\Gamma)$.

(11.26) $K_1(R\Gamma)$.

This is constructed as follows: Take the free abelian group over the set of isomorphisms $f\colon F \to F$ of finitely generated free $R\Gamma$-modules. Divide by the subgroup generated by elements of the following form
(i) $[gf]-[f]-[g]+[\mathrm{id}(F)]$ whenever f and g are automorphisms of F.
(ii) $[f_1]-[f]+[f_2]$ whenever one has a commutative diagram

$$\begin{array}{ccccccccc} 0 & \to & F_1 & \to & F & \to & F_2 & \to & 0 \\ & & \downarrow f_1 & & \downarrow f & & \downarrow f_2 & & \\ 0 & \to & F_1 & \to & F & \to & F_2 & \to & 0 \end{array}$$

with exact rows and automorphisms f, f_1, and f_2.

(11.27) $Wh(R\Gamma)$.

This is the quotient of $K_1(R\Gamma)$ by the subgroup generated by elements of the form

$$[\pm R\mathrm{Hom}(?,f)]-[\mathrm{id}(R\mathrm{Hom}(?,x))]$$

whenever f is an automorphism of x and

$$\pm R\mathrm{Hom}(?,f)\colon R\mathrm{Hom}(?,x) \to R\mathrm{Hom}(?,x),\ g \mapsto \pm fg.$$

(11.28) $\tilde{K}_1(R\Gamma)$ and $\tilde{W}h(R\Gamma)$.

These are the quotients of $K_1(R\Gamma)$ and $Wh(R\Gamma)$ by the subgroup which is generated by all elements $[\mathrm{id}(F)]$, F a finitely generated free $R\Gamma$-module.

For generalities about such constructions in algebraic K-theory, see Bass [1968]. For the relation to classical algebraic K-theory of group rings, see the exercises; the identification with the standard K-theoretic groups is necessary if one wants to use the splitting results.

We write K for one of the groups K_0, $K_0(\text{free})$, $\tilde{K}_0$, K_1, Wh, $\tilde{K}_1$, $\tilde{W}h$. If $\Gamma_x \subset \Gamma$ is the full subcategory with the single object x, we write $K(R[x])$ for $K(R\Gamma_x)$. Of course, $R[x]$ is a group ring and one has to be careful to use our definitions (11.23)–(11.28) also in this case. We have previously defined functors S_x and E_x. They preserve the relations needed to define the K-groups and therefore induce homomorphisms $K(S_x)$: $K(R\Gamma) \to K(R[x])$ and $K(E_x)$: $K(R[x]) \to K(R\Gamma)$. Taking sums of the $K(S_x)$ and $K(E_x)$ over $x \in \mathrm{Is}(\Gamma)$, we obtain

$$K(S)\colon\; K(R\Gamma) \to \bigoplus_{x \in \mathrm{Is}(\Gamma)} K(R[x])$$

$$K(E)\colon\; \bigoplus_{x \in \mathrm{Is}(\Gamma)} K(R[x]) \to K(R\Gamma).$$

(11.29) Proposition. *The maps $K(S)$ and $K(E)$ are isomorphisms inverse to each other.*

Proof. Proposition (11.16, i) implies $K(S)K(E) = \mathrm{id}$. It remains to show that $K(E)K(S) = \mathrm{id}$. Let M be a finitely generated projective $R\Gamma$-module resp. $f\colon M \to M$ an automorphism of a finitely generated free $R\Gamma$-module M. We write M as a quotient of (11.17) and show by induction on n that $K(E)K(S)[M] = [M]$ resp. $K(E)K(S)[f] = [f]$ (compare proof of (11.18)). Let $x = x_n$ be maximal among the $x_1, \ldots, x_n$ in (11.17). Then x is M- resp. F-maximal and from (11.22) we obtain an exact sequence

$$0 \to E_x S_x M \to M \to \mathrm{Cok}_x M \to 0$$

and similarly for (M, f) in place of M. Thus we have the K-group identities

$$[M] = K(E_x)K(S_x)[M] + [\mathrm{Cok}_x M]$$
$$[f] = K(E_x)K(S_x)[f] + [\mathrm{Cok}_x f]$$

As shown in the proof of (11.18), $\mathrm{Cok}_x M$ is a quotient of the module (11.17) corresponding to $x_1, \ldots, x_{n-1}$ so that the induction hypothesis can be applied to it. We now usw $S_y(\mathrm{Cok}_x M) = 0$ if $x \cong y$ and $S_y(\mathrm{Cok}_x M) \cong \mathrm{Cok}_x M$ if $x \ncong y$ (exercise 9). □

C. Further properties.

Let Γ_1 and Γ_2 be EI-categories. Let

$$M \in [\Gamma_1^{op}, R\text{-Mod}] \text{ and } N \in [\Gamma_2^{op}, R\text{-Mod}].$$

We have the tensor product over R

$$M \otimes_R N \in [\Gamma_1^{op} \times \Gamma_2^{op}, R\text{-Mod}]$$

defined on objects by

$$(M \otimes_R N)(x, y) = M(x) \otimes_R N(y).$$

We want to describe the behaviour of $\otimes_R$ with respect to the S- and E-functors.

(11.30) Suppose $A \in R[x]$-Mod and $B \in R[y]$-Mod. Then

$$\begin{aligned}
&E_x(A) \otimes_R E_y(B)\\
&= (A \otimes_{R[x]} R\operatorname{Hom}(?, x)) \otimes_R (B \otimes_{R[y]} R\operatorname{Hom}(?, y))\\
&\cong (A \otimes_R B) \otimes_{R[x] \otimes R[y]} (R\operatorname{Hom}(?, x) \otimes_R R\operatorname{Hom}(?, y))\\
&\cong (A \otimes_R B) \otimes_{R[(x, y)]} R\operatorname{Hom}(?, (x, y))\\
&= E_{(x,y)}(A \otimes_R B).
\end{aligned}$$

This isomorphism is natural in A and B.

(11.31) We have defining exact sequences

$$0 \to M(x)_S \to M(x) \to S_x(M) \to 0$$
$$0 \to N(y)_S \to N(y) \to S_y(N) \to 0.$$

They yield an exact sequence

$$M(x)_S \otimes_R N(y) + M(x) \otimes_R N(y)_S \xrightarrow{j} M(x) \otimes_R N(y) \to S_x(M) \otimes_R S_y(N) \to 0$$

which may not be injective on the left. We look at $(M \otimes_R N)(x, y)_S$. This is generated by images

(11.32) $M(u) \otimes_R N(v) \to M(x) \otimes_R N(y)$

for non-isomorphisms $(\varphi, \psi)\colon (x, y) \to (u, v)$. This means that φ or ψ (or both) are non-isomorphisms. If φ is a non-isomorphism, then

$$M(u) \to M(x)_S \subset M(x)$$

so that (11.32) factors over $M(x)_S \otimes_R N(y) \to M(x) \otimes_R N(y)$. Hence the image of j contains $(M \otimes_R N)(x, y)_S$. On the other hand, each element in this image is in $(M \otimes_R N)(x, y)_S$ so that the image of j is precisely the S-part. Therefore,

$$S_x(M) \otimes_R S_y(N) \cong S_{(x,y)}(M \otimes_R N).$$

This isomorphism is natural in M and N.

(11.33) Next we study the behaviour of Ind_F with respect to E. Here, $F\colon \Gamma_1 \to \Gamma_2$ is a covariant functor between EI-categories. We claim that

$$\mathrm{Ind}_F E_x(A) \cong E_{F(x)}(\mathrm{Ind}_{F_x} S_x(A)).$$

The right hand side has the following meaning. Given $x \in \mathrm{Ob}(\Gamma_1)$,

$$F_x\colon \mathrm{Aut}(x) \to \mathrm{Aut}(F(x)),\ f \mapsto F(f)$$

is a group homomorphism which makes $R[F(x)]$ into a left (or right) $R[x]$-module. This induces the natural transformation

$$\begin{aligned} R[x]^{\mathrm{op}} - \mathrm{Mod} &\to R[F(x)]^{\mathrm{op}} - \mathrm{Mod} \\ A &\mapsto A \otimes_{R[x]} R[F(x)]. \end{aligned}$$

Proof of the claim. We know that $E_x = \mathrm{Ind}_i$ for $i\colon \Gamma_x \to \Gamma$. Moreover, $\mathrm{Ind}_F \mathrm{Ind}_i(A) = \mathrm{Ind}_{Fi}(A)$. Now use the definition of Ind_{Fi}; see (11.9). □

(11.34) We determine the behaviour of Ind_F with respect to S_y. We claim that

$$S_y(\mathrm{Ind}_F P) \cong \oplus\, \mathrm{Ind}_{F_x} S_x(P)$$

for a projective $R\Gamma_1$-module P of finite type; the direct sum is taken over a complete set of non-isomorphic objects $x \in \mathrm{Ob}(\Gamma_1)$ such that $F(x) \cong y$.

Proof. We write $P \cong \oplus\, E_x S_x(P)$ (see (11.18)) and obtain

$$\mathrm{Ind}_F P \cong \oplus\, \mathrm{Ind}_F E_x S_y(P).$$

Then we use $\mathrm{Ind}_F E_x = E_{F(x)} \mathrm{Ind}_{F_x}$ (see (11.33)) and arrive at

$$\mathrm{Ind}_F(P) \cong \oplus\, E_{F(x)} \mathrm{Ind}_{F_x} S_x(P).$$

If we apply S_y to this relation, then, by (11.16), only the summands with $y \cong F(x)$ remain. □

(11.35) We now look at the restriction Res_F for $F\colon \Gamma_1 \to \Gamma_2$. We are interested in the following question: When does Res_F map finitely generated free resp. projective modules to modules with the same property? Since Res_F is compatible with direct sums, this reduces to the question: Is $\mathrm{Res}_F R\Gamma_2(?, y)$ a finitely-generated free $R\Gamma_1$-module for all $y \in \mathrm{Ob}(\Gamma_2)$?

Since S_x preserves the properties of being finitely generated or free, we must ensure that, for all $x \in \mathrm{Ob}(\Gamma_1)$, the $R[x]$-module $S_x \mathrm{Res}_F R\Gamma_2(?, y)$ is a finitely generated free $R[x]$-module. We have

$$\operatorname{Res}_F R\Gamma_2(?, y) = R\operatorname{Hom}(F(?), y)$$
$$\operatorname{Res}_F E_y(A) = A \otimes_{R[y]} R\operatorname{Hom}(F(?), y).$$

In order to form S_x, look at the singular part. Let $g\colon x \to x'$ be a non-isomorphism. Then the singular part is generated by the images

$$R\operatorname{Hom}(F(x'), y) \to R\operatorname{Hom}(F(x), y)$$

and similarly with $A \otimes_{R[y]}$ in front. The image is thus generated by basis elements of the form

$$F(x) \xrightarrow[F(g)]{} F(x') \xrightarrow[h]{} y.$$

When we factor out these, we want to obtain something finitely generated. We therefore use the following definition.

A morphism $f\colon F(x) \to y$ in Γ_2 is called **irreducible** if the following holds: Whenever f factors as $f = hF(g)$, then g is necessarily an isomorphism.

Let $\operatorname{Irr}(x, y) \subset \operatorname{Hom}_{\Gamma_2}(F(x), y)$ be the set of irreducible morphisms. We have a right action of $\operatorname{Aut}(x)$ on this set by composition. If the quotient $\operatorname{Irr}(x, y)/\operatorname{Aut}(x)$ is finite, then we set

$$\operatorname{index}(x, y) = |\operatorname{Irr}(x, y)/\operatorname{Aut}(x)|,$$

the cardinality of this set. The set $\operatorname{Irr}(x, y)$ has also a left $\operatorname{Aut}(y)$-action.

Altogether, $R\operatorname{Irr}(x, y)$ is a $R[y]$-$R[x]$-bimodule. With these definitions, we can state the

(11.36) Lemma. $S_x \operatorname{Res}_F R\Gamma_2(?, y) \cong R\operatorname{Irr}(x, y)$ *as a bimodule.*

Proof. $S_x \operatorname{Res}_F R\Gamma_2(?, y) = R\operatorname{Hom}(F(x), y)/R\operatorname{Hom}(F(x), y)_S$ and this is the free R-module generated by the irreducible elements in $\operatorname{Hom}(F(x), y)$. □

We apply (11.36) in the following chain of isomorphisms of $R[y]$-modules

(11.37)
$$\begin{aligned} S_x \operatorname{Res}_F E_y(A) &= S_x(A \otimes_{R[y]} R\operatorname{Hom}(F(?), y)) \\ &\cong A \otimes_{R[y]} S_x R\operatorname{Hom}(F(?), y) \\ &\cong A \otimes_{R[y]} R\operatorname{Irr}(x, y). \end{aligned}$$

If P is an $R\Gamma_2$-module of finite type, we can apply (11.37) to (11.18) for P and obtain the isomorphism

$$S_x \operatorname{Res}_F(P) \cong \bigoplus_{y \in \operatorname{Is}(\Gamma_2)} S_y(P) \otimes_{R[y]} R\operatorname{Irr}(x, y).$$

In the sequel, we use the following definition. A functor $F\colon \Gamma_1 \to \Gamma_2$ is called admissible for K-theory, or ***K*-admissible** for short, if it has the following properties:

(11.38)

(i) For each $y \in \mathrm{Ob}(\Gamma_2)$ we have $\mathrm{Irr}(x, y) = \emptyset$ for all $x \in \mathrm{Ob}(\Gamma_1)$ except for finitely many isomorphism classes.

(ii) For all $x \in \mathrm{Ob}(\Gamma_1)$, the $R[x]$-module $R\mathrm{Irr}(x, y)$ is finitely generated and free.

(iii) Let $h: F(z) \to y$ be in Γ_2. There exists $g: z \to x$ in Γ_1 and $f: F(x) \to y$ in Γ_2 such that $f \circ F(g) = h$ and $f \in \mathrm{Irr}(x, y)$. Suppose $g': z \to x'$ and $f': F(x') \to y$ give another such factorization of h. Then there exists an isomorphism $k: x \to x'$ in Γ_1 such that $k \circ g = g'$ and $f' \circ F(k) = f$.

(11.39) Proposition. (i) Res_F *preserves the property of being finitely generated free if and only if F is K-admissible.*
(ii) *Suppose F is K-admissible. Then*

$$\mathrm{Res}_F R\Gamma_2(?, x) \cong \bigoplus_{x \in \mathrm{Is}(\Gamma_1)} \mathrm{index}(x, y) R\Gamma_1(?, x).$$

(Here the notation nM stands for the direct sum of n copies of the module M.)

Proof. Suppose Res_F preserves the property of being finitely generated free. Then we can apply (11.18) and (11.37) to the module $\mathrm{Res}_F R\mathrm{Hom}(?, y) = R\mathrm{Hom}(F(?), y)$ and see that the natural transformation

$$T: \bigoplus_{x \in \mathrm{Is}(\Gamma_1)} R\mathrm{Irr}(x, y) \otimes_{R[x]} R\mathrm{Hom}(?, x) \to R\mathrm{Hom}(F(?), y),$$

which maps $f \otimes g$ to $fF(g)$, is an isomorphism. It is easy to see that K-admissibility is equivalent to T being an isomorphism. The assertion (ii) is just another way of writing this isomorphism. □

We now apply the preceding general considerations to the discrete fundamental group category; see (10.9).

Let $i: H \to G$ be an inclusion of compact Lie groups. There is an associated functor

$$F = \pi_1^d(i, X): \pi_1^d(H, \mathrm{res}_H X) \to \pi_1^d(G, X)$$

for each G-space X; see (10.14), Ex. 8. We want to determine the irreducible morphisms from $F(\alpha)$ to β where $\alpha: H/K \to X$ and $\beta: G/L \to X$ are objects in the respective categories. Since each morphism $\sigma: H/K \to \mathrm{res}_H G/L$ factors as $H/K \to \mathrm{Image}\,\sigma \to \mathrm{res}_H G/L$, one sees that a morphism $(\sigma, \phi): \alpha \to \beta$ is irreducible if and only if $H/K \to \mathrm{res}_H G/L$ is injective. Using this, one shows

(11.40) Proposition. *Let $i: H \to G$ be an inclusion of compact Lie groups of the same dimension. Then the following holds:*
(i) *$\pi_1^d(i, X)$ is K-admissible.*

(ii) Index(α, β) *is the number of orbits* c *of type* H/K *in* $\mathrm{res}_H G/L$ *such that there exists a path in* X^K *from* α *to* $\beta|c$. □

Finally, we consider functors induced by Ind_F and Res_F in K-groups and their behaviour with respect to the splittings. Let $F\colon \Gamma_1 \to \Gamma_2$ be a functor between EI-categories. Then we have induced homomorphisms

(11.41) $K(\mathrm{Ind}_F)\colon\ K(\Gamma_1) \to K(\Gamma_2)$

if K stands for one of the symbols K_0, K_0 (free), $\tilde{K}_0$, K_1, Wh, $\tilde{K}_1$, $\tilde{W}h$. If F is K-admissible, then, similarly, we have

(11.42) $K(\mathrm{Res}_F)\colon\ K(\Gamma_2) \to K(\Gamma_1)$.

The next proposition describes the splitting behaviour. We leave the verification to the reader. Let $\mathrm{Ind}_F(\bar{x}, \bar{y})\colon K(R[x]) \to K(R[y])$ be the homomorphism induced by induction along F_x if $F\bar{x} = \bar{y}$ and 0 otherwise. (Here, $\bar{x}$ is the isomorphism class of x)

(11.43) Proposition. *The following diagrams are commutative (the second if* F *is* K*-admissible).*

(i)

$$\begin{array}{ccc} K(R\Gamma_1) & \xrightarrow{\quad K(\mathrm{Ind}_F)\quad} & K(R\Gamma_2) \\ {\scriptstyle K(S)}\Big\downarrow\Big\uparrow{\scriptstyle K(E)} & & \Big\downarrow\Big\uparrow \\ \bigoplus\limits_{\bar{x}} K(R[x]) & \xrightarrow[(\mathrm{ind}_F(\bar{x},\bar{y}))]{\qquad\qquad} & \bigoplus\limits_{\bar{y}} K(R[y]) \end{array}$$

(ii)

$$\begin{array}{ccc} K(R\Gamma_2) & \xrightarrow{\quad K(\mathrm{Res}_F)\quad} & K(R\Gamma_1) \\ {\scriptstyle K(S)}\Big\downarrow\Big\uparrow{\scriptstyle K(E)} & & \Big\downarrow\Big\uparrow \\ \bigoplus\limits_{\bar{y}} K(R[y]) & \xrightarrow[(?\,\otimes_{R[y]} R\,\mathrm{Irr}(x,y))]{\qquad\qquad} & \bigoplus\limits_{\bar{x}} K(R[x]) \end{array}$$

The horizontal maps at the bottom are to be interpreted as matrices. □

The results of this section are based on tom Dieck [1981] and unpublished work of Lück. Compare also Triantafillou [1982].

(11.44) Exercises.

1. For $M \in [\Gamma_1^{op}, R\text{-Mod}]$ and $N \in [\Gamma_2^{op}, R\text{-Mod}]$, define $M \otimes_R N$ by $(x, y) \mapsto M(x) \otimes_R N(y)$. Show that $\oplus$ and $\otimes_R$ are associative and commutative up to natural isomorphism. Show that $\otimes_R$ and $\otimes_\Gamma$ are distributive over $\oplus$, i.e. $(M \oplus N) \otimes_\Gamma P \cong (M \otimes_\Gamma P) \oplus (N \otimes_\Gamma P)$. Verify (1.7).
2. Show that the following assertions about $R\Gamma$-modules P are equivalent: (i) P is projective. (ii) Each exact sequence $0 \to M \to N \to P \to 0$ splits. (iii) $\mathrm{Hom}_\Gamma(-, P)$ is exact. (iv) P is direct summand of a free $R\Gamma$-module. Show that an $R\Gamma$-module P is finitely generated and projective if and only if it is a direct summand of a finitely generated free $R\Gamma$-module. Show that $\oplus$ and $\otimes_R$ preserve the properties of being finitely generated, free, or projective.
3. Show that Ind_F is compatible with the properties of being finitely generated, free, or projective.
4. Verify (11.40) in detail.
5. Suppose X is a G-space with simply connected fixed point sets X^K. Let α: $H/K \to \mathrm{res}_H X$ resp. β: $G/L \to X$ be objects in $\pi_1^d(H, \mathrm{res}_H X)$ resp. $\pi_1^d(G, X)$. Let $\mathrm{Mono}_H(H/K, \mathrm{res}_H G/L)$ be the set of H-homotopy classes $H/K \to \mathrm{res}_H G/L$ which have injective representatives. Show in addition to (11.39) that the sets $\mathrm{Irr}(\alpha, \beta)$ and $\mathrm{Mono}_H(H/K, \mathrm{res}_H G/L)$ are isomorphic as left $\mathrm{Aut}_1^d(\beta) = \pi_0(\mathrm{Aut}_G(G/L))$-sets and as right $\mathrm{Aut}_1^d(\alpha) = \pi_0(\mathrm{Aut}_H(H/K))$-sets. Moreover, index (α, β) is the number of orbits of type H/K in $\mathrm{res}_H(G/L)$.
6. Establish the existence of (11.41) and (11.42). Give a proof of (11.43).
7. Show that $\otimes_R$ induces pairings

$$K(R\Gamma_1) \otimes L(R\Gamma_2) \;\to\; M(R(\Gamma_1 \times \Gamma_2))$$

if (K, L, M) is one of the following triples:
$(K_0(\text{free}), K_0(\text{free}), K_0(\text{free}))$
$(K_0(\text{free}), K_0, K_0)$
$(K_0(\text{free}), \tilde{K}_0, \tilde{K}_0)$
(K_1, K_1, K_1)
$(K_0(\text{free}), K_1, K_1)$
$(K_0(\text{free}), \tilde{K}_1, \tilde{K}_1)$
(Wh, Wh, Wh)
$(K_0(\text{free}), Wh, Wh)$
$(K_0(\text{free}), \widetilde{Wh}, \widetilde{Wh})$.
Use (11.30) and (11.31) to establish compatibility with splittings.
8. Let G be a finite group. Show that $K_0(\mathbb{Z}\mathrm{Or}(G), \text{free})$ is isomorphic to the additive group of the Burnside ring. What does the splitting (11.29) mean in this case?

9. Verify the statements at the end of the proof of (11.29).
10. Show that our $K_0(R[x])$ is the usual K_0-group of finitely generated projective $R[x]$-modules. Show that our $\tilde{K}_1(R[x])$ is the group which is usually denoted by $K_1(R[x])$.

Chapter II: Algebraic Topology

This chapter comprises the following sections.
1. Introduction and basic properties of equivariant CW-complexes.
2. Characterization of equivariant homotopy equivalences in terms of fixed point data. The equivariant version of the classical suspension theorem of Freudenthal.
3. Generalization of elementary classical obstruction theory to an equivariant situation.
4. Obstruction theory as a tool to classify equivariant homotopy classes by mapping degrees in certain cases. Existence of congruences between mapping degrees. Applications to lens spaces.
5. Determination of the congruences of section 4 for maps between complex representation spheres.
6. Definition of equivariant (co-) homology theories. Development from (naive) spectra to such theories. Theorems about changing the group in this context.
7. A splitting theorem for equivariant stable homotopy.
8. The Burnside ring as stable homotopy group. Lefschetz-Dold index.
9. Mackey functors for finite groups. Introduction of Bredon (co-) homology. Elementary theorems about the transfer associated to an orbit map.
10. The notion of a homotopy representation. Dimension function and homotopy representation group as basic invariants.

1. Equivariant CW-complexes.

Cell complexes are constructed by iterated attaching of cells. We remind the reader of a few general topological and categorical facts about the attaching process. In this section, G is a locally compact Hausdorff group. Subgroups are closed if not specified otherwise.

A diagram

$$\begin{array}{ccc} A & \xrightarrow{\;f\;} & Y \\ {\scriptstyle j}\downarrow & & \downarrow{\scriptstyle J} \\ X & \xrightarrow{\;F\;} & Z \end{array}$$

of G-spaces and G-maps is called **pushout** (in the category of G-spaces) if for each pair of G-maps $f'\colon Y \to U$, $j'\colon X \to U$ with $f'f = j'j$, there exists a unique G-map $u\colon Z \to U$ with $uJ = f'$, $uF = j'$.

The pushouts which are most important for us are obtained via attaching of spaces. This is the special case that $j\colon A \subset X$ is a *closed embedding*. In this case, a pushout is constructed as follows. Suppose X and Y are disjoint. Consider the following equivalence relation R on $X \cup Y$:

$$(z_1, z_2) \in R \Leftrightarrow \begin{cases} z_1, z_2 \in A \text{ and } f(z_1) = f(z_2); & \text{or} \\ z_1 \in A,\ z_2 \in f(A) \text{ and } f(z_1) = z_2; & \text{or} \\ z_2 \in A,\ z_1 \in f(A) \text{ and } f(z_2) = z_1; & \text{or} \\ z_1 \notin A \text{ and } z_1 = z_2. \end{cases}$$

The quotient space $(X \cup Y)/R = Z$ is denoted by $Y \cup_f X$. The canonical maps $X \to X \cup Y \leftarrow Y$ induce $F\colon X \to Y \cup_f X$ and $J\colon Y \to Y \cup_f X$. We say that $Y \cup_f X$ is obtained from Y by **attaching** X via f. As an exercise, the reader may prove the following.

(1.1) Proposition. *J is a closed embedding. The morphism $(F, f)\colon (X, A) \to (Y \cup_f X, Y)$ is a relative homeomorphism.* □

The closed inclusion $j\colon A \to X$ is called a **G-cofibration** if it has the **homotopy extension property** for all G-maps $f\colon X \to Y$ and for all G-homotopies $\varphi\colon A \times I \to Y$ with $\varphi(a, 0) = f(a)$ for $a \in A$. That means, given f and φ, there exists $\phi\colon X \times I \to Y$ such that $\phi | A \times I = \varphi$ and $\phi(x, 0) = f(x)$. For an exhaustive treatment of cofibrations, see tom Dieck-Kamps-Puppe [1970]; most parts of this book are easily extended to an equivariant situation and there is no need to discuss such matters here. As another exercise, we state

(1.2) Proposition. *If $j\colon A \to X$ is a G-cofibration, then $J\colon Y \to Y \cup_f X$ is a G-cofibration.* □

The ordinary theory of CW-complexes (see J.H.C. Whitehead [1949], G.W. Whitehead [1978] Chapter II, Lundell-Weingram [1969]) is easily extended to an equivariant setting. This has been carried out by Matumoto [1971] and Illman [1972], [1972a].

We collect the basic definitions and results. We use the notation $S^{n-1} = \{x \in \mathbb{R}^n \mid |x| = 1\}$ and $D^n = \{x \in \mathbb{R}^n \mid |x| \leqq 1\}$ for the unit $(n-1)$-sphere and the unit n-ball in $\mathbb{R}^n$. We set $S^{-1} = \emptyset$ and consider D^0 as a point-space. The spaces S^{n-1} and D^n carry the trivial G-action. We set $\mathring{D}^n = D^n \setminus S^{n-1}$. We refer the reader to G. W. Whitehead [1978] for standard homotopy theory.

Let $n \geqq 0$ be an integer. Let A be a G-space. Given a family $(H_j | j \in J)$ of closed subgroup H_j of G and G-maps

$$\varphi_j\colon G/H_j \times S^{n-1} \to A,\ j \in J,$$

we consider pushouts of G-spaces

$$\begin{array}{ccc} \coprod_{j\in J} G/H_j \times S^{n-1} & \xrightarrow[\varphi]{} & A \\ \cap & & \downarrow i \\ \coprod_{j\in J} G/H_j \times D^n & \xrightarrow[\phi]{} & X. \end{array} \tag{1.3}$$

We put $\varphi | G/H_j \times S^{n-1} = \varphi_j$ and $\phi | G/H_j \times D^n = \phi_j$. Such pushouts exist and i is a closed embedding. In this situation, we use the following terminology: X is obtained from A by (simultaneously) **attaching** the family **of (equivariant) n-cells** $(G/H_j \times D^n | j \in J)$ of **type** $(G/H_j | j \in J)$. We call $\phi(G/H_j \times D^n)$ resp. $\phi(G/H_j \times \mathring{D}^n)$ a **closed** resp. an **open** n-cell of type G/H_j; moreover, $\varphi(G/H_j \times S^{n-1})$ is called the (combinatorial) **boundary** of $\phi(G/H_j \times D^n)$. The map

$$(\phi_j, \varphi_j)\colon (G/H_j \times D^n, G/H_j \times S^{n-1}) \to (X, A)$$

is called the **characteristic map** of the corresponding n-cell, whereas φ_j is called the **attaching map**.

We shall also use the terms characteristic map and attaching map in a slightly different context. Since equivariant maps $\psi\colon G/H \times D^n \to X$ correspond bijectively to non-equivariant maps $\psi'\colon D^n \to X^H$, via the assignment $\psi(gH, x) = g\psi'(x)$, we also call the map $(\phi'_j, \varphi'_j)\colon (D^n, S^{n-1}) \to (X^H, A^H)$ a **characteristic** map of the n-cell $\phi(G/H_j \times D^n)$. We use the following fact:

(1.4) If A is a Hausdorff space, then X is a Hausdorff space.

If X is obtained from A by attaching n-cells, we consider $i\colon A \to X$ as inclusion. The space $X \backslash A$ is then homeomorphic to $\coprod_{j\in J} G/H_j \times \mathring{D}^n$ and is the topological sum of the open n-cells. The open n-cells are uniquely determined by the pair (X, A) being the G-components of $X \backslash A$, i.e. the components of $(X \backslash A)/G$ lifted to $X \backslash A$.

A reformulation of the pushout property shows that maps and homotopies may be extended cellularly from A to X. We spell this out for homotopies.

(1.5) Proposition. *Suppose X is obtained from A by attaching n-cells using characteristic maps ϕ_j as above. Let $f\colon (X, A) \to (Y, B)$ be a G-map and suppose that, for each $j \in J$, a G-homotopy*

$$g_j\colon G/H_j \times (D^n, S^{n-1}) \times I \to (Y, B)$$

of $f\phi_j$ rel $G/H_j \times S^{n-1}$ is given. Then there exists a unique G-homotopy

$$H\colon (X, A) \times I \to (Y, B)$$

rel A such that $H(\phi_j \times \mathrm{id}) = g_j$.

Proof. One shows first that

$$\begin{array}{ccc} \coprod_j G/H_j \times S^{n-1} \times I & \xrightarrow[\varphi \times \mathrm{id}]{} & A \times I \\ \downarrow & & \downarrow \\ \coprod_j G/H_j \times D^n \times I & \xrightarrow[\phi \times \mathrm{id}]{} & X \times I \end{array}$$

is again a pushout. The homotopy $A \times I \to Y$, which constantly equals $f|A$, and the homotopies g_j yield H via the universal property of the pushout. □

Suppose (X, A) is a pair of G-spaces with A being a Hausdorff space.

An **equivariant CW-decomposition** of (X, A) consists of a filtration $(X_n | n \in \mathbb{Z})$ of X such that the following holds:

(i) $A \subset X_0$; $A = X_n$ for $n < 0$; $X = \cup X_n$.

(ii) For each $n \geqq 0$, the space X_n is obtained from X_{n-1} by attaching equivariant n-cells.

(iii) X carries the colimit topology with respect to (X_n), i.e. $B \subset X$ is closed if and only if, for all n, $B \cap X_n$ is closed in X_n.

If (X_n) is an equivariant CW-decomposition of (X, A), then (X, A) is called a **relative equivariant CW-complex with respect to the filtration** (X_n). If $A = \emptyset$, then X is called an **equivariant CW-complex**. The subspace X_n is called the ***n*-skeleton** of (X, A). The n-cells of (X_n, X_{n-1}) are called **the *n*-cells of** (X, A). We call (X, A) **finite (countable** etc.) if $X \backslash A$ consists of a finite (countable ect.) number of equivariant cells. If $X = X_n$ and $X \neq X_{n-1}$, then n is called the **cellular dimension** of (X, A).

Since we only deal with the type of cell complexes defined above, we use the term ***G*-complex** instead of G-equivariant CW-complex. We point out that the filtration (X_n) is part of the structure whereas the characteristic maps are not. Proposition (1.15) shows clearly why characteristic maps should not be part of the structure.

Now we collect the elementary properties of G-complexes (compare G.W. Whitehead [1978], II).

(1.6) Proposition. *Let (X, A) be a relative G-complex. Then*

(i) *X is a Hausdorff space.*

(ii) *X_n is closed in X.*

(iii) *C is closed in X if and only if, for each closed cell E of (X, A), the subset $C \cap E$ is closed in E, and $C \cap A$ is closed in A.*

(iv) *A compact subset of X is contained in the union of A with finitely many cells of (X, A).* □

We consider subcomplexes. Let (Y, B) be a relative G-complex. A **G-subcomplex of** (Y, B) is a pair (X, A) with the following properties:

(i) X is a G-subspace of Y.

(ii) A is a closed G-subspace of B.

(iii) X is the union of A and a collection of open cells of Y, whose boundaries are all contained in X.

The term subcomplex is justified by

(1.7) Proposition. *If (X, A) is a G-subcomplex of (Y, B), then (X, A) is a G-complex with filtration $(X_n = X \cap Y_n \mid n \in \mathbb{Z})$. Moreover, X is closed in Y.* □

(1.8) Example. Let (X, A) be a relative G-complex and $n \geqq 0$ an integer. Then (X, X_n) and (X_n, A) are relative G-complexes and (X_n, A) is a subcomplex of (X, A). In this sense, the skeleta are subcomplexes.

(1.9) Example. Let $((X(j), A(j)) \mid j \in J)$ be a family of subcomplexes of (X, A). Then $(\bigcap_j X(j), \bigcap_j A(j))$ is a subcomplex. Moreover, if $\bigcup_j A(j)$ is closed in X, then $(\bigcup_j X(j), \bigcup_j A(j))$ is a subcomplex.

We often have to attach cells to G-complexes. We want to obtain again G-complexes (and not just relative ones). The next proposition shows when this is possible.

(1.10) Proposition. *Let A be a G-complex. Suppose X is obtained from A by attaching n-cells. Suppose the image of the attaching map $\varphi\colon \coprod G/H_j \times S^{n-1} \to A$ is contained in A_{n-1}. Then X is a G-complex with skeleta $X_q = A_q$ for $q < n$ and $X_q = A_q \bigcup_\varphi (\coprod G/H_j \times D^n)$ for $q \geqq n$.* □

(1.11) Proposition. *Let X be a G-complex and A a subcomplex. Then the quotient space X/A is a G-complex with n-skeleton X_n/A_n.* □

More generally, one can investigate equivalence relations on complexes which yield complexes as quotient spaces. See Lundell-Weingram [1969], II. Prop. 5.7. Further properties are formulated as exercises.

We investigate certain standard notions of the theory of transformation groups in connection with G-complexes.

Let (X, A) be a relative G-complex and let $\mathfrak{F}$ be a closed family of subgroups of G (see I.6). Set $X(\mathfrak{F}) = \{x \in X \mid G_x \in \mathfrak{F}\}$.

(1.12) Proposition. *Suppose $A(\mathfrak{F})$ is closed in A. Then $(X(\mathfrak{F}), A(\mathfrak{F}))$ is a subcomplex of (X, A).*

Proof. If $x \in X(\mathfrak{F}) \backslash A(\mathfrak{F})$, then x is contained in an open cell of type G/G_x with $G_x \in \mathfrak{F}$. The boundary of such a cell is necessarily contained in $X(\mathfrak{F})$. □

If H is a subgroup of G, set

$$\mathfrak{F}_H = \{K \subset G \mid H \text{ subconjugate } K\}$$
$$\mathfrak{F}'_H = \{K \in \mathfrak{F}_H \mid (K) \neq (H)\}.$$

Given a G-complex X, we can then form

$$X(H) = X(\mathfrak{F}_H)/X(\mathfrak{F}'_H)$$

and this, in a canonical way, is a G-complex (by (1.11) and (1.12)). This G-complex contains the (H)-orbit bundle of X and an extra base point.

Let H be a subgroup of G and let Y be an H-complex.

(1.13) Proposition. *The space $X = G \times_H Y$ is a G-complex with n-skeleton $X_n = G \times_H Y_n$. If E is an (open, closed) n-cell of type H/K of Y, then $G \times_H E$ is an (open, closed) n-cell of type G/K of X.*

Proof. Consider the canonical commutative diagram

$$\begin{array}{ccc} \coprod_n (G \times Y_n) & \xrightarrow{(2)} & G \times Y \\ \downarrow (1) & & \downarrow (3) \\ \coprod_n (G \times_H Y_n) & \xrightarrow{(4)} & G \times_H Y. \end{array}$$

The map (2) is an identification because it is the product of an identification with the identity of the locally compact space G. Hence the composition $(3) \circ (2) = (4) \circ (1)$ is an identification. Since (1) is an identification, (4) is one, too. One checks that $G \times_H Y_n \to G \times_H Y$ is an embedding. Hence X carries the colimit topology with respect to the X_n.

If

$$\begin{array}{ccc} A & \longrightarrow & B \\ \downarrow & & \downarrow \\ C & \longrightarrow & D \end{array}$$

is a pushout of H-spaces, then we obtain a pushout of G-spaces by applying the functor $Y \mapsto G \times_H Y$ (exercise 6). Using this, we see that $G \times_H Y_n$ is obtained from $G \times_H Y_{n-1}$ by attaching G-equivariant n-cells. □

(1.14) Proposition. *Let G be a compact Lie group, H a closed subgroup and X a G-complex. Then the space X^H is a WH-complex with n-skeleton X_n^H.*

Proof. The space X_n^H is closed in X_n because X_n is Hausdorff. If $C \subset X^H$ is given and $C \cap X_n^H$ is closed in X_n^H, then $C \cap X_n^H$ is closed in X_n. One sees that X^H carries the correct colimit topology.
Let

$$\begin{array}{ccc} \coprod_{j \in J} G/H_j \times S^{n-1} & \longrightarrow & X_{n-1} \\ \downarrow & & \downarrow \\ \coprod_{j \in J} G/H_j \times D^n & \longrightarrow & X_n \end{array}$$

be a pushout. Pass to H-fixed points. Then one obtains again a pushout (exercise 5). Decompose G/H_j^H into WH-orbits; by I(5.10), there are only finitely many WH-orbits. □

Now let G be an arbitrary discrete group. For discrete groups, there is a close connection between G-complexes and ordinary CW-complexes. (For the following, compare also Bredon [1972], III.1.)

Let X be a G-space and an ordinary CW-complex. We say that G **acts cellularly** on X if the following holds:

(i) For each $g \in G$ and each open cell E of X, the left translation gE is again an open cell of X.

(ii) If $gE = E$, then the induced map $E \to E$, $x \mapsto gx$ is the identity.

(1.15) Proposition. *Let X be a* CW*-complex with cellular action of G. Then X is a G-complex with n-skeleton X_n.*

Proof. X_n is a G-subspace of X and X has the colimit topology with respect to the X_n. It remains to be shown that X_n is obtained from X_{n-1} by attaching equivariant n-cells.
By definition of a CW-complex, we have a pushout diagram

$$\begin{array}{ccc} J \times S^{n-1} & \xrightarrow{\varphi} & X^{n-1} \\ \downarrow & & \downarrow \\ J \times D^n & \xrightarrow{\phi} & X^n \end{array}$$

with a discrete set J. (One has $J \times S^{n-1} = \coprod_{j \in J} S^{n-1}$.) The group G acts on the set of n-cells and hence on J, by definition of a cellular action. We write J as disjoint union of orbits J_α, $\alpha \in A$, and choose G-isomorphisms

$$G/H_\alpha \cong J_\alpha, \; gH_\alpha \mapsto gj_\alpha, j_\alpha \in J_\alpha.$$

For $s \in \mathring{D}^n$ and $h \in H_\alpha$, we have $h\phi(j_\alpha, s) = \phi(j_\alpha, s)$ by (ii). Thus given $h \in H_\alpha$,

$$h\phi(j_\alpha, s) = \phi(hj_\alpha, s)$$

holds for all $s \in \mathring{D}^n$ and, by continuity, for all $s \in D^n$. Therefore,

$$\phi_\alpha\colon\ G/H_\alpha \times D^n \to X^n,\ (gH_\alpha, s) \mapsto g\phi(j_\alpha, s)$$

is a well-defined continuous G-map. Analogously, we obtain $\varphi_\alpha\colon G/H_\alpha \times S^{n-1} \to X^{n-1}$. Collecting the φ_α and ϕ_α to build maps φ' and ϕ', we obtain a commutative diagram of G-maps

$$\begin{array}{ccc} \coprod_{\alpha \in A} G/H_\alpha \times S^{n-1} & \xrightarrow{\varphi'} & X^{n-1} \\ \downarrow & & \downarrow \\ \coprod_{\alpha \in A} G/H_\alpha \times D^n & \xrightarrow{\phi'} & X^n. \end{array}$$

We have to show that this diagram is a pushout. Assume $f\colon X^n \to Z$ is given and the maps $f|X^{n-1}$ and

$$\{gH_\alpha\} \times D^n \xrightarrow{\phi'} X^n \xrightarrow{f} Z$$

are continuous. Since D^n is compact, the map $\phi'|\{gH_\alpha\} \times D^n$ is an identification and therefore f is continuous on $\phi'(\{gH_\alpha\} \times D^n)$. But these are just the closed n-cells of X^n and therefore f is continuous. □

(1.16) Proposition. *Let G be a discrete group. Let X be a G-complex and H a subgroup of G. Then X, considered as an H-space, is an H-complex with the same skeleta.*

Proof. Start with a pushout

$$\begin{array}{ccc} \coprod_{j \in J} G/H_j \times S^{n-1} & \longrightarrow & X^{n-1} \\ \downarrow & & \downarrow \\ \coprod_{j \in J} G/H_j \times D^n & \longrightarrow & X^n. \end{array}$$

Decompose the G/H_j as H-sets into a disjoint union of orbits. The same diagram then shows that X^n is obtained from X^{n-1} by attaching H-equivariant n-cells. □

If G is discrete and X is a G-complex, then X is in a canonical way a CW-complex with cellular G-action. Hence (for discrete groups) we have two equivalent definitions of G-complexes.

There are G-actions on CW-complexes which satisfy (i) but not (ii) in the definition of a cellular action and have a certain relevance. Usually, for a suitable subdivision, (ii) is then satisfied. For instance, if G acts cellularly on a triangulated manifold, then for the dual cell decomposition (ii) is almost never satisfied: look at fixed points. Therefore, it is sensible to consider a more general notion of an equivariant cell complex where a nontrivial action on D^n is allowed. In this case, one would attach $G/H \times D(V)$ by an attaching map $G/H \times S(V) \to A$ where $D(V)$ and $S(V)$ are the unit ball and unit sphere in an orthogonal representation space V. In this book, G-complexes are used as a convenient class of spaces; the geometry of the cell decomposition will not be relevant.

Actually, for compact Lie groups, G-complexes may be rather inconvenient. If H is a closed subgroup of G, then the restriction $\operatorname{res}_H X$ of a G-complex X is in general not an H-complex in a canonical way. This is caused by the fact that $\operatorname{res}_H G/K$ has no canonical cell decomposition. The existence and uniqueness of cell decompositions and triangulations of G-manifolds is treated thoroughly in Illman [1978], [1983]. We also refer to the work of Illman for later use of CW-complex methods for manifolds, e.g. in obstruction theory. The category of CW-complexes is technically very convenient; but usually it is difficult to see whether naturally occuring spaces have the structure of a CW-complex or at least the homotopy type of a CW-complex.

On the other hand, G-complexes also cause some technical problems. Sometimes one has to retopologize, i.e. to work in a category of compactly generated spaces as described in Steenrod [1967] or R. Vogt [1971]. One problem is that, for compact Lie groups, the product of two G-complexes has no natural cell structure. For instance, a product of two zero-cells is a space of the form $G/H \times G/K$ and such products have to be decomposed further in order to obtain a cellular structure. For some general results about G-complexes, see Waner [1980]. For triangulation problems, see also Verona [1980].

(1.17) Exercises.

1. If (X, A) is a relative G-complex, then X/A is a G-complex with n-skeleton X_n/A.
2. Let (X, A) be a relative G-complex. Then the orbit space $(X/G, A/G)$ is a relative CW-complex with n-skeleton X_n/G.
3. Let (X, A) be a relative G-complex. Then $A \subset X$ is a G-cofibration. If A is normal, then X is normal. There exists an open G-neighbourhood U of A in X such that A is a strong G-deformation retract of U. (See Lundell-Weingram [1969], II. Theorem 6.1.)
4. If $S \to D$ is an ordinary cofibration and Z a G-space, then $Z \times S \to Z \times D$ is a G-cofibration.
5. Let $H \subset G$. Then $(Y \cup_f X)^H$ is obtained from Y^H by attaching X^H via f^H.

6. Let
$$\begin{array}{ccc} A & \longrightarrow & B \\ \downarrow & & \downarrow \\ C & \longrightarrow & D \end{array}$$
be a pushout of H-spaces, $H \subset G$. By application of the functor $Y \mapsto G \times_H Y$, one obtains a pushout of G-spaces.
7. Prove (1.4).
8. Let G be a compact Lie group. Let X and Y be G-complexes. Show that $X \times Y$ with diagonal action has the G-homotopy type of a G-complex. Let H be a subgroup of G. Show that $\operatorname{res}_H X$ has the H-homotopy type of an H-complex.
9. Let G and H be compact Lie groups. Let X resp. Y be a G-resp. H-complex. Show that $X \times Y$ with $G \times H$-action is in a canonical way a $G \times H$-complex if furnished with the compactly generated topology. (Products of cells are cells.)
10. Let SV be a representation sphere of a finite group G. Show directly that SV admits a G-equivariant triangulation by looking at the boundary of the convex hull of $\{\pm g e_i \mid g \in G;\ e_1, \ldots, e_m \in SV \text{ basis of } V\}$.

2. Maps between complexes.

Let X and Y be G-complexes. A G-map $f\colon X \to Y$ ist called **cellular** if $f(X_n) \subset Y_n$ for all n. The following cellular approximation theorem is an important technical tool.

(2.1) Theorem. *Let $f\colon X \to Y$ be a G-map. Then there exists a G-homotopy $H\colon X \times I \to Y$ such that $H_0 = f$ and H_1 is cellular.*

Before we prove the theorem, we recall some facts from homotopy theory. A pair (Y, B) of topological spaces is said to be ***n*-connected** if for every relative CW-complex (X, A) with $\dim(X, A) \leqq n$, any map $f\colon (X, A) \to (Y, B)$ is homotopic rel A to a map of X into B. A necessary and sufficient condition for (Y, B) to be n-connected is that every map $f\colon (D^q, S^{q-1}) \to (Y, B)$ is homotopic rel S^{q-1} to a map of D^q into B $(q = 0, 1, \ldots, n)$. Let $h\colon B \to Y$ be any map. Let Z_h be the mapping cylinder of h, that is $Z_h = B \times I \cup Y/\sim$, $(b, 1) \sim h(b)$ and $B \to B \times 0 \subset Z_h$ be the inclusion. Then h is said to be ***n*-connected** if the pair (Z_h, B) is n-connected. We call h ∞-connected if it is n-connected for all n. The map h is 0-connected if and only if it induces an epimorphism $\pi_0(h)\colon \pi_0(B) \to \pi_0(Y)$ of the sets π_0 of path components. It is n-connected, $n \geqq 1$, if and only if $\pi_0(h)$ is bijective and $\pi_q(h)\colon \pi_q(B, b) \to \pi_q(Y, h(b))$ is an isomorphism for $q < n$ and an epimorphism for $q = n$ (for each base point $b \in B$). Using the exact homotopy sequence

$$\ldots \to \pi_q(B, b) \to \pi_q(Y, b) \to \pi_q(h, b) \to \ldots,$$

one sees that a 0-connected map h is n-connected, $n \geqq 1$, if and only if $\pi_q(h, b) = 0$ for $1 \leqq q \leqq n$ and any b. Recall that elements of $\pi_q(h, b)$ are represented by commutative diagrams

$$\text{(2.2)} \qquad \begin{array}{ccc} S^{q-1} & \longrightarrow & B \\ \cap\big\downarrow & & \big\downarrow h \\ D^q & \longrightarrow & Y \end{array}$$

of pointed maps.

Let

$$\begin{array}{ccc} A & \longrightarrow & Y \\ \cap\big\downarrow & & \big\downarrow\cap \\ X & \longrightarrow & Z \end{array}$$

be a pushout and assume that A is a deformation retract of a neighbourhood in X. Then

(2.3) Lemma. *If (X, A) is n-connected, then (Z, Y) is n-connected.*

Proof. This is a special case of the Blakers-Massey excision theorem. The most elementary proof seems to be in tom Dieck-Kamps-Puppe [1970] § 15. □

We apply this lemma to schow

(2.4) Lemma. *For each G-complex Y, the pair (Y^H, Y_n^H) is n-connected.*

Proof. Generally, if (C, B) and (B, A) are n-connected, then (C, A) is n-connected. If we show that (Y_{n+1}^H, Y_n^H) is n-connected, then it follows that (Y_{n+k}^H, Y_n^H) is n-connected for $k \geqq 1$ and since Y^H carries the colimit topology with respect to the Y_n^H, the result follows. By applying (2.3) to a pushout of the type

$$\begin{array}{ccc} \coprod_\alpha G/H_\alpha^H \times S^n & \longrightarrow & Y_n^H \\ \big\downarrow & & \big\downarrow \\ \coprod_\alpha G/H_\alpha^H \times D^{n+1} & \longrightarrow & Y_{n+1}^H, \end{array}$$

we see that (Y_{n+1}^H, Y_n^H) is n-connected. □

Proof of (2.1). We show inductively the existence of homotopies $H^n: X \times I \to Y$ such that $H_0^0 = f$, $H_1^{n-1} = H_0^n$ for $n \geqq 1$, $H_1^n(X_i) \subset Y_i$ for $i \leqq n$, H^n constant homotopy on X_{n-1}. The desired homotopy H is then given by

$$H(x, t) = \begin{cases} H^i(x, 2^{i+1}(t - 1 + 2^{-i})) & \text{for} \quad 1 - 2^{-i} \leqq t \leqq 1 - 2^{-i-1} \\ H^i(x, 1) & \text{for} \quad x \in X_i,\ t = 1. \end{cases}$$

For the induction step, we assume that $f(X_i) \subset Y_i$ holds for $i < n$. Let ϕ_j: $G/H_j \times (D^n, S^{n-1}) \to (X_n, X_{n-1})$ be a characteristic map for an n-cell. By Proposition (1.5), it suffices to find a G-homotopy g_j of $f\phi_j$ rel $G/H_j \times S^{n-1}$ such that, at time $t = 1$, g_j maps X_n into Y_n. Such G-homotopies correspond to ordinary homotopies $(D^n, S^{n-1}) \to (Y^H, Y_{n-1}^H)$. Therefore, it suffices to deform $f\phi_j | D^n$ rel S^{n-1} into Y_n^H. This is possible since (Y^H, Y_n^H) is n-connected by (2.4). □

Let ν be a function from the set of closed subgroups of G into $\{0, 1, 2, \ldots, \infty\}$ such that $\nu(H) = \nu(gHg^{-1})$. A G-map f: $X \to Y$ is called **ν-connected** if f^H: $X^H \to Y^H$ is an ordinary $\nu(H)$-equivalence for all H. A relative G-complex (X, A) is said to have dimension less than or equal to ν, in symbols $\mathrm{Dim}(X, A) \leq \nu$, if the cells in $X \backslash A$ are of the form $G/H \times D^k$ with $k \leq \nu(H)$. If $A = \emptyset$, we write $\mathrm{Dim}\, X \leq \nu$. The maximal k such that cells of the form $G/H \times D^k$ exist in $X \backslash A$ is denoted by $\mathrm{Dim}(X, A)(H)$. We set $\mathrm{Dim}(X, A)(H) = \infty$ if no such k exists.

Thus this cellular dimension $\mathrm{Dim}(X, A)(H)$ coincides with the topological dimension $\dim(X \backslash A)^H/NH$ of the space $(X \backslash A)^H/NH$. We call $H \mapsto \mathrm{Dim}(X, A)(H)$ the **cellular dimension function** of the relative G-complex (X, A). Note that in (2.6) this function is only relevant when applied to isotropy subgroups.

It is a typical phenomenon in transformation group theory that classical invariants yield functions on conjugacy classes of subgroups.

(2.5) Proposition. *Let (X, A) be a relative G-complex satisfying* $\mathrm{Dim}(X, A) \leq \nu$ *and let f: $Y \to Z$ be a ν-equivalence. Suppose G-maps h: $X \times \{0\} \cup A \times I \to Z$ and k: $A \times \{1\} \to Y$ are given such that $fk(a, 1) = h(a, 1)$ for $a \in A$. Then there exist G-maps H: $X \times I \to Z$ extending h and K: $X \times \{1\} \to Y$ extending k such that $fK(x, 1) = H(x, 1)$ for $x \in X$.*

Proof. The maps H and K are constructed inductively over $X_n \times I$ resp. $X_n \times \{1\}$. Thus it suffices to consider the case that X is obtained from A by attaching a cell of type $G/H \times D^k$, $k \leq \nu(H)$ (compare (1.5)). Then the assertion essentially amounts to insert the dotted arrows in a commutative diagram of the type

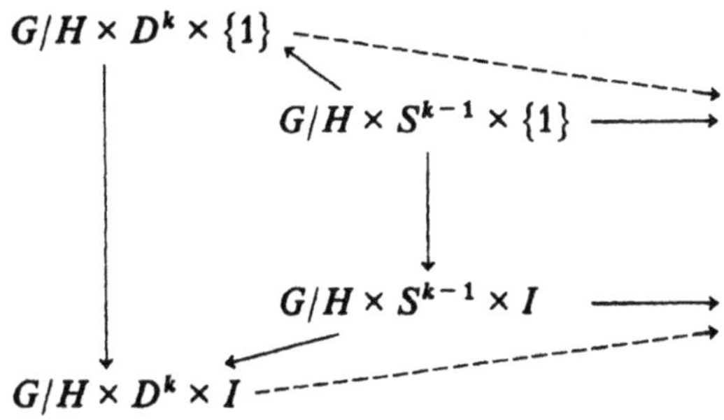

by I(4.3), this is translated into a corresponding non-equivariant statement about f^H: $Y^H \to Z^H$ and this follows from $\nu(H)$-connectivity (exercise 1). □

(2.6) Proposition. *Let f: $Y \to Z$ be a ν-equivalence and X a G-complex. Then f_*: $[X, Y]_G \to [X, Z]_G$ is surjective (resp. bijective) if* $\operatorname{Dim} X \leq \nu$ *(*resp. $\operatorname{Dim} X < \nu$*).*

Proof. Apply (2.5) to $(X, \emptyset)$ in order to obtain surjectivity and to $(X \times I, X \times \partial I)$ in order to obtain injectivity. □

The next result is a very useful special case.

(2.7) Proposition. *Let f: $Y \to Z$ be a G-map between G-complexes such that f^H is a homotopy equivalence for all H. Then f is a G-homotopy equivalence.*

Proof. f is a ν-equivalence where $\nu(H) = \infty$ for all H. Thus f_*: $[X, Y]_G \to [X, Z]_G$ is bijective for all G-complexes X. The assertion follows from category theory. □

A variant of (2.7) for G-ANR's is proved in James-Segal [1978]. A variant for spaces with suitable cofibration properties is indicated in H. Hauschild [1978]; the result of Hauschild is based on Boardman-Vogt [1973], Appendix 3: see exercises 3–6.

We now turn our attention to the suspension construction of homotopy theory.

The suspension theorem of Freudenthal and the excision theorem of Blakers and Massey are easily extended to an equivariant situation. If the reader has seen one such generalization, he will be able to apply similar methods to other topics in homotopy theory.

Let V be a G-module (= a finite-dimensional real representation of G). We write $S^V = V \cup \{\infty\}$ for the one-point-compactification of V with base-point ∞. If n denotes the trivial n-dimensional representation, then S^n is indeed a model for the n-sphere.

Let X and Y be pointed G-spaces. We set $\Sigma^V X = S^V \wedge X$ and consider the suspension map between pointed G-homotopy sets

$$\text{(2.8)} \qquad \Sigma^V\colon\ [X, Y]_G^0 \to [\Sigma^V X, \Sigma^V Y]_G^0, \quad [f] \mapsto [\mathrm{id} \wedge f].$$

Note that there exist many different suspension maps due to the different group actions on the suspension sphere S^V.

The generalization of the Freudenthal suspension theorem is concerned with the following problem: Under which conditions is Σ^V in (2.8) an isomorphism or epimorphism? This problem can be rephrased in the following way. For a pointed G-space B, let $\Omega^V B$ denote the space of all pointed maps $f\colon S^V \to B$ with compact-open-topology and G-action $(g \cdot f)(x) = gf(g^{-1}x)$. As in ordinary homotopy theory, one has the adjointness for pointed G-homotopy sets

$$[\Sigma^V A, B]_G^0 \cong [A, \Omega^V B]_G^0.$$

One obtains a commutative diagram

$$\begin{array}{ccc} [X, Y]_G^0 & \xrightarrow{\ \Sigma^V\ } & [\Sigma^V X, \Sigma^V Y]_G^0 \\ & \searrow^{\eta_*} & \downarrow \cong \\ & & [X, \Omega^V \Sigma^V Y]_G^0 \end{array}$$

where $\eta\colon Y \to \Omega^V \Sigma^V Y$ is adjoint to the identity of $\Sigma^V Y$. By (2.6), it therefore suffices to study the connectivity of the map η. The next result is about this connectivity.

Let $\mathrm{Hur}(B)$, the **Hurewicz number** of B, denote the largest element k of $\{1, 2, 3, \ldots, \infty\}$ such that $\pi_i(B) = 0$ for all $i < k$. Let n be a function from conjugacy classes of subgroups to non-negative integers with the following properties:

(2.9) (i) For $H \subset G$ with $\dim V^H > 0$, one has $n(H) \leq 2\,\mathrm{Hur}(Y^H) - 1$.
(ii) For all pairs of subgroups $K \subset H \subset G$ such that $\dim V^K > \dim V^H$, one has $n(H) \leq \mathrm{Hur}(Y^K) - 1$.

Then we can state

(2.10) Theorem. *Let n be a function satisfying* (2.9). *Then $\eta\colon Y \to \Omega^V \Sigma^V Y$ is an n-equivalence.*

Proof. We have to show that η^H is an $n(H)$-equivalence. In order to show this, we consider the following diagram

$$\begin{array}{ccc}
\pi_r(Y^H) & \xrightarrow{\pi_r(\eta^H)} & \pi_r((\Omega^V \Sigma^V Y)^H) \\
\downarrow \cong & & \downarrow \cong \\
[S^r, Y]^0_H & \xrightarrow{\eta_*} & [S^r, \Omega^V \Sigma^V Y]^0_H \\
& \searrow \Sigma^V & \downarrow \cong \\
& & [S^{r+V}, \Sigma^V Y]^0_H
\end{array}$$

Thus η^H is an $n(H)$-equivalence if, in this diagram, Σ^V is an isomorphism for $r < n(H)$ and an epimorphism for $r = n(H)$. We use the next diagram in which R_1 and R_2 indicate restrictions to H-fixed point sets.

$$\begin{array}{ccc}
[S^r, Y]^0_H & \xrightarrow{\Sigma^V} & [S^{r+V}, \Sigma^V Y]^0_H \\
\cong \downarrow R_1 & & \downarrow R_2 \\
[S^r, Y^H]^0 & \xrightarrow{\Sigma^{VH}} & [S^{r+VH}, \Sigma^{VH} Y^H]^0
\end{array}$$

By the classical Freudenthal suspension theorem (e. g. tom Dieck-Kamps-Puppe [1970], §16), Σ^{VH} is bijective (resp. surjective) if $r \leq 2\,\mathrm{Hur}(Y^H) - 2$ (resp. $r \leq 2\,\mathrm{Hur}(Y^H) - 1$). Since, by hypothesis (2.9, i), $n(H) \leq 2\,\mathrm{Hur}(Y^H) - 1$, it suffices to show that R_2 is injective for $r \leq n(H)$. This will be done by obstruction theory.

Suppose $g: S^{r+V} \to \Sigma^V Y$ is an H-map such that $R_2(g) = 0$. Then g extends to $\tilde{g}: C(i) \to \Sigma^V Y$ where $C(i)$ denotes the mapping cone of the inclusion $i: S^{r+VH} \to S^{r+V}$. We show that g is nullhomotopic by extending g to the cone CS^{r+V} on S^{r+V}. We look at the obstructions of this extension problem. The cells of CS^{r+V} not contained in $C(i)$ are of the form $H/K \times D^{j+1}$, $K \subset H$, $\dim V^K > \dim V^H$, $j \leq r + \dim V^K$. Therefore, the obstructions to extending g over such cells lie in $\pi_j((\Sigma^V Y)^K) = \pi_j(\Sigma^{VK} Y^K)$. We want these groups to be zero for $n \leq r + \dim V^K$ because then all obstructions vanish. They are zero for $j \leq \dim V^K + \mathrm{Hur}(Y^K) - 1$. But, by hypothesis (2.9, ii), $r \leq n(H) \leq \mathrm{Hur}\, Y^K - 1$. □

Further references: Illman [1972], [1972a], Matumoto [1971], Hauschild [1977] for exercises 7–9, Namboodiri [1983] for (2.10), Adams [1984].

(2.11) Exercises.

1. Make sure that you can prove (2.5) in the non-equivariant situation.
2. Give a proof of (2.3).
3. Let X and Y be finite dimensional pointed G-complexes of finite orbit type. Show that the suspension

$$[\Sigma^V X, \Sigma^V Y]_G^0 \to [\Sigma^U \Sigma^V X, \Sigma^U \Sigma^V Y]_G^0$$

is bijective for all U provided that the representation V is large enough. Show that such V exist.

4. Given a diagram of G-spaces and G-maps

$$\begin{array}{ccc} Y & \xrightarrow[p]{} & Z \\ \uparrow{\scriptstyle f_A} & & \uparrow{\scriptstyle h} \\ A & \xrightarrow[\subset]{} & X \end{array}$$

and a G-homotopy H_A: $h|A \simeq pf_A$. Assume that $A \subset X$ is a G-cofibration. Then there exists a G-map f: $X \to Y$ extending f_A and a G-homotopy H: $h \simeq pf$ extending H_A provided that
(i) p is an equivariant homotopy equivalence;
or
(ii) p is an ordinary homotopy equivalence and $X \backslash A$ is a numerable free G-space.

5. Let p: $(X, A) \to (Y, B)$ be a G-map such that $p_A = p|A$: $A \to B$ is a G-homotopy equivalence and p is an ordinary homotopy equivalence. Suppose that $X \backslash A$ and $Y \backslash B$ are numerable free G-spaces and $A \subset X$, $B \subset Y$ are G-cofibrations. Then any G-homotopy inverse q_B of p_A can be extended to a G-homotopy inverse q of p and any G-homotopy H_B: $\mathrm{id}_B \simeq p_A q_B$ to a G-homotopy H: $\mathrm{id}_Y \simeq pq$.
6. Let f: $X \to Y$ be a G-map such that for all $H \subset G$ the map f^H is an ordinary homotopy equivalence. Suppose that for all $H \subset G$ the spaces X_H and Y_H are numerable free NH/H-spaces and $G(X^H \backslash X_H) \subset GX^H$, $G(Y^H \backslash Y_H) \subset GY^H$ are G-cofibrations. Suppose that X and Y have finite orbit type. Then f is a G-homotopy equivalence.
7. Let V be a real representation of G. Let $I = [0, 1]$ with basepoint 0. Define

$$\pi_{i+V}^G(X) = [S^{i+V}, X]_G^0 \quad \text{for} \quad i \geq 0$$
$$\pi_{i+V}^G(X, A) = [I \wedge S^{i-1+V}, \partial I \wedge S^{i-1+V}; X, A]_G^0 \quad \text{for} \quad i \geq 0$$

for a pointed G-pair (X, A). The sets $\pi_{i+V}^G(X)$ resp. $\pi_{i+V}^G(X, A)$ are a group (abelian group) for $i \geq 1$ resp. $i \geq 2$ ($i \geq 2$ resp. $i \geq 3$) using the i-part of $i + V$ as suspension coordinate. The group structure is defined as in the non-equivariant case. There is a long exact homotopy sequence

$$\pi_V^G(X) \leftarrow \pi_V^G(A) \leftarrow \pi_{1+V}^G(X, A) \leftarrow \pi_{1+V}^G(X) \leftarrow \dots$$

8. Let X_1, X_2 be subcomplexes of the G-complex X and put $X_0 = X_1 \cap X_2$. Suppose that

$$\pi_i(X_1^H, X_0^H) = 0 \quad \text{for} \quad 0 < i < p_H$$
$$\pi_i(X_2^H, X_0^H) = 0 \quad \text{for} \quad 0 < i < q_H$$

and $\dim V^H < p_H + q_H - 2$. Then the map (induced by inclusion)

$$i_V\colon \pi_V^G(X_2, X_0) \to \pi_V^G(X, X_1) \text{ is an isomorphism}$$

and

$$i_{1+V}\colon \pi_{1+V}^G(X_2, X_0) \to \pi_{1+V}^G(X, X_1) \text{ is an epimorphism.}$$

(Hint: Write maps as induced by a map and apply the classical Blakers-Massey excision theorem and the reasoning of (2.6).)

9. Let V be a G-representation. Show that the suspension

$$[S^{n+rV}, S^{rV}]_G^0 \to [S^{n+rV} \wedge S^V, S^{rV} \wedge S^V]_G^0$$

is an isomorphism for $n \leq r - 2$.

3. Obstruction theory.

In this section, G denotes a compact Lie group. This section is entirely pragmatic. We translate the classical obstruction theory into an equivariant setting. For our applications, greatest theoretical generality is not needed.

We begin with a description of suitable equivariant cohomology groups which are derived in a simple-minded manner from ordinary cohomology groups. They can be interpreted as cohomology groups with local coefficients.

Let (X, A) be a relative G-complex with free G-action on $X \backslash A$, i.e. X_n is obtained from X_{n-1} by attaching n-cells of type G (also called free n-cells). The filtration (X_n) leads to a cellular chain complex $C_*(X, A)$

$$\text{(3.1)} \qquad \dots \to H_{n+1}(X_{n+1}, X_n) \xrightarrow{d} H_n(X_n, X_{n-1}) \to \dots$$

Homology is singular homology with coefficients in $\mathbb{Z}$ and d is the boundary morphism in the exact sequence for the triple (X_{n+1}, X_n, X_{n-1}); see Dold [1972], V.1., for this and cellular homology in general.

The G-action on X_n induces a G-action on $H_n(X_n, X_{n-1})$. Let G_0 denote the component of e in G. Then this action factors over G/G_0 by homotopy invariance of homology. Therefore, $H_n(X_n, X_{n-1})$ becomes a module over the integral group ring $\mathbb{Z}(G/G_0)$ of the finite group G/G_0 and then (3.1) is a chain complex of such modules.

Let M be another $\mathbb{Z}(G/G_0)$-module. The cochain complex

(3.2) $\quad \mathrm{Hom}_{\mathbb{Z}(G/G_0)}(C_*(X, A), M) = C_G^*(X, A; M)$

yields cohomology groups which we denote by

(3.3) $\quad \mathfrak{H}_G^*(X, A; M)$.

Modules M which are relevant for our applications are given as follows: Let Y be a path-connected, n-simple space, i.e. $\pi_1(Y, y)$ acts trivially on $\pi_n(Y, y)$. Then the canonical map $\pi_n(Y, y) \to [S^n, Y]$ from pointed to free homotopy classes is bijective. The action of G on Y induces therefore a well defined action of G/G_0 on $\pi_n Y$.

The cohomology groups $\mathfrak{H}_G^*(X, A; M)$ are actually ordinary cohomology groups with local coefficients. We are going to explain this.

To begin with, one observes that the projection $X \to X/G_0$ induces an isomorphism

$$C_*(X, A) \to C_*(X/G_0, A/G_0)$$

of chain complexes. The proof runs as follows. Let

$\phi\colon \coprod_j G \times (D^n, S^{n-1}) \to (X_n, X_{n-1})$ be a characteristic map. It gives an isomorphism

$$\bigoplus_j H_n(G \times (D^n, S^{n-1})) \to H_n(X_n, X_{n-1}).$$

In a similar fashion one obtains an isomorphism

$$\bigoplus_j H_n(G_0 \backslash G \times (D^n, S^{n-1})) \to H_n(G_0 \backslash X_n, G_0 \backslash X_{n-1});$$

factoring out G_0 yields an isomorphism

$$H_n(G \times (D^n, S^{n-1})) \to H_n(G_0 \backslash G \times (D^n, S^{n-1}))$$

because (using suspension) we have to look at H_0 where only path-components matter.

The pair $(X/G_0, A/G_0)$ is a relative free G/G_0-complex with n-skeleton $(X/G_0)_n = X_n/G_0$. Therefore,

(3.4) $\quad \mathfrak{H}_G^n(X, A; M) \cong \mathfrak{H}_{G/G_0}^n(X/G_0, A/G_0; M)$.

If the $\mathbb{Z}(G/G_0)$-module M is interpreted as a local coefficient system $\tilde{M}$ on $X/G \backslash A/G$, then (3.4) can be thought of as cohomology $H^n(X/G, A/G; \tilde{M})$ with local coefficients. Usually, these groups are only treated for coefficient systems on X/G; compare G. W. Whitehead [1978], VI. It is for this reason that we rewrite the theory. Elements of (3.2) have a description as cellular cochains: One should first note that $H_n(X_n, X_{n-1})$ is a free $\mathbb{Z}(G/G_0)$-module. The characteristic map ϕ given above yields a canonical basis: the images of the canonical generator $H_n(D^n, S^{n-1})$ under the j-th component of ϕ

$$\{e\} \times (D^n, S^{n-1}) \to G \times (D^n, S^{n-1}) \xrightarrow[\phi_j]{} (X_n, X_{n-1}).$$

An element of $C_G^n(X, A; M)$ may thus be identified with a function on this basis with values in M. The homological description is needed to give a convenient definition of the boundary operator.

Another fact to be kept in mind is that the groups $\mathfrak{H}_G^n(X, A; M)$ are zero if n is larger than the cellular dimension of (X, A) as a relative complex. This is the dimension of $(X \backslash A)/G$ as a space; so one has to observe that

$$\dim(X \backslash A) = \dim(X \backslash A)/G + \dim G$$

for the dimensions of the spaces involved.

The cohomology groups above are defined using the cellular structure. A cellular map $f\colon (X, A) \to (Y, B)$ induces homomorphisms

$$\text{(3.5)} \quad \begin{aligned} f^*&\colon C_G^*(Y, B; M) \to C_G^*(X, A; M) \\ f^*&\colon \mathfrak{H}_G^*(Y, B; M) \to \mathfrak{H}_G^*(X, A; M). \end{aligned}$$

There is also a suspension isomorphism. In its construction and also elsewhere we need properties of the external homology product

$$H_m(X, A) \otimes H_n(Y, B) \to H_{m+n}((X, A) \times (Y, B))$$
$$(X, A) \times (Y, B) := (X \times Y, X \times B \cup A \times Y).$$

The following diagram is commutative (see Dold [1972], VII, 2.1):

$$\text{(3.6)} \quad \begin{array}{ccc} H_m(X, A) \otimes H_n(Y, B) & \longrightarrow & H_m(X \times Y, A \times Y \cup X \times B) \\ & & \downarrow \partial \\ \downarrow (\partial \otimes \mathrm{id}, (-1)^m \mathrm{id} \otimes \partial) & & H_{m+n-1}(A \times Y \cup X \times B, A \times B) \\ & & \uparrow (i_{1*}, i_{2*}) \\ (H_{m-1}(A) \otimes H_n(Y, B)) \oplus (H_m(X, A) \otimes H_{n-1} B) & \to & H_{m+n-1}(A \times Y, A \times B) \\ & & \oplus\, H_{m+n-1}(X \times B, A \times B) \end{array}$$

where i_1, i_2 are inclusions and ∂ the appropriate boundary maps of the exact homology sequences. Another commutative diagram we need is (3.7). It follows from (3.6).

$$(3.7)\qquad \begin{array}{ccc}
H_m(X,A)\otimes H_n(Y,B) & \xrightarrow{(-1)^m\mathrm{id}\otimes\partial} & H_m(X,A)\otimes H_{n-1}(B,C)\\
\downarrow & & \downarrow\\
H_{m+n}((X,A)\times(Y,B)) & \xrightarrow{\partial'} & H_{m+n-1}((X,A)\times(B,C))\\
\Delta\downarrow & & \uparrow i_*\\
H_{m+n-1}(X\times B\cup A\times Y, A\times Y) & \xleftarrow[\cong]{} & H_{m+n-1}((X,A)\times B).
\end{array}$$

Here ∂ and Δ are boundary maps for the appropriate triples, i is the inclusion, the isomorphism is excision, and ∂' is defined so as to make the bottom rectangle commutative.

Let $z \in H_1(I, \partial I)$ be the canonical generator and let $\sigma_z\colon u \mapsto z \times u$ denote the product with z. If (X, A) is a relative complex, denote $(I, \partial I) \times (X, A)$ by (X^*, A^*). By (3.7), the following diagram is commutative

$$(3.8)\qquad \begin{array}{ccc}
H_n(X_n, X_{n-1}) & \xleftarrow{\partial} & H_{n+1}(X_{n+1}, X_n)\\
\downarrow -\sigma_z & & \downarrow \sigma_z\\
H_{n+1}((I,\partial I)\times(X_n, X_{n-1})) & \xleftarrow{\partial'} & H_{n+2}((I,\partial I)\times(X_{n+1}, X_n))\\
i_*\downarrow\cong & & i_*\downarrow\cong\\
H_{n+1}(X^*_{n+1}, X^*_n) & \xleftarrow{\partial} & H_{n+2}(X^*_{n+2}, X^*_{n+1})
\end{array}$$

where the isomorphisms i_* are excision maps and ∂' is defined as in (3.7). Of course, σ_z is just the ordinary suspension isomorphism. The homomorphisms $(-1)^n i_* \circ \sigma_z\colon H_n(X_n, X_{n-1}) \to H_{n+1}(X^*_{n+1}, X^*_n)$ therefore define a chain map $C_*(X, A) \to C_{*+1}(X^*, A^*)$ and an isomorphism

(3.9) $\quad \sigma\colon\ \mathfrak{H}^n_G(X, A; M) \to \mathfrak{H}^{n+1}_G(X^*, A^*; M),$

called **suspension isomorphism**.

Our next task is to develop the exact obstruction sequence. For this purpose, we use the following **conventions**: We fix an integer $n \geqq 1$ and assume that the space Y is n-simple. The pair (X, A) is a relative G-complex with free action on $X\backslash A$ and k-skeleton X_k. Obstruction theory is concerned with finding maps $X \to Y$ by extending inductively a given map $X_k \to Y$ to X_{k+1}. It is sufficient to deal with path-connected spaces Y. We thus assume Y to be path-connected. Then it is clear that any G-map $A \to Y$ has an extension to X_1 (see proof of

(3.15)). If X_1 is path-connected, then so is X_k for $k \geq 1$. Then the first main result of obstruction theory is

(3.10) Theorem. *For each integer $n \geq 1$ there exists an exact obstruction sequence*

$$[X_{n+1}, Y]_G \to \operatorname{Im}([X_n, Y]_G \to [X_{n-1}, Y]_G) \xrightarrow{c^{n+1}} \mathfrak{H}_G^{n+1}(X, A; \pi_n Y)$$

which is natural in (X, A) and Y.

The exactness of this sequence means that each homotopy class $X_{n-1} \to Y$ which is extendable over X_n has an associated obstruction element in the cohomology group $\mathfrak{H}_G^{n+1}(X, A; \pi_n Y)$ (as defined in (3.3)); this obstruction element is zero if and only if the homotopy class $X_{n-1} \to Y$ is extendable over X_{n+1}.

In analogy to the non-equivariant situation (G. W. Whitehead [1978], V. 5), we construct the exact sequence in several steps.

Let us start with a characteristic map

$$\phi\colon \coprod_j G \times (D_j^{n+1}, S_j^n) \to (X_{n+1}, X_n).$$

Here, (D_j^{n+1}, S_j^n) is just a copy of (D^{n+1}, S^n). We first describe an exact sequence

(3.11) $$[X_{n+1}, Y]_G \to [X_n, Y]_G \xrightarrow[c^{n+1}]{} C_G^{n+1}(X, A; \pi_n Y).$$

The first map is induced by the inclusion $X_n \subset X_{n+1}$. Let $[h] \in [X_n, Y]_G$ be given. We obtain the diagram

$$H_{n+1}(X_{n+1}, X_n) \xleftarrow[\varrho]{} \pi_{n+1}(X_{n+1}, X_n) \xrightarrow[\partial]{} \pi_n(X_n) \xrightarrow[h_*]{} \pi_n(Y) = [S^n, Y]$$

where ϱ is the Hurewicz homomorphism and ∂ the boundary operator. Suppose X_n and X_{n+1} are path-connected. Then (X_{n+1}, X_n) is n-connected (2.4). By the relative Hurewicz theorem (G. W. Whitehead [1978], IV.7), the map ϱ is surjective and the kernel is generated by elements of the form $x - \alpha x$, where $\alpha \in \pi_1(X_n)$ and αx denotes the action of this element α on x. Since Y is n-simple, the elements $x - \alpha x$ are mapped into zero by $h_* \partial$. Thus

$$h_* \partial \varrho^{-1} =: c^{n+1}(h)$$

is a well-defined element in $C_G^{n+1}(X, A; \pi_n Y)$ which only depends on the G-homotopy class of h. The G-equivariance of this homomorphism is immediate. If X_n is not path-connected, we treat each path-component in a similar way. For simplicity, we only deal with path-connected X_n for the rest of the proof. In general, the sequence in (3.10) is the product of the corresponding sequences for the path-components of X_n.

The element $c^{n+1}(h)$ has the following description as a cellular cochain. Let

$\phi = (\phi_j)$: $\coprod_j G \times (D^{n+1}, S^n) \to (X_{n+1}, X_n)$ be a characteristic map and $e_j \in C_{n+1}(X, A)$ be the basis element corresponding to ϕ_j. Then $c^{n+1}(h)(e_j) \in [S^n, Y]$ is the homotopy class represented by $h\phi_j$.

The element $c^{n+1}(h)$ has the following naturality property. Let f: $(X', A') \to (X, A)$ be a cellular map. Then

$$c^{n+1}(hf) = f^* c^{n+1}(h).$$

From the exact homotopy sequence of the pair (X_{n+1}, X_n) it follows immediately that $c^{n+1}(h) = 0$ if h is extendable over X_{n+1}. Conversely, assume $c^{n+1}(h) = 0$. Then, in particular, for each $j \in J$, the composition

$$S_j^n \to G \times S_j^n \to X_n \xrightarrow[h]{} Y$$

is nullhomotopic and therefore h can be extended (equivariantly!) over the $(n+1)$-cells.

Now we need

(3.12) Lemma. *For all $h \in [X_n, Y]_G$, the element $c^{n+1}(h)$ is a cocycle (called the obstruction cocycle).*

Proof. This follows from the commutative diagram

$$\begin{array}{ccccccc} H_{n+2}(X_{n+2}, X_{n+1}) & \xleftarrow[\varrho]{} & \pi_{n+2}(X_{n+2}, X_{n+1}) & & & & \\ \downarrow \partial & & \downarrow \partial & & & & \\ H_{n+1}(X_{n+1}) & \xleftarrow[\varrho]{} & \pi_{n+1}(X_{n+1}) & & & & \\ \downarrow & & \downarrow & & & & \\ H_{n+1}(X_{n+1}, X_n) & \xleftarrow[\varrho]{} & \pi_{n+1}(X_{n+1}, X_n) & \xrightarrow[\partial]{} & \pi_n(X_n) & \xrightarrow[h_*]{} & \pi_n(Y) \end{array}$$

in which the homomorphisms ϱ are Hurewicz maps and the vertical maps part of long exact sequences. □

Now assume that $[h_0]$ and $[h_1]$ are elements of $[X_n, Y]_G$ with the same image in $[X_{n-1}, Y]_G$. We want to show that $c^{n+1}(h_0)$ and $c^{n+1}(h_1)$ differ by a coboundary.

Choose a G-homotopy k: $I \times X_{n-1} \to Y$ with $k_i = h_i | X_{n-1}$ for $i = 0,1$. Consider the relative G-complex $(I \times X_n, I \times A)$ with n-skeleton

$\hat{X}_n = I \times X_{n-1} \cup \partial I \times X_n$ and $(n+1)$-skeleton $\hat{X}_{n+1} = I \times X_n$. Together, the maps h_0, k, h_1 define a map $F\colon \hat{X}_n \to Y$. We obtain an obstruction element

$$c^{n+1}(F) \in \operatorname{Hom}(H_{n+1}(\hat{X}_{n+1}, \hat{X}_n), \pi_n Y).$$

By composing $c^{n+1}(F)$ with the suspension isomorphism σ_z introduced earlier, we finally obtain an element

$$\begin{aligned} c^{n+1}(F) \circ \sigma_z =: d(h_0, k, h_1) &\in \operatorname{Hom}(H_n(X_n, X_{n-1}), \pi_n Y) \\ &= C^n_G(X, A; \pi_n Y). \end{aligned}$$

This is called the **difference cochain**; it is denoted $d(h_0, h_1)$ if k is the constant homotopy.

The element $d(h_0, k, h_1)$ has the following description as a cellular cochain. Suppose $\phi\colon (D^n, S^{n-1}) \to (X_n, X_{n-1})$ is the characteristic map of an n-cell defining a basis element $e \in C_n(X, A)$. Together, the maps h_0, k, h_1 define a map

$$\{0\} \times D^n \cup I \times S^{n-1} \cup \{1\} \times D^n \xrightarrow[\phi \times \mathrm{id}]{} \{0\} \times X_n \cup I \times X_{n-1} \cup \{1\} \times X_n \xrightarrow[(h_0, k, h_1)]{} Y.$$

Composing with a standard homeomorphism

$$S^n \cong \partial I \times D^n \cup I \times S^{n-1}$$

yields a homotopy class $x \in [S^n, Y]$ and

$$d(h_0, k, h_1)(e) = x.$$

We list a few properties of the difference cochain.

(3.13) Proposition.
(i) $d(f_0, \mathrm{const}, f_0) = 0$.
(ii) $d(f_0, k, f_1) + d(f_1, h, f_2) = d(f_0, k + h, f_2)$.
(iii) $d(f_0, k, f_1) = -d(f_1, k^-, f_0)$ *where* k^- *is the inverse homotopy to* k.
(iv) *For each* f_0, k, *and* $d \in C^n_G(X, A; \pi_n Y)$, *there exists* $f_1\colon X_n \to Y$ *with* $f_1 | X_{n-1} = k_1$ *such that* $d(f_0, k, f_1) = d$.

Proof. All these assertions follow from standard homotopy theory and are best proved by using the description above of the difference cochain as a cellular cochain. □

(3.14) Lemma. $\delta d(h_0, k, h_1) = c^{n+1}(h_0) - c^{n+1}(h_1)$.

Proof. We use the following diagram

$$\begin{array}{ccc}
H_n(X_n, X_{n-1}) & \xleftarrow{\quad\partial\quad} & H_{n+1}(X_{n+1}, X_n) \\
{\scriptstyle -\sigma}\big\downarrow \cong & (1) & {\scriptstyle \sigma}\big\downarrow \cong \\
H_{n+1}(I\times X_n, I\times X_{n-1}\cup \partial I\times X_n) & \xleftarrow{\quad\partial'\quad} & H_{n+2}(I\times X_{n+1}, I\times X_n\cup \partial I\times X_{n+1}) \\
\big\downarrow {\scriptstyle c^{n+1}(F)} \quad (2) & \searrow^{j_*} \quad (3) & \big\downarrow {\scriptstyle \partial''} \\
\pi_n(Y) & \xleftarrow{\quad c(F)\quad} & H_{n+1}(\partial I\times X_{n+1}\cup I\times X_n, \partial I\times X_n\cup I\times X_{n-1})
\end{array}$$

The maps ∂ and ∂'' are boundary maps of triples and ∂' is defined as in (3.7). The homomorphism $c(F)$ is the obstruction cocycle for F and the relative complex $(I\times X, I\times A)$. The map j is an inclusion. The triangle (2) is commutative by naturality of the obstruction cocycle, (1) is commutative by (3.7), and (3) is commutative by naturality of homology sequences. We have $c(F)\partial'' = 0$ because the composition of $c(F)$ with

$$H_{n+2}(\partial I\times X_{n+2}\cup I\times X_{n+1}, \partial I\times X_{n+1}\cup I\times X_n) \xrightarrow{\partial}$$
$$H_{n+1}(\partial I\times X_{n+1}\cup I\times X_n, \partial I\times X_n\cup I\times X_{n-1})$$

is already zero by (3.12).

Let

$$j_0, j_1\colon (X_{n+1}, X_n) \to (\partial I\times X_{n+1}\cup I\times X_n, \partial I\times X_n\cup I\times X_{n-1})$$

be the canonical inclusions. Then, for $c\in H_{n+1}(X_{n+1}, X_n)$, we have

$$\partial''\sigma(c) = j_{1*}(c) - j_{0*}(c) - j_*\partial'\sigma(c)$$

in the diagram above. This follows from (3.6).

Using the properties of the large diagram established so far, we obtain

$$\begin{aligned}
-c^{n+1}(F)\sigma\partial(c) &= c(F)j_*\partial'\sigma(c)\\
&= c(F)(j_{1*} - j_{2*} - \partial''\sigma)(c)\\
&= c(F)j_{1*}(c) - c(F)j_{0*}(c)\\
&= c^{n+1}(h_1)(c) - c^{n+1}(h_0)(c).
\end{aligned}$$

The last equality follows again from the naturality of the obstruction cocycle. □

By now, we have established the maps in the sequence appearing in (3.10). Exactness remains to be shown. Let $f\colon X_n \to Y$ be given such that $c^{n+1}(f)$ is cohomologous to zero. Then f might not be extendable to X_{n+1} without alteration. We want to modify f relative to X_{n-1} to obtain a map g such that the resulting difference cochain $d(f, g)$ satisfies $\delta d(f, g) = c^{n+1}(f)$. Using (3.14), we see that g is extendable over X^{n+1}. By (3.13, iv), such g exists.

This completes the proof of (3.10). □

For the rest of this section, we use the following **conventions**: We fix an integer $n \geqq 1$. The space Y is n-simple and $(n-1)$-connected. The pair (X, A) is a relative G-complex with free action on $X \backslash A$.

Let $f\colon A \to Y$ be a G-map.

(3.15) Proposition.

(i) *The map $f\colon A \to Y$ is extendable over $X_n \to Y$. Any two extensions are homotopic* rel A *on X_{n-1}.*

(ii) *Let $k\colon f_0 \simeq f_1\colon A \to Y$ be a homotopy and let $F_0, F_1\colon X_n \to Y$ be extensions of f_0, f_1. There exists a homotopy K between $F_0 | X_{n-1}$ and $F_1 | X_{n-1}$ extending k.*

Proof.

(i) f can be extended to X_0: The space X_0 is a disjoint union of A and homogeneous spaces G. Since Y is path-connected, any two extensions are homotopic rel A. An extension to X_1 is possible: One has to show that each attaching map $\varphi\colon G \times S^0 \to X_0$ composed with $X_0 \to Y$ is null-homotopic. This is the case because Y is path-connected. This proves the assertion for $n = 1$. Now assume $n \geqq 2$. For $1 \leqq r \leqq n-1$, let $h\colon X_r \to Y$ be an extension. Since $\pi_r Y = 0$, and therefore $C_G^{r+1}(X, A; \pi_r Y) = 0$, (3.5) gives an extension of h to X_{r+1}. Now suppose that $g_0, g_1\colon X_n \to Y$ are extensions of f. Consider the map

$$F\colon \partial I \times X_n \cup I \times A \to Y$$

which is the constant homotopy on $I \times A$ and equals g_i on $\{i\} \times X_n$. Consider the relative complex $(I \times X_n, \partial I \times X_n \cup I \times A)$. By the first part of the proof, F has an extension to the n-skeleton and in particular to $I \times X_{n-1}$. This yields the desired homotopy rel A.

(ii) is obtained by applying (i) to $(I \times X_{n-1}, \partial I \times X_{n-1} \cup I \times A)$. □

We now consider maps $f\colon A \to Y$ and their extensions $\bar{f}\colon X_n \to Y$. By (3.15), $\bar{f}$ determines uniquely an element in

$$\operatorname{Im}([X_n, Y]_G \to [X_{n-1}, Y]_G).$$

Using (3.10), we obtain a well-defined $c^{n+1}(\bar{f}) \in \mathfrak{H}_G^{n+1}(X, A; \pi_n Y)$. By (3.15), homotopic maps f_0 and f_1 have extensions $\bar{f}_0$ and $\bar{f}_1$ which are homotopic on X_{n-1}. Therefore, $c^{n+1}(\bar{f}_0) = c^{n+1}(\bar{f}_1)$ and we obtain a well-defined map

$$\gamma\colon [A, Y]_G \to \mathfrak{H}_G^{n+1}(X, A; \pi_n Y)$$

by $\gamma[f] = c^{n+1}(\bar{f})$. We have $\gamma[f] = 0$ if and only if f has an extension to X_{n+1}, i.e. the sequence

$$[X_{n+1}, Y]_G \to [A, Y]_G \to \mathfrak{H}_G^{n+1}(X, A; \pi_n Y)$$

is exact. We call $\gamma[f]$ the **primary obstruction to extending** f. The map γ is again natural in Y and (X, A); the reader should make explicit what this means.

We are now going to classify the extensions $X_n \to Y$ up to homotopy rel A. We put $X = X_n$. We apply the preceding construction to $(I \times X, I \times A \cup \partial I \times X)$ and obtain a map

$$[I \times A \cup \partial I \times X, Y]_G \xrightarrow{\gamma} \mathfrak{H}_G^{n+1}(I \times X, I \times A \cup \partial I \times X; \pi_n Y)$$

which we compose with the suspension isomorphism σ^{-1} of (3.9). Since $\gamma(F) = 0$ precisely when F has an extension to $I \times X$, we regard γ as an obstruction to the extension of a homotopy. We want to show that this obstruction element can be used to classify homotopy classes $X_n \to Y$ rel A.

To begin with, we collect the properties of this construction. A map $F\colon I \times A \cup \partial I \times X \to Y$ will be considered as a homotopy k between $f_i = F|\{i\} \times X$ and $\sigma^{-1}\gamma(F)$ will be denoted by $\gamma(f_0, k, f_1)$ or simply by $\gamma(f_0, f_1)$ if k is the constant homotopy. The next proposition follows essentially from (3.13).

(3.16) Proposition.
(i) $\gamma(f_0, f_0) = 0$.
(ii) $\gamma(f_0, k, f_1) + \gamma(f_1, h, f_2) = \gamma(f_0, k + h, f_2)$.
(iii) $\gamma(f_0, k, f_1) = -\gamma(f_1, k^-, f_0)$ *where* k^- *is the inverse homotopy to* k.
(iv) *For each* f_0, k, *and* $d \in \mathfrak{H}_G^n(X, A; \pi_n Y)$, *there exists* f_1 *such that* $d = \gamma(f_0, k, f_1)$.
(v) *For* $\alpha\colon Y \to Y'$, *one has*
$\alpha_*\gamma(f_0, k, f_1) = \gamma(\alpha f_0, \alpha k, \alpha f_1)$.
(vi) *For* $\beta\colon (X', A') \to (X, A)$, *one has*
$\beta^*\gamma(f_0, f_1) = \gamma(f_0\beta, f_1\beta)$. □

Using (3.16), we can prove the desired homotopy classification theorem.

Let $f_0\colon X_n \to Y$ be a fixed G-map. Suppose $f\colon X_n \to Y$ is any other map with $f|A = f_0|A$. Then $\gamma(f_0, f) \in \mathfrak{H}_G^n(X, A; \pi_n Y)$. Let $[X, Y]_G^A$ be the set of G-homotopy classes rel A of maps $f\colon X \to Y$ with $f|A = f_0|A$.

(3.17) Theorem. *Let Y be an n-simple and $(n-1)$-connected space. Then $\gamma(f_0, f)$ only depends on the class $[f]^A \in [X_n, Y]_G^A$ of f. The assignment $[f]^A \mapsto \gamma(f_0, f)$ gives a bijection*

$$\gamma(f_0)\colon [X_n, Y]_G^A \to \mathfrak{H}_G^n(X, A; \pi_n Y).$$

Proof. $\gamma(f_0, f) = 0$ if and only if there exists a homotopy rel A between f_0 and f_1. If f and f' are homotopic rel A, then $\gamma(f, f') = 0$ and, by (3.16),

$$\gamma(f_0,f) = \gamma(f_0,f) + \gamma(f,f') = \gamma(f_0,f').$$

Therefore, $\gamma(f_0)$ is well-defined. If $\gamma(f_0,f) = \gamma(f_0,f')$, then, by (3.16), $\gamma(f,f') = 0$ and hence f and f' are homotopic rel A. Finally, (3.16, iv) shows that $\gamma(f_0)$ is surjective. □

In general, the set $[X_n, Y]_G^A$ has no natural group structure. One can interpret (3.11) and (3.12) to the effect that there is a natural transitive action of $\mathfrak{H}_G^n(X, A; \pi_n Y)$ on this set. For other presentations of equivariant obstruction theory, see Bredon [1967] and Lashof-Rothenberg [1978]. A sensible translation of the classical obstruction theory to equivariant fibrations uses equivariant cohomology with local coefficients; a local coefficient system in this case is a contravariant functor from the fundamental group category I.10 to the category of abelian groups.

Another application of obstruction theory to transformation groups is described in Lashof [1979].

Finally, we quote a result from non-equivariant obstruction theory for later use. Let again Y be n-simple and $(n-1)$-connected. By the universal coefficient theorem, we have, with $\pi = \pi_n Y$,

$$H^n(Y;\pi) \cong \operatorname{Hom}(H_n(Y), \pi).$$

Let $\iota(Y) \in H^n(Y;\pi)$ be the class corresponding to the inverse of the Hurewicz isomorphism.

(3.18) Proposition. *Let (X, A) be a relative CW-complex and $f\colon A \to Y$ be a given map. The primary obstruction $\gamma(f) \in H^{n+1}(X, A;\pi)$ is the image of $\iota(Y)$ under*

$$H^n(Y;\pi) \xrightarrow{f^*} H^n(A;\pi) \xrightarrow{\delta} H^{n+1}(X, A;\pi).$$

Proof. G. W. Whitehead [1978], V (6.2). □

Using this proposition, one shows that, for $f_0, f_1\colon X_n \to Y$, the element $\gamma(f_0, f_1) \in H^n(X;\pi)$ is given by

(3.19) $(-1)^n(f_1^*\iota(Y) - f_0^*\iota(Y))$.

If $\pi_n(Y) \cong \mathbb{Z}$, then $f_1^*\iota(Y)$ is usually called the degree of $f_1\colon X_n \to Y$. Thus we see that in this case

(3.20) f_0 and f_1 are homotopic if and only if they have the same degree.

(3.21) Exercises.

1. Let $p\colon E \to B$ be a G-map and B a numerably locally trivial free G-space.
   (i) p is a G-fibration if and only if $p/G\colon E/G \to B/G$ is a fibration.

(ii) p has a G-section if and only if p/G has a section.
Thus, if p is a G-fibration over the connected G-complex B with typical fibre F, then the obstruction to finding an equivariant section are the non-equivariant obstructions for sections of p/G and therefore lie in groups $H^n(B/G;\ \pi_{n-1}(F))$; here, of course, local coefficients have to be used in general. Note that for finite G the sequence

$$1 \to \pi_1(B) \to \pi_1(B/G) \to \pi_1(BG) \to 1$$

is exact ($\pi_1(BG) \cong G$).

2. If $p\colon E \to B$ is a G-fibration over the G-complex B and $A \subset B$ is a subcomplex such that G acts freely on $B\backslash A$, then the obstructions to extending a section over A to a section over B lie in suitable groups $H^n(B/G, A/G; \pi_{n-1}(F))$. Carry this out by using the method of exercise 1.

4. The classification theorem of Hopf.

We recall a classical theorem of H. Hopf, see (3.20). Let M^nbe a closed connected orientable n-manifold. Then $H^n(M^n; \mathbb{Z}) \cong \mathbb{Z}$. Choose generators (= orientations) $z(S^n)$ and $z(M^n)$ of $H^n(S^n; \mathbb{Z})$ and $H^n(M^n; \mathbb{Z})$. Given $f\colon M^n \to S^n$, its degree $d(f) \in \mathbb{Z}$ is defined by $f^* z(S^n) = d(f) z(M^n)$. Hopf's theorem then asserts that $[M^n, S^n] \to \mathbb{Z}, [f] \mapsto d(f)$ is a bijection; in particular, homotopy classes of maps are characterized by their degrees. We generalize this result to transformation groups.

In this section, G denotes a compact Lie group.

Let X be a G-complex of finite orbit type and let Y be a G-space. We are looking for conditions on X and Y which allow a classification of homotopy classes $X \to Y$ in terms of suitable degrees. In general, there need not exist G-maps $X \to Y$ at all, e.g. if X has trivial and Y free G-action. For the construction of G-maps $X \to Y$, we want to use the inductive procedure of I (7.4). Let $\operatorname{Iso}(X)$ consist of the conjugacy classes $(H_1), \ldots, (H_t)$. The indexing is chosen in such away that H_i subconjugate to H_j implies $j \leqq i$. Let $\mathfrak{F} = (H_1) \cup \ldots \cup (H_k)$, $H = H_{k+1}$, and $\mathfrak{F}' = \mathfrak{F} \cup (H)$. Then $X(\mathfrak{F}') = X_{(H)} \cup X(\mathfrak{F})$. Suppose we have a G-map $f(\mathfrak{F})\colon X(\mathfrak{F}) \to Y$. The G-extensions of $f(\mathfrak{F})$ to $X(\mathfrak{F}')$ correspond bijectively to the WH-extensions of $f(\mathfrak{F})^H\colon X(\mathfrak{F})^H = X^{>H} \to Y^H$ to $X(\mathfrak{F}')^H = X^H$. Our aim is to compute the homotopy set $[X, Y]_G$ and to describe invariants for its elements. The invariants shall be mapping degrees. Suppose $f, h\colon X \to Y$ are given and assume that their restrictions to $X(\mathfrak{F})$ are G-homotopic. Since $X(\mathfrak{F}) \subset X$ is a G-cofibration (1.12), we can assume that $f|X(\mathfrak{F}) = h|X(\mathfrak{F})$. The restrictions to $X(\mathfrak{F}') = X_{(H)} \cup X(\mathfrak{F})$ are G-homotopic rel $X(\mathfrak{F})$ if and only if f^H and h^H are WH-homotopic rel $X^{>H}$. Now suppose

(4.1) The space Y^H is $n(H)$-simple and $(n(H)-1)$-connected for an integer $n(H) \geq 1$.

Then we have a primary obstruction (3.17)

(4.2) $\gamma(f^H, h^H) \in \mathfrak{H}^{n(H)}_{W(H)}(X^H, X^{>H}; \pi_{n(H)}(Y^H))$

against a WH-homotopy between f^H and h^H rel $X^{>H}$. There are no more obstructions if we assume

(4.3) X^H is a finite-dimensional WH-complex which has free WH-cells $WH \times D^k$ of dimension k at most $n(H)$.

We look for conditions under which elements (4.2) can be identified as expressions in mapping degrees. We begin by analysing the cohomology group appearing in (4.2).

Let W be a finite group. If M is a $\mathbb{Z}W$-module which is isomorphic to $\mathbb{Z}$ as abelian group, then the $\mathbb{Z}W$-module structure can be specified by a homomorphism

$$e_M: W \to \mathrm{Aut}(\mathbb{Z}) = \{+1, -1\}$$

such that $w \in W$ acts as multiplication by $e_M(w)$. We call e_M the **orientation behaviour** of M.

Let U be a W-complex and $B \subset U$ a subcomplex. Let M be a $\mathbb{Z}W$-module. We make the following assumptions.

(4.4) (i) $\dim U = n > 0$.
(ii) W acts freely on $U \backslash B$.
(iii) $P := H^n(U, \mathbb{Z}) \cong \mathbb{Z}, \quad M \cong \mathbb{Z}$.
(iv) $e_P = e_M$.

Let $t: \mathfrak{H}^n_W(U, B; M) \to H^n(U, B; M)$ be the map which forgets the W-action.

We need the following algebraic device. Let M_1 and M_2 be $\mathbb{Z}W$-modules. We have a **norm homomorphism**

$$N: \mathrm{Hom}(M_1, M_2) \to \mathrm{Hom}_W(M_1, M_2)$$

from the $\mathbb{Z}$-linear into the $\mathbb{Z}W$-linear homomorphisms which is defined by

$$N(f)(x) = \sum_{w \in W} wf(w^{-1}x).$$

The following is easily checked

(4.5) Lemma.
(i) *If u and v are $\mathbb{Z}W$-linear, then $N(ufv) = uN(f)v$.*
(ii) *If M_1 is a free $\mathbb{Z}W$-module, then N is surjective.* □

We have the defining exact sequences for cohomology groups

$$(4.6)\quad 0 \leftarrow \mathfrak{H}^n_W(U,B;M) \leftarrow \mathrm{Hom}_W(H_n(U_n,U_{n-1}),M) \xleftarrow[d_W]{} \mathrm{Hom}_W(H_{n-1}(U_{n-1},U_{n-2}),M)$$

$$0 \leftarrow H^n(U,B;M) \leftarrow \mathrm{Hom}(H_n(U_n,U_{n-1}),M) \xleftarrow[d]{} \mathrm{Hom}(H_{n-1}(U_{n-1},U_{n-2}),M).$$

The map N satisfies $d_W N = Nd$, (4.5, i), and therefore induces a map, also called N, between the cohomology groups which is surjective by (4.5, ii). The composition Nt is multiplication by $|W|$.

The exact sequence of $\mathbb{Z}W$-modules

$$(4.7)\quad 0 \to H_n(U,B) \xrightarrow[i]{} H_n(U_n,U_{n-1}) \xrightarrow[\partial]{} H_{n-1}(U_{n-1},U_{n-2})$$

induces a sequence

$$(4.8)\quad \mathrm{Hom}_W(H_n(U,B),M) \xleftarrow{\delta_0} \mathrm{Hom}_W(H_n(U_n,U_{n-1}),M) \xleftarrow{\delta_1} \mathrm{Hom}_W(H_{n-1}(U_{n-1},U_{n-2}),M).$$

We show exactness, i.e. $\ker\delta_0 = \mathrm{im}\,\delta_1$. If we apply $\mathrm{Hom}(-,M)$ to (4.7), we obtain an exact sequence, analogous to (4.8), which is exact even if we continue with zero on the left (universal coefficient formula). Now suppose that $\varphi \in \ker\delta_0$. Choose $\psi\colon H_n(U_n,U_{n-1}) \to H_n(U_n,U_{n-1})$ so that $N(\psi) = \mathrm{id}$; this is possible by (4.5, ii). The composition $\varphi\psi i$ is W-equivariant because all self-maps of $H_n(U;\mathbb{Z}) \cong \mathbb{Z}$ are. Therefore,

$$|W|\varphi\psi i = N(\varphi\psi i) = \varphi N(\psi) i = \varphi i = 0;$$

the last equation holds since $\varphi \in \ker\delta_0$. Hence $\varphi\psi i = 0$ and, therefore, there exists $\varphi' \in \mathrm{Hom}(H_{n-1}(U_{n-1},U_{n-2}),M)$ such that $\varphi\psi = \varphi'\partial$. This implies

$$N(\varphi')\partial = N(\varphi'\partial) = N(\varphi\psi) = \varphi N(\psi) = \varphi$$

and hence $\varphi \in \mathrm{im}\,\delta_1$.

If we now compare the sequences (4.6) and (4.8), we obtain an induced injective homomorphism

$$j\colon \mathfrak{H}^n_W(U,B;M) \to \mathrm{Hom}_W(H_n(U,B),M).$$

The following diagram is commutative

$$\begin{array}{ccccc}
H^n(U,B;M) & \xrightarrow{N} & \mathfrak{H}^n_W(U,B;M) & \xrightarrow{t} & H^n(U,B;M) \\
\downarrow \cong & & \downarrow j & & \downarrow \cong \\
\mathrm{Hom}(H_n(U,B),M) & \xrightarrow{N} & \mathrm{Hom}_W(H_n(U,B),M) & \xrightarrow{t} & \mathrm{Hom}(H_n(U,B),M) \\
\downarrow i & & \downarrow & & \downarrow i \\
\mathrm{Hom}(H_n(U),M) & \xrightarrow{N} & \mathrm{Hom}_W(H_n(U),M) & \xrightarrow{t} & \mathrm{Hom}(H_n(U),M).
\end{array}$$

(4.9) Proposition. *Suppose* (U, B) *satisfies* (4.4).
(i) *If* $\dim B \leq n-1$, *then the image of*

$$\mathfrak{H}^n_W(U,B;M) \xrightarrow{t} H^n(U,B;M) \to H^n(U;M) = \mathbb{Z}$$

consists of the multiples of $|W|$.
(ii) *If* $\dim B \leq n-2$, *then both* $\mathfrak{H}^n_W(U,B;M)$ *and* $H^n(U,B;M)$ *are isomorphic to* $\mathbb{Z}$.

Proof.
(i) If $\dim B \leq n-1$, then, in the diagram above, i is surjective. Now use diagram chasing and the fact that N, in the top row, is surjective.
(ii) In this case $H_n(U,B) = H_n(U)$; therefore, j maps injectively into $\mathrm{Hom}_W(H_n(U);\mathbb{Z}) = \mathbb{Z}$. □

We now return to the investigation of $[X, Y]_G$ and make the following assumptions.

(4.10) (i) X is a G-complex of finite orbit type and Y a G-space.
(ii) For $K \in \mathrm{Iso}(X)$, the space X^K is a finite dimensional WK-complex of topological dimension $n(K)$ and Y^K is $(n(K)-1)$-connected if $n(K) > 0$. (The dimension of the empty set is -1.)
(iii) Let $\phi(G, X, Y) = \{(K) \mid K \in \mathrm{Iso}(X),\ WK \text{ finite},\ \pi_{n(K)}(Y^K) \neq 0\}$.
For $(K) \in \phi(G, X, Y)$, we assume that

$$P(K) := \tilde{H}^{n(K)}(X^K;\mathbb{Z}) \cong \mathbb{Z}$$

and $$M(K) := \pi_{n(K)}(Y^K) \cong \begin{cases} \mathbb{Z} & n(K) > 0 \\ \mathbb{Z}/2 & n(K) = 0 \end{cases}$$

with $\mathbb{Z}/2$ standing for a 2-element set. Then, by the Hurewicz and universal coefficient theorems,

$$Q(K) := \tilde{H}^{n(K)}(Y^K;\mathbb{Z}) \cong \mathbb{Z}.$$

(iv) For $(K) \in \phi(G, X, Y)$, the WK-modules $P(K)$ and $Q(K)$ are isomorphic.

If (ii), (iii), or (iv) hold for a group K, then also for each conjugate group. An orientation for X consists of a choice of generators

$$z(K, X) \in \tilde{H}^{n(K)}(X^K; \mathbb{Z})$$

for each $(K) \in \phi(G, X, Y)$ and similarly for Y. Having X and Y oriented, we can associate to each G-map and each $(K) \in \phi(G, X, Y)$ a degree $d(f^K) \in \mathbb{Z}$ defined by

$$(f^K)^* z(K, Y) = d(f^K) z(K, X).$$

Assumption (4.10, iv) allows a choice of orientations in such a manner that $d(f^K) = d(f^L)$ whenever $(K) = (L)$.

(4.11) Theorem. *Suppose X and Y satisfy* (4.10)(i)–(iv) *and are oriented. Then the following holds:*

(i) *$[X, Y]_G$ is not empty.*

(ii) *Let $f: X \to Y$ be a G-map. Suppose that $(K) \in \phi(G, X, Y)$, $n(K) > 0$, $\dim X^{>K} < n(K)$. Suppose that $d = d(f^K) \bmod |WK|$. Then there exists a G-map $h: X \to Y$ such that $d(h^K) = d$ and $h|X^{>K} = f|X^{>K}$.*

(iii) *Let K be as in* (ii). *Suppose that $f_0, f_1: X \to Y$ are G-maps such that $f_0|X^{>K}$ and $f_1|X^{>K}$ are homotopic as WK-maps. Then*

$$d(f_0^K) = d(f_1^K) \bmod |WK|.$$

(iv) *If $1 + \dim X^{>K} < n(K)$ for all $(K) \in \phi(G, X, Y)$, then two G-maps $f_0, f_1: X \to Y$ are G-homotopic if and only if $d(f_0^K) = d(f_1^K)$ for all $(K) \in \phi(G, X, Y)$.*

Remarks.

(i) In case $\phi(G, X, Y)$ is empty, all G-maps $X \to Y$ are G-homotopic.

(ii) If WH is finite and $H \subset K$, then WK is finite. This is seen as follows: $\emptyset \neq (G/K)^H$ carries a free WK-action and consists of a finite number of WH-orbits I(5.10) and hence is a finite set.

(iii) It is an interesting problem to determine in (iii) the exact relationship between $d(f^H) \bmod |WH|$ and the $d(f^K)$. We shall return to this problem.

Proof of (4.11). Using the hypothesis (4.10, ii), (i) follows immediately from (3.15).

(iv) Suppose f and h are given such that $d(f^H) = d(h^H)$ for all $H \in \operatorname{Iso}(X)$, WH finite. We want to show that f and g are G-homotopic. We do this by induction over orbit bundles, assuming that their restrictions to $X(\mathfrak{F})$ are G-homotopic. Since $X(\mathfrak{F}) \subset X$ is a G-cofibration, we can and shall assume that $f|X(\mathfrak{F}) = h|X(\mathfrak{F})$. We show that in this case the restrictions to $X(\mathfrak{F}') = X_{(H)} \cup X(\mathfrak{F})$ are G-homotopic relative $X(\mathfrak{F})$. There is a single obstruction

$\gamma(f^H, h^H) \in \mathfrak{H}_W^{n(H)}(X^H, X^{>H}; \pi_{n(H)}(Y^H))$; see (4.2). If WH is infinite, then this cohomology group is zero: Since the topological dimension of X^H is $n(H)$, the cellular dimension of $(X^H, X^{>H})$ is less than $n(H)$. For finite WH, we use (4.5) and the map t. The element $t\gamma(f^H, h^H) \in H^{n(H)}(X^H, X^{>H}; \pi_{n(H)}(Y^H))$ is the obstruction against a non-equivariant homotopy $f^H \simeq h^H$ rel $X^{>H}$. Since, by assumption, $\dim X^{>H} < n(H) - 1$, the canonical map

$$j\colon H^{n(H)}(X^H, X^{>H}; \pi_{n(H)}(Y^H)) \to H^{n(H)}(X^H; \pi_{n(H)}(Y^H)) \cong \mathbb{Z}$$

is an isomorphism. The element $jt\gamma(f^H, h^H)$ is the obstruction against a homotopy $f^H \cong h^H$.

It follows from (3.19) that, up to sign, $jt\gamma(f^H, h^H)$ is the difference of degrees $d(f^H) - d(h^H)$. Thus $\gamma(f^H, h^H) = 0$. This completes the proof of (iv).

(ii) The WK-extensions $h^K\colon X^K \to Y^K$ of $f^K|X^{>K}$ are classified by $\gamma(h^K, f^K) \in \mathfrak{H}_{WK}^{n(K)}(X^K, X^{>K}; \pi_{n(K)}(Y^K))$. By (4.9, i), we can find h^K with degree d. Each WK-map $X^K \to Y^K$ has an extension to a G-map $X \to Y$; there are no obstructions.

(iii) follows as (ii) by using (4.9, i) and (3.17). □

As a first application of (4.11), we derive the classical homotopy classification of lens spaces. Let $G = \mathbb{Z}/m$ be the cyclic group of order m. Given positive integers $r_1, \dots, r_n$ prime to m, we have the representation $V(r_1, \dots, r_n)$

$$\mathbb{Z}/m \times \mathbb{C}^n \to \mathbb{C}^n, (k, (z_1, \dots, z_n)) \mapsto (\zeta^{r_1} z_1, \dots, \zeta^{r_n} z_n)$$

($\zeta = \exp(2\pi i/m)$) with unit sphere $SV(r_1, \dots, r_n) = S(r_i)$. We orient $S(r_i)$ in a canonical manner using the standard orientation of $\mathbb{C}^n$. A G-map f between two such spheres is a G-homotopy equivalence if and only if it is an ordinary homotopy equivalence (2.5), and this is the case if and only if f has degree ± 1 (why?). We call f an **oriented homotopy equivalence** if it has degree 1. By (4.11), a G-map $f\colon SV(r_1, \dots, r_n) \to SV(s_1, \dots, s_n)$ has a degree $d(f)$ whose congruence class mod m is uniquely determined and each element in this congruence class can be realized by a suitable map. It thus suffices to find G-maps f such that $d(f) \equiv 1 \bmod m$.

(4.12) Proposition. *For a G-map $f\colon SV(r_1, \dots, r_n) \to SV(s_1, \dots, s_n)$, we have*

$$d(f) r_1 \cdot \ldots \cdot r_n \equiv s_1 \cdot \ldots \cdot s_n \bmod m.$$

Thus $SV(r_1, \dots, r_n)$ and $SV(s_1, \dots, s_n)$ are oriented G-homotopy equivalent if and only if $\prod r_i \equiv \prod s_i \bmod m$.

Proof. For the proof, it suffices to compute the degree of one particular map. The map $f\colon S^1 \to S^1, z \mapsto z^t$ satisfies $f(\zeta^r z) = \zeta^{rt} f(z)$. Therefore, this is an equivariant map $SV(r) \to SV(l)$ if $rt \equiv l \bmod m$; the degree of this map is t. Let V and W be orthogonal representations of any compact Lie group G. Then

$S(V \oplus W)$ is G-homeomorphic to the join $SV * SW$. Employing the join construction to maps like f above, we obtain an equivariant map

$$h\colon SV(r_1) * \ldots * SV(r_n) \to SV(s_1) * \ldots * SV(s_n)$$

such that its degree satisfies the congruence of (4.12). □

The orbit space $S(r_i)/G = L(r_i)$ is called a **lens space**. We want to investigate when two such lens spaces are homotopy equivalent. Let $h\colon L(r_i) \to L(s_i)$ be a homotopy equivalence. The orbit map $p\colon S(r_i) \to L(r_i)$ is a universal covering. Therefore, we find $H\colon S(r_i) \to S(s_i)$ covering h. The map H is equivariant in the following sense: Choose base points $x_0 \in L(r_i)$, $h(x_0) = y_0 \in L(s_i)$, $a_0 \in p^{-1}(x_0)$, $t_0 \in p^{-1}(y_0)$. We can find H such that $H(a_0) = t_0$. We obtain an isomorphism $\pi_1(L(r_i)) \cong G$ by lifting $[w] \in \pi_1$ to $\tilde{w}$ with $\tilde{w}(0) = a_0$, $\tilde{w}(1) = g(a_0)$ and mapping $[w]$ to g. Similarly, $\pi_1(L(s_i)) \cong G$. The isomorphism $h_*\colon \pi_1(L(r_i), x_0) \to \pi_1(L(s_i), y_0)$ corresponds to an automorphism $\alpha\colon G \to G$. We compare the maps $H_g\colon s \mapsto H(gs)$ and ${}_gH\colon s \mapsto gH(s)$. Both are liftings of h. We have $H_g(a_0) = \alpha(g)t_0$. Uniqueness of liftings yields $\alpha(g)H(s) = H(gs)$, which is the desired equivariance property of H.

(4.13) Proposition. *$L(r_i)$ and $L(s_i)$ are homotopy equivalent if and only if there exists $u \in \mathbb{Z}/m^*$ such that*

$$\prod r_i \equiv \pm u^n \prod s_i \bmod m.$$

Proof. Given a homotopy equivalence h, the argument above shows the existence of an equivariant map H with degree $d(H)$ satisfying

$$d(H)\prod r_i \equiv u^n \prod s_i \bmod m$$

if α is given as $\alpha(t) = ut$, $(u, m) = 1$. It follows from covering space theory that H is a homotopy equivalence. Therefore, H has degree ± 1.

Conversely, if the congruence is satisfied, we can find an equivariant map

$$H\colon S(r_i) \to S(us_i)$$

of degree ± 1. The quotient map is a homotopy equivalence $L(r_i) \to L(s_i)$. □

Let M be a connected, oriented, closed manifold of dimension $2n-1$ with smooth free orientation preserving action of $\mathbb{Z}/m$. Let V be a complex representation of $\mathbb{Z}/m$ of complex dimension n with free action on $S(V)$. Then there exist equivariant maps $f\colon M \to S(V)$ and the degree $d(f) \bmod m$ is uniquely determined. Of course, given M, we have a choice of several V. By the preceding discussion of lens spaces, we can alter the class $d(f) \in \mathbb{Z}/m$ by an arbitrary unit $a \in \mathbb{Z}/m^*$. Thus M alone determines uniquely a class in the set $\mathbb{Z}/m \bmod \mathbb{Z}/m^*$. If $m = p^t$ is a power of the prime p, then this class can be specified by the element of the form p^a, $0 \le a \le t$, in it. This degree class is an important invariant of the group action on M. It is an unsolved problem to determine this degree class from other geometric properties of the manifold. As an example, consider a free action of $\mathbb{Z}/2$ on the closed n-manifold M. Assume that the equivariant tangent bundle $TM \oplus \mathbb{R}$ is isomorphic to $M \times V$ where $V = \mathbb{R}^{n+1}$ with antipodal action. Then the degree $d(M)$ of a $\mathbb{Z}/2$-map $M \to S(V)$ is determined modulo 2. One can show (see Löffler [1980], (2.3))

(4.14) $$d(M) \equiv \tfrac{1}{2}\chi(M) \quad \text{for } n \text{ even}$$
$$d(M) \equiv \chi_{\frac{1}{2}}(M) \quad \text{for } n \text{ odd}.$$

($\chi(M)$ Euler characteristic, $\chi_{\frac{1}{2}}(M)$ the mod 2 semi-characteristic of Kervaire.)

We now return to the general problem mentioned in Remark (iii) after (4.11). We show that quite generally there exist **linear** congruences between degrees of fixed point mappings. In order to do so, we have to recall a few facts from homotopy theory.

If X and Y are pointed G-spaces, let $[X, Y]_G^0$ denote the set of pointed G-homotopy classes $X \to Y$. Let $\Sigma X = S^1 \wedge X$ denote the suspension of X. As in ordinary homotopy theory, $[\Sigma X, Y]_G^0$ carries a natural group structure and $[\Sigma^2 X, Y]_G^0$ is abelian. (The construction of the group structure uses only the „suspension coordinate“ and has nothing to do with the group action.) If X is any G-space, let σX denote the unreduced suspension, i.e. $I \times X$ with both $0 \times X$ and $1 \times X$ identified to a point (in particular $\sigma X = S^0$ if $X = \emptyset$). These two points are G-fixed and $0 \times X$ can be taken as a base point for σX. A G-map $f: X \to Y$ induces a pointed G-map $\sigma f: \sigma X \to \sigma Y$, $[t, x] \mapsto [t, fx]$. This construction is compatible with G-homotopies. Taking the n-fold suspension, we obtain a map

$$[X, Y]_G \to [\Sigma^n \sigma X, \Sigma^n \sigma Y]_G^0$$

for G-spaces X and Y.
The forgetful map $[\Sigma^n \sigma X, \Sigma^n \sigma Y]_G^0 \to [\Sigma^n \sigma X, \Sigma^n \sigma Y]_G$ is bijective for $n \geq 2$ because then $\Sigma^n \sigma Y$ has a simply connected fixed point set (4.15).

If X and Y satisfy the hypotheses (4.10), then $\Sigma^n \sigma X$ and $\Sigma^n \sigma Y$, $n \geq 2$, again satisfy (4.10). The only additional isotropy group appearing may be G itself, namely in case $X^G = \emptyset$, and then

$$(\Sigma^n \sigma X)^G = \Sigma^n \sigma(X^G) = S^n.$$

We now suppose that $X = \Sigma A$ for a pointed G-complex A and that Y^G is simply connected. Then

(4.15) Lemma. *The forgetful map* $[\Sigma A, Y]_G^0 \to [\Sigma A, Y]_G$ *is bijective.* □

If, moreover, $X = \Sigma A$ and Y satisfy (4.10), then the assignment of the mapping degree $[f] \mapsto d(f^H)$ is a homomorphism $[\Sigma A, Y]_G^0 \to \mathbb{Z}$ and from (4.11) we obtain an injective homomorphism

(4.16) $\varphi\colon [\Sigma A, Y]^0_G \to \prod_{(H)} \mathbb{Z}, f \mapsto (d(f^H))$

where $(H) \in \phi(G, X) := \{(H) \mid H \in \mathrm{Iso}(X),\ WH \text{ finite}\}$. By (4.11, ii), the image of φ has maximal rank (i.e. cokernel φ is finite). Therefore, this image must be describable by a set of congruence relations among the $d(f^H)$, $(H) \in \phi(G, X)$. In order to deal with such congruences, it suffices to consider the case of a finite group G and to ask for the dependence of $d(f)$ on the $d(f^H)$ for $H \neq 1$.

Let A_s be the singular part of A so that ΣA_s is the singular part of ΣA. The homomorphism $[\Sigma A, Y]^0_G \to \mathbb{Z}/|G|$, $[f] \mapsto d(f) \bmod |G|$ factors, by (4.11, iii), over the restriction $[\Sigma A, Y]^0_G \to [\Sigma A_s, Y]^0_G$ and, together with (4.11, iv), we see that we have a pull-back diagram of abelian groups

(4.17)
$$\begin{array}{ccc} [\Sigma A, Y]^0_G & \xrightarrow{\ r\ } & [\Sigma A_s, Y]^0_G \\ \downarrow{\scriptstyle d} & & \downarrow{\scriptstyle \alpha} \\ \mathbb{Z} & \longrightarrow & \mathbb{Z}/|G|. \end{array}$$

By induction over orbit bundles we obtain from this

(4.18) Proposition. *The cokernel of* φ *in* (4.16) *has order* $\prod |WH|$ *where the product runs over all* $(H) \in \phi(G, X)$. □

Moreover, by linear algebra, we see that there exist rational numbers a_K such that in the situation (4.17), $\sum_{K \neq 1} a_K d(f^K) \in \mathbb{Z}$ and

$$d(f) \equiv \sum_{K \neq 1} a_K d(f^K) \bmod |G| \quad \text{for all} \quad f \in [\Sigma A_s, y].$$

We will show in a moment that the a_K can be chosen to be integers.

In the general case of the homotopy set $[X, Y]_G$, one can first pass by suspension to $[\Sigma^2 \sigma X, \Sigma^2 \sigma Y]^0_G$ and apply the preceding considerations in order to obtain congruences between mapping degrees. Note that suspension does not change degrees.

We write (4.16) in the form

$$\varphi\colon [X, Y]^0_G \to C(G, X)$$

with $C(G, X) = C(\phi(G, X), \mathbb{Z})$ being the group of all functions $\phi(G, X) \to \mathbb{Z}$. Of course, $\varphi[f](H) = d(f^H)$ is the degree of f^H. We call $\varphi[f]$ the **degree function** of f.

For the next theorem, we make the following assumptions:

(i) $X = \Sigma A$ for a finite dimensional pointed G-complex A of finite orbit type.
(ii) For $H \subset G$, the topological dimension of X^H is $n(H)$, and Y^H is $(n(H)-1)$-connected if $n(H) > 0$.
(iii) For $(H) \in \phi(G)$, we have $n(H) > 0$, $P(H) = H^{n(H)}(X^H) \cong \mathbb{Z}$, and $\pi_{n(H)}(Y^H) \cong \mathbb{Z}$. Then $Q(H) = H^{n(H)}(Y^H) \cong \mathbb{Z}$.
(iv) For $(H) \in \phi(G)$, the WH-modules $P(H)$ and $Q(H)$ are isomorphic.
(v) For $(K) \in \phi(G, X)$, we have $\dim X^K > 1 + \dim X^{>K}$.
(vi) Y^G is simply connected.
(vii) For $(H) \in \phi(G)$, the space X^H is the fixed point set of an isotropy group.

Under these hypotheses, theorem (4.11) holds. One has to choose orientations for all $(H) \in \phi(G)$ and then one has degrees $d(f^H)$ for G-maps $f\colon X \to Y$. For the proof of (4.19), the reader should keep in mind the following facts. Suppose $(H) \in \phi(G)$ and X, Y satisfy (i)–(vii) above. Consider the WH-spaces X^H and Y^H. Let V be a WH-representation and S^V its one-point-compactification. Then we can choose some V such that the WH-spaces $S^V X^H$ and $S^V Y^H$ (the suspensions by S^V, compare section 6) satisfy (i)–(vii).

(4.19) Theorem. *Suppose the pointed G-spaces $X = \Sigma A$ and Y satisfy the hypotheses* (i)–(vii) *above. Then:*

(i) *There exists a set of integers $n(H, K)$ defined for $(H), (K) \in \phi(G, X)$, $(H) \leq (K)$ with $n(H, H) = 1$ such that a function $z \in C(G, X)$ is contained in the image of φ if and only if the congruences*

$$\text{(4.20, H)} \qquad \sum_{\substack{(K) \in \phi(G,X) \\ (K) \geqq (H)}} n(H, K) z(K) \equiv 0 \bmod |WH|$$

are satisfied for all $(H) \in \phi(G, X)$.

(ii) *The image of φ has a $\mathbb{Z}$-basis $e(H)$, $(H) \in \phi(G, X)$, with the following properties:*

$$e(H)(K) \equiv 0 \bmod |WH| \text{ for all } (K) \in \phi(G, X);$$
$$e(H)(K) = 0 \text{ if } (K) \nleqq (H);\ (K) \in \phi(G, X);$$
$$e(H)(H) = |WH|.$$

Proof. It suffices to find a congruence (4.20, 1) for finite G which is satisfied by all z in the image of φ. The congruence (4.20, H) is then obtained by looking at X^H and Y^H as WH-spaces.

Then, by (4.18), the congruences characterize the image of φ. Also by (4.18), if we have found elements $e(H)$ with properties as stated in (ii), they constitute a $\mathbb{Z}$-basis.

The theorem is proved by induction on G: Suppose the theorem is true for all groups of order less than n. If G is not finite, suppose that $|WK| < n$ for all $(K) \in \phi(G, X)$. If G is finite, suppose that $|G| = n$.

We first show the existence of elements $e(H)$ represented by a map $f(H)\colon X \to Y$ with suitable degree function. The map is constructed inductively over orbit bundles, using (4.11). Given $K \in \mathrm{Iso}(X)$, we want to construct a map $f(H)^K\colon X^K \to Y^K$ as WK-map which is already given on $X^{>K}$ as $h\colon X^{>K} \to Y^K$ say. For $1 \neq L/K \subset WK$, the mapping degrees of h satisfy the conditions

$$\begin{aligned} d(h^L) &= 0 && \text{for} \quad (L) \nleq (H) \\ d(h^L) &= |WH| && \text{for} \quad (L) = (H) \\ d(h^L) &\equiv 0 \bmod |WH| && \text{for} \quad (L) \lneqq (H) \end{aligned}$$

by induction. We distinguish three cases.
Case 1: $(K) \nleq (H)$. Then $f(H)^K$ is the constant map, up to homotopy.
Case 2: $(K) = (H)$. The constant map $X^K \to Y^K$ has the correct form on $X^{>K}$. Thus, by (4.11), there exists $X^K \to Y^K$ of degree $|WH|$ being constant on $X^{>K}$.
Case 3: $(K) \leq (H)$. Then $d(h^L) \equiv 0 \bmod |WH|$ for all L as above. If $|WK| < n$, then, by (4.19, i), we have linear congruences at our disposal; it follows that we can find $f(H)^K$ with degree zero $\bmod |WH|$. If $|WK| = |G| = n$, then we know by (4.11) that the degree of $f(H)^K$ extending h lies in a well-defined class $a \bmod |G|$. We aim to show that $a \equiv 0 \bmod |WH|$. If $|WH| = |G|$, we are in case 2. Thus suppose $|WH| < |G|$. Then there exists a subgroup $U_p \subset G$ whose order equals that of a p-Sylow subgroup $WH(p)$ of WH. If we consider any extension $\tilde{h}$ of h as U_p-map, we know by induction that $a \equiv 0 \bmod |U_p|$. We conclude that $a \equiv 0 \bmod |WH|$. This completes the inductive proof of (ii).

Now let G be finite. We must show the existence of a congruence

$$z(1) \equiv \sum_{(H) \neq 1} n(H) z(H) \bmod |G|.$$

We use the pullback (4.17). We have an injective map

$$\varphi_s\colon [X_s, Y]_G^0 \to C(G, X_s).$$

The elements $r[f(H)] =: y(H)$, $(H) \neq 1$, form a $\mathbb{Z}$-basis of $[X_s, Y]$. We want to show that there exists a map $\beta\colon C(G, X_s) \to \mathbb{Z}/|G|$ such that $\alpha = \beta\varphi_s$. Notice that the elements $|WH|^{-1} \varphi_s y(H)$ form a basis of $C(G, X_s)$. Thus it suffices to show that we can find $x_H \in \mathbb{Z}/|G|$ such that $\alpha y(H) = |WH| x_H$. Then we can define β by $\beta(|WH|^{-1} \varphi_s y(H)) = x_H$. But x_H can be taken as $|WH|^{-1} d(f(H))$, by construction of $f(H)$ according to (ii). This completes the proof. □

We add a few remarks to (4.19). The numbers $n(H, K)$, of course, depend on X and Y. Given X and Y, the $n(H, K) \bmod |WH|$ are not uniquely determined. As a corollary of (4.19), one sees that the exponent of the cokernel of φ is the least common multiple of the $|WH|$, $(H) \in \phi(G, X)$.

In the next section, we study an important special case where the congruences of (4.19) can be determined explicitly.

5. Maps between complex representation spheres.

Let G be a compact Lie group. We consider G-maps $f\colon S(V) \to S(W)$ between unit spheres of complex representations V and W. In order to talk about mapping degrees, we need specific orientations. We recall some standard definitions. An orientation of a real vector space U is given by an equivalence class $[x_1, \dots, x_m]$ of a basis $(x_1, \dots, x_m)$, two bases being equivalent if their transition matrix has positive determinant. An orientation of U induces an orientation of the unit sphere $S(U)$ as follows: For each $x \in S(U)$, let $z(x)$ be the outward pointing unit vector at x. The tangent space $T_x S(U) \subset T_x U \cong U$ shall carry the orientation defined by any basis $b_1(x), \dots, b_{m-1}(x)$ such that $[b_1(x), \dots, b_{m-1}(x), z(x)]$ is the given orientation of U. The orientations of the $T_x S(U)$ define an orientation of the manifold $S(U)$. The one-point-compactification S^U of U carries the orientation which restricts to the given orientation of U. If U is the scalar-restriction of a complex vector space V with $\mathbb{C}$-basis $x_1, \dots, x_n$, then the orientation $[x_1, ix_1, \dots, x_n, ix_n]$ is independent of the choice of the basis and is called the canonical orientation of V. We apply this procedure to all the fixed point sets V^H of a complex representation and obtain definite orientations of the $S(V^H)$ and $(S^V)^H$. Let $\phi(G) = \{(H) \mid H \subset G,\ WH \text{ finite}\}$. Let $(H) \mapsto \dim V^H$ be the **dimension function** of V or $S(V)$.

Thus if $\dim V^H = \dim W^H$, then f^H has a uniquely defined degree $d(f^H) \in \mathbb{Z}$. If V and W have the same dimension function, then the G-homotopy class of f is determined by the totality of the degrees $d(f^H)$ for $(H) \in \phi(G)$, i.e. for (H) such that NH/H is finite; this is a consequence of (4.11). We want to determine which functions $z\colon \phi(G) \to \mathbb{Z}$ arise as the degree function $(H) \mapsto d(f^H)$ of maps $f\colon S(V) \to S(W)$ for given V, W with the same dimension function. The description uses congruences between values of functions, as in (4.19). We begin to derive specific congruences by studying the degree of f in equivariant K-theory.

Let $K_G(X)$ be the Grothendieck group of complex vector bundles over the G-space X. If X is pointed, then $\tilde{K}_G(X) := \operatorname{kernel}(K_G(X) \to K_G(*))$. The groups $K_G(X)$ and $\tilde{K}_G(X)$ are modules over the complex representation ring $R(G)$. A basic and deep fact of K_G-theory is the Bott-isomorphism. We quote the result.

(5.1) Theorem. *Let V be a complex representation. Then $\tilde{K}_G(S^V)$ is a free $R(G)$-module on a single basis element $b(V) \in \tilde{K}_G(S^V)$, called the* **Bott class**.

Proof. Atiyah [1968]. Segal [1968] for abelian G. □

Formally, (5.1) is a suspension isomorphism for K_G-theory and serves to make K_G-theory into an equivariant cohomology theory (Atiyah [1967]).

We use (5.1) in the following situation. If V and W are complex representations of G and $f\colon S^V \to S^W$ is a pointed map, then we define the **K_G-theory degree** of f to be the element $a_G(f) = a(f) \in R(G)$ satisfying $f^* b(W) = a(f) b(V)$ where $f^*\colon \tilde{K}_G(S^W) \to \tilde{K}_G(S^V)$ denotes the $R(G)$-linear map induced by f. If $h\colon S(V) \to S(W)$ is a G-map, then we obtain an induced pointed G-map $f\colon S^V \to S^W$ by radial extension (i.e. suspension) of h. We consider $a(f)$ as a character, i.e. as a function on G. A natural problem is now to determine the value $a(f)(g) \in \mathbb{C}$ for $g \in G$ by topological properties of f.

We abbreviate $\tilde{K}_G(S^V)$ by $K_G(V)$. Let C denote the closed subgroup of G generated by g. Decompose $V = V^C \oplus V_C$, i.e. V_C is the (orthogonal) complement of the C-fixed point set. The maps f, f^C and the inclusions $V^C \subset V$ and $W^C \subset W$ induce a commutative diagram

(5.2)
$$\begin{array}{ccc} K_C(W) & \xrightarrow{\ f^*\ } & K_C(V) \\ \downarrow i_W & & \downarrow i_V \\ K_C(W^C) & \xrightarrow{\ (f^C)^*\ } & K_C(V^C). \end{array}$$

We write $i_W b(W) = \lambda(W_C) b(W^C)$ and similarly for V. Now suppose that $\dim V^C = \dim W^C$ so that we can talk about the degree $d(f^C)$ of f^C. The restriction homomorphism $R(G) \to R(C)$ maps $a_G(f)$ to $a_C(f)$. Commutativity of (5.2) yields the equation

$$\lambda(W_C) d(f^C) = a_C(f) \lambda(V_C)$$

because $(f^C)^*$ is multiplication by the degree of f^C. Evaluating characters at $g \in C$ gives

(5.3) $$\lambda(W_C)(g) \cdot d(f^C) = a_C(f)(g) \cdot \lambda(V_C)(g).$$

We have to recall the determination of $\lambda(V_C)$. Let V be any complex G-representation of dimension n and $\Lambda^i(V)$ its i-th exterior power. Then let

(5.4) $$\lambda_{-1}(V) = \sum_{i=0}^{n} (-1)^i \Lambda^i(V) \in R(G).$$

Here, $\Lambda^0(V)$ stands for the trivial one-dimensional representation. The element $\lambda_{-1}(V)$ may be called the **K_G-theoretic Euler class** of V. The basic property of exterior powers

$$\Lambda^k(V \oplus W) \cong \sum_{i+j=k} \Lambda^i(V) \otimes \Lambda^j(W)$$

yields the equality

(5.5) $$\lambda_{-1}(V \oplus W) = \lambda_{-1}(V) \lambda_{-1}(W).$$

In our situation, we have (see Segal [1968], p. 140)

(5.6) Proposition. $\lambda(V_C) = \lambda_{-1}(V_C)$. □

This proposition enables us to compute the character value $\lambda(V_C)(g)$. Write V_C as a direct sum of irreducible one-dimensional representations

$$V_C = \bigoplus_{i=1}^{k} V_i.$$

Then $\lambda_{-1}(V_C) = \prod_{i=1}^{k} \lambda_{-1}(V_i) = \prod_{i=1}^{k} (1 - V_i)$. Since the V_i are nontrivial, $1 \neq V_i(g)$, and therefore $\lambda_{-1}(V_C)(g) \neq 0$. We thus can divide by $\lambda_{-1}(V_C)(g)$ in (5.3) and, using the notation $\lambda_{-1}(W_C)(g)/\lambda_{-1}(V_C)(g) =: \lambda_{-1}(W_C - V_C)(g)$, we can rewrite (5.3) as follows

(5.7) $\quad a_C(f)(g) = \lambda_{-1}(W_C - V_C)(g)\, d(f^C)$.

Let us suppose for the moment that G is finite. Then for any $a \in R(G)$, by the orthogonality relations,

$$\sum_{g \in G} a(g) = |G| \dim a^G \equiv 0 \bmod |G|.$$

If $C \subset G$ is cyclic, we let $C^* \subset C$ be the set of its generators and put $a^*(C) = \sum_{h \in C^*} a(h)$. Then

(5.8) $\quad \sum_{g \in G} a(g) = \sum_{(C)} |G|\,|NC|^{-1} a^*(C)$,

the right hand sum being taken over conjugacy classes of cyclic subgroups. We write

$$n(V - W, C) = \sum_{h \in C^*} \lambda_{-1}(W_C - V_C)(g)$$

and obtain from (5.7)

$$a_C(f)^* = n(V - W, C)\, d(f^C).$$

If $\dim V^C \neq \dim W^C$, then $(f^C)^*$ is the zero map. This follows from the fact that the homotopy group $\pi_{2n}(S^{2m})$ for $n \neq m$ is a finite group and $f \mapsto f_*$ is a homomorphism $\pi_{2n}(S^{2m}) \to \operatorname{Hom}(\tilde{K}(S^{2m}), \tilde{K}(S^{2n})) \cong \operatorname{Hom}(\mathbb{Z}, \mathbb{Z}) \cong \mathbb{Z}$. Thus by defining the degree $d(f^C) = 0$ if $\dim V^C \neq \dim W^C$, (5.7) becomes a true statement in this case, too.

Using these remarks and (5.8), we obtain

(5.9) $\quad \sum_{(C)} |G|\,|NC|^{-1} n(V - W, C)\, d(f^C) \equiv 0 \bmod |G|$.

Note that for $V = W$ this congruence will reappear in section IV.5.

In order to read (5.9) correctly, we need

(5.10) Lemma. *$n(V-W, C)$ is a rational number. If $\dim V_C \leq \dim W_C$, then it is an integer. If C is the trivial group, then $n(V-W, C)=1$.*

Proof. Let $c = |C|$. For each generator $g \in C$ and each one-dimensional representation (= character) U of C, the character value $U(g)$ is a c-th root of unity. Therefore, $\lambda_{-1}(W_C - V_C)(g)$ is contained in the cyclotomic field $\mathbb{Q}(\zeta)$, $\zeta = \exp(2\pi i/c)$. The sum $\sum_{g \in C^*} \lambda_{-1}(W_C - V_C)(g)$ is then obtained by applying to $\lambda_{-1}(W_C - V_C)(g)$ all Galois automorphisms of $\mathbb{Q}(\zeta)$ over $\mathbb{Q}$ and then adding the results. The resulting sum is Galois invariant hence contained in $\mathbb{Q}$, by Galois theory. If $\dim V_C \leq \dim W_C$, then $\lambda_{-1}(W_C - V_C)(g)$ is a cyclotomic integer; therefore $n(V-W, C)$ is a rational integer. (Note that $\lambda_{-1}(W_C - V_C)(g)$ is a cyclotomic unit if $\dim V_C = \dim W_C$.) □

(5.11) Remark. Using (5.10), we see that the sum in (5.9) is a priori a rational number. But it equals $|G| \dim a_G(f)^G$ and is therefore actually an integer. This fact leads to non-trivial integrality conditions as the following example shows.

(5.12) Example. Let $G = \mathbb{Z}/2$. Let W be a trivial k-dimensional representation, U an l-dimensional representation with antipodal action and put $V = U \oplus W$. Let $f\colon S^V \to S^W$ be a pointed G-map. Then $\lambda(V-W, G) = 2^{-l}$ and (5.9) yields $d(f^G) \equiv 0 \bmod 2^{l+1}$.

We come back to the case of arbitrary compact Lie groups. If NH/H is finite, we can apply the preceding considerations to the WH-map f^H and obtain a congruence

$$\text{(5.13)} \quad \sum_{(K)} |NH/NH \cap NK| n(V^H - W^H, K/H) d(f^K) \equiv 0 \bmod |WH|,$$

the sum being taken over all NH-conjugacy classes (K) with H normal in K and K/H cyclic. Rewriting this in terms of G-conjugacy classes (K) with $H \lhd K$, K/H cyclic, we obtain a congruence of the form

$$\text{(5.14)} \quad \sum_{(K)} n(H, K) d(f^K) \equiv 0 \bmod |WH|.$$

In case $\dim V^H = \dim W^H$ for all $H \subset G$, the $n(H, K)$ are integers by (5.10) and, moreover, $n(H, H) = 1$. We now show that the congruences (5.14) essentially characterize the degree functions. Before doing this, we add a couple of remarks. If $(H) \in \phi(G)$ and $H \subset K$, then $(K) \in \phi(G)$ (use I (5.10)) so that (5.14) involves only degrees $d(f^K)$ for $(K) \in \phi(G)$.

We know from the equivariant Hopf Theorem (4.11) that a homotopy class $[f] \in [SV, SW]$ is determined by the degrees $d(f^H)$ for $(H) \in \phi(G, SV) = \{(K) \mid K \in \operatorname{Iso} SV, WK \text{ finite}\}$. All other degrees $d(f^K)$ must therefore be determined by those special degrees. First look at $(K) \in \phi(G)$ which is not an isotropy group. Then $SV^K = \cup \{SV^H \mid H \in \operatorname{Iso} SV, H \supset K\}$. There are only finitely many conjugacy classes entering this union, I (5.11). Suppose that $gHg^{-1} \supset K$ and $H \supset K$. Then $gH \in G/H^K$. Since WK is finite, the set $(G/H)^K$ is finite, I (5.10), and there is only a finite number of groups gHg^{-1} containing K. Thus SV^K is a union of finitely many SV^H and therefore $SV^K = SV^H$ for a suitable isotropy group. Hence $d(f^H) = d(f^K)$.

Now suppose (K) is not in $\phi(G)$. Then there exists a subgroup L of G such that $K \lhd L$, L/K is a torus and $L \in \phi(G)$; see IV (4.5). We show that $d(f^K) = \pm d(f^L)$ in this case; the sign is seen to depend on the complex structure of the representations involved, i.e. on the orientations involved.

Let us look at complex representations V and W of a circle $C = S^1$ such that $\dim V = \dim W$ and $\dim V^C = \dim W^C$. Let $f\colon SV \to SW$ be a C-map. We want to use (5.3) in order to compare the degrees $d(f)$ and $d(f^C)$. Let g be a generator of C, i.e. the powers g^n are dense in C. Then

$$a_C(f)(g) = \lambda_{-1}(W_C - V_C)(g)\, d(f^C).$$

Passing to the limit, we obtain

$$d(f) = a_C(f)(1) = \lim_{g \to 1} \lambda_{-1}(W_C - V_C)(g)\, d(f^C).$$

Let $V(k)$ be the representation $S^1 \times \mathbb{C} \to \mathbb{C}$, $(u, z) \mapsto u^k z$. Then $\lambda_{-1} V(k)(g) = 1 - g^k$. If

$$V_C = \bigoplus_{i=1}^{n} V(k_i) \quad \text{and} \quad W_C = \bigoplus_{i=1}^{n} V(l_i),$$

then $\lambda_{-1}(W_C - V_C)(g) = \prod_{i=1}^{n} (1 - g^{l_i})(1 - g^{k_i})^{-1}$. The limit therefore equals $\prod l_i k_i^{-1}$ and hence

(5.15) $$d(f) = \prod \frac{l_i}{k_i}\, d(f^C).$$

In order to apply this equality to our problem about congruences, we use

(5.16) Proposition. *Let V and W be real representations of a torus T. Suppose the dimensions of V^H and W^H are equal for all $H \subset T$. Then V and W are isomorphic.*

Proof. Let (S, d) be the subgroup of T with d components such that $(S, d)_0$, the component of e, is equal to S where S is a torus of codimension one. Let $V(S, d)$ be the irreducible representation with kernel (S, d). Let V_0 be the trivial representation. If

$$V = \sum_{(S,d)} m(S, d)\, V(S, d) + m_0 V_0$$

is the decomposition of V into irreducible representations, then

$$m_0 = \dim V^T \quad \text{and} \quad m(S, d) = \sum_k \mu(k) \tfrac{1}{2}(\dim V^{(S, dk)} - \dim V^T),$$

with μ being the Möbius function of elementary number theory. Hence $m(S, d)$ and m_0 are computable from the fixed point dimensions. □

If V and W are complex S^1-representations as above with the same fixed point dimensions for all subgroups, then (5.16) shows that $\{i \mid |k_i| = a\}$ and $\{i \mid |l_i| = a\}$ have the same cardinality; therefore, $\prod l_i / \prod k_i = \pm 1$ in this case.

The equality (5.15) has other applications: Since the degrees are integers, it implies certain divisibility properties for the degrees or the l_i, k_i.

For further information on congruences the reader should consult IV.5.

We call $\phi(G) \to \mathbb{Z}, (H) \mapsto d(f^H)$ the **degree function** of f. It is a continuous function (see IV.3). Let $C(G)$ denote the group of continuous functions $\phi(G) \to \mathbb{Z}$. The next result is a variant of (4.19).

(5.17) Theorem. *Let V and W be complex representations of the compact Lie group G such that* $\dim V^H = \dim W^H$ *for all $H \subset G$. Then a function $z \in C(G)$ is the degree function of a G-map $f\colon S(V) \to S(W)$ if and only if the following conditions hold:*

(i) *For each $(H) \in \phi(G)$, the congruence* (5.14)
$\sum_{(K)} n(H, K) z(K) \equiv 0 \bmod |WH|$ *holds.*

(ii) *If* $\dim V^H = 0$, *then* $z(H) = 1$.

(iii) *If* $V^H = V^K$, *then* $z(H) = z(K)$.

Proof. We have seen that the degrees of G-maps satisfy these conditions. Suppose $z \in C(G)$ is given satisfying (i)–(iii). We construct f inductively over the orbit bundles. Let $X = S(V)$, $Y = S(W)$, and $\mathrm{Iso}(X) = \{(H_1), \ldots, (H_t)\}$ such that $(H_i) > (H_j)$ implies $i < j$. Let $X_r = \{x \in X \mid (G_x) = (H_j) \text{ for some } j \le r\}$. We construct inductively G-maps $f_r\colon X_r \to Y$ such that

(5.18) degree $f_r^L = z(L)$ for $L \in T_r$ where
$T_r = \{K \subset G \mid (K) \in \phi(G); (K) \ge (H_i) \text{ some } i \le r \text{ or } (K) > (H_{r+1})\}$.

Note that $X_r^L = X^L$ for L as in (5.18). Put $H = H_{r+1}$. The G-extensions $f_{r+1}\colon X_{r+1} \to Y$ of f_r correspond via restriction bijectively to the WH-extensions $h\colon X^H \to Y^H$ of $f_r' = f_r | X^{>H}$ where $X^{>H} = X^H \cap X_r$. By (3.10), the obstructions to the existence of h lie in $\mathfrak{H}_{NH}^*(X^H, X^{>H}; \pi_{*-1}(Y^H))$ and these groups are zero because $\dim X^H = \dim Y^H$. Let f_{r+1}' be a WH-extension of f_r'. Let $f\colon X \to Y$ be a G-map with $f^H = f_{r+1}'$ and $f | X_r = f_r$, which exists by the same obstruction argument. Suppose $(H) \in \phi(G)$. Then for the fixed point degrees we have

$$d(f^H) + \sum_{K \neq H} n(H, K) d(f^K) \equiv 0 \bmod |WH|.$$

By induction, $d(f^K) = z(K)$ so that in this case $d(f^H) \equiv z(H) \bmod |WH|$. Since $\dim X^H \ge \dim X^{>H} + 2$, we can, as in II.4, alter f_{r+1}' rel $X^{>H}$ to obtain an NH-

map f'_{r+1} so that $d(f'_{r+1}) = z(H)$. Let $f_{r+1}: X_{r+1} \to Y$ be a G-map with $f_{r+1}|X^H = f'_{r+1}$ and $f_{r+1}|X_r = f_r$. If $(H) \notin \phi(G)$, we let f_{r+1} be any G-map extending f_r. We have to verify (5.18) for f_{r+1}. If $L \in T_r \subset T_{r+1}$, then $f^L_{r+1} = f^L_r$ and f^L_{r+1} has the correct degree. If $L \in T_{r+1}$, $(L) = (H)$, then, by construction of f_{r+1}, we have $d(f^H_{r+1}) = z(L)$. If $L \in T_{r+1}$, $(L) > (H_{r+2})$, there exists an isotropy group P, $(P) = (H_s)$, such that $P \supseteq L$. Then $s \leq r+1$, $(P) \in \phi(G)$, $X^L_{r+1} = X^L = X^P = X^P_{r+1}$ and therefore $d(f^L_{r+1}) = d(f^P_{r+1}) = z(P)$. But $z(L) = z(P)$ by assumption (iii) so that f^L_{r+1} has the correct degree.

In order to find f_1, we proceed as above by taking $X_0 = \emptyset$. □

The description of maps between real representations is a more subtle problem; see Tornehave [1982] for an elegant treatment. For other results on mapping degrees of $f: S(V) \to S(W)$, see Atiyah-Tall [1969], Meyerhoff-Petrie [1976], Chung-Nim Lee and Wasserman [1975], Rubinsztein [1973], Snaith [1971], tom Dieck [1979]. Homotopy equivalences between representations are studied in tom Dieck [1978], Traczyk [1978], [1982], Kawakubo [1980], [1980a], Morimoto [1982]. Madsen-Rothenberg [1985] and W.C. Hsiang-Pardon [1982] show that for finite groups G of odd order homeomorphic G-representations are linearly isomorphic; see also Schultz [1977a] for p-groups. For non-isomorphic but homeomorphic representations, see Cappell-Shaneson [1981], [1982]. Earlier de Rham [1940], [1950], [1964] has shown that G-diffeomorphic SV, SW imply isomorphic V, W; see also Rothenberg [1978].

(5.19) Exercises.

1. Let G be a finite abelian group. For $H \subset G$ such that G/H is cyclic, let $V(H)$ be that direct summand of a G-representation V which contains the irreducible submodules with kernel H. Show that SV and SW are G-homotopy equivalent if and only if for each H such that G/H is cyclic, the spheres $SV(H)$ and $SW(H)$ are G-homotopy equivalent.
2. Let V and W be two complex representations of the compact Lie group G. Call SV and SW oriented homotopy equivalent if there exists a G-map $f: SV \to SW$ such that for each $H \subset G$ the map f^H has degree one with respect to the canonical orientations. Let U be another complex representation and suppose that $S(V \oplus U)$ and $S(W \oplus U)$ are oriented homotopy equivalent. Show that the same holds for SV and SW.

6. Stable homotopy. Homology. Cohomology.

Direct limits of homotopy sets over suspension maps are called stable homotopy sets. We have to use the beginnings of the theory in an equivariant setting.

Let G be a compact Lie group. We consider the directed category of complex

G-modules $U, V, W, \ldots$ and write $V < W$ if there exists U and an isomorphism $U \oplus V \cong W$. For homotopy considerations we need

(6.1) Lemma. *Any two isomorphisms $f_0, f_1\colon U \to V$ between complex G-modules are G-homotopic as isomorphisms.*

Proof. It suffices to show that automorphisms are homotopic to the identity. Let $U = U_1 \oplus \ldots \oplus U_r$ be the isotypical decomposition of U, i.e. each U_i is the direct sum of isomorphic irreducible submodules and for $i \neq j$ the corresponding irreducible submodules are non-isomorphic. By Schur's lemma of representation theory,

$$\mathrm{Aut}(U) \cong \mathrm{Aut}(U_1) \times \ldots \times \mathrm{Aut}(U_r),$$

and $\mathrm{Aut}(U_i) = \mathrm{GL}(n_i, \mathbb{C})$ if U_i is a direct sum of n_i irreducible modules. Now use the fact that $\mathrm{GL}(n, \mathbb{C})$ is path-connected. □

Let $S^V = V \cup \{\infty\}$ denote the one-point-compactification of the representation space V with ∞ as base-point. Let X and Y be pointed G-spaces. Suppose $V < W$ and choose an isomorphism $\varphi\colon U \oplus V \cong W$. Note that there is a canonical pointed G-homeomorphism $S^{U \oplus V} \cong S^U \wedge S^V$. We use the notation $S^V \wedge X = S^V X$. We consider the suspension map $b_{W,V}$ between pointed G-homotopy sets

$$\begin{array}{ccc} [S^V X, S^V Y]_G^0 & \xrightarrow{(1)} & [S^U S^V X, S^U S^V Y]_G^0 \\ \Big\downarrow b_{W,V} & & (2) \Big\downarrow \cong \\ [S^W X, S^W Y]_G^0 & \xleftarrow[(3)]{\cong} & [S^{U\oplus V} X, S^{U\oplus V} Y]_G^0 . \end{array}$$

The map (1) is suspension with $\mathrm{id}(S^U)$, the map (2) is induced by the canonical homeomorphism above, and (3) by the isomorphism φ. By (6.1), $b_{W,V}$ is independent of the choice of φ. If $U < V < W$, then $b_{W,V} b_{V,U} = b_{W,U}$. The direct limit of this system of mappings is denoted by $\omega_0^G(X; Y)$. The meaning of a direct limit is the following: Elements in $[S^V X, S^V Y]_G^0$ represent elements in $\omega_0^G(X; Y)$ and if $y = b_{W,V} x$, then y and x represent the same element. Since each complex G-module is isomorphic to one of the type $G \times \mathbb{C}^n \to \mathbb{C}^n$, one can take the direct limit over an honest directed set of G-modules. If $\dim_{\mathbb{R}} V^G \geq 2$, then $[S^V X, S^V Y]_G^0$ carries a natural structure of an abelian group, as does any homotopy set $[S^V X, Z]_G^0$ in this case, and $b_{W,V}$ is a homomorphism. This induces the structure of an abelian group in $\omega_0^G(X; Y)$. We also use the notations

$$\omega_n^G(X; Y) = \omega_0^G(S^n X; Y), \quad n \geq 0$$

$$\omega_{-n}^{G}(X; Y) = \omega_0^G(X; S^n Y), \quad n \geq 0$$
$$\omega_n^G(X; Y) = \omega_G^{-n}(X; Y).$$

Moreover, we set

$$\omega_n^G(S^0; Y) = \omega_n^G(Y), \quad \omega_G^n(X; S^0) = \omega_G^n(X)$$

and $\quad \omega_n^G(S^0; S^0) = \omega_n^G = \omega_G^{-n}.$

It is obvious how to make $\omega_n^G(X; Y)$ into a functor, contravariant in X and covariant in Y. Actually, these functors form a cohomology theory in X and a homology theory in Y as we shall see in a moment. But first let us generalize this construction slightly by using a naive notion of spectrum.

(6.2) Definition. A **spectrum** consists of the following data:
(a) A pointed G-space $X(U)$, one for each complex G-module U;
(b) a pointed G-map $\varphi_{W,V}\colon S^U X(V) \to X(W)$ whenever $U \oplus V \cong W$;
(c) a pointed G-homotopy equivalence $h(W, V)\colon X(V) \to X(W)$ whenever V and W are isomorphic. The following axioms are assumed to hold:
(i) $h(W_3, W_2)h(W_2, W_1) = h(W_3, W_1)$. So $X(W)$ depends only on the isomorphism class of W.
(ii) If U and U' are isomorphic and $k\colon S^U \to S^{U'}$ is the map induced by an isomorphism, then the following diagram is commutative up to pointed G-homotopy

$$\begin{array}{ccc} S^U \wedge X(V_1) & \xrightarrow{\varphi_{W_1,V_1}} & X(W_1) \\ \downarrow{\scriptstyle k \wedge h(V_2, V_1)} & & \downarrow{\scriptstyle h(W_2, W_1)} \\ S^{U'} \wedge X(V_2) & \xrightarrow{\varphi_{W_2,V_2}} & X(W_2). \end{array}$$

(iii) The following diagram is commutative up to pointed G-homotopy

$$\begin{array}{ccc} S^U \wedge S^V \wedge X(W) & \xrightarrow{\mathrm{id}(S^U) \wedge \varphi_{V \oplus W, W}} & S^U \wedge X(V \oplus W) \\ \downarrow{\scriptstyle \cong} & & \downarrow{\scriptstyle \varphi_{U \oplus V \oplus W, V \oplus W}} \\ S^{U \oplus V} \wedge X(W) & \xrightarrow{\varphi_{U \oplus V \oplus W, W}} & X(U \oplus V \oplus W). \end{array}$$

Given a spectrum as above and pointed G-spaces X and Y, we consider suspension maps $b_{W,V}$

$$\begin{array}{ccc} [S^V X, X(V) \wedge Y]^0_G & \xrightarrow[(1)]{} & [S^U S^V X, S^U \wedge X(V) \wedge Y]^0_G \\ \Big\downarrow b_{W,V} & & \Big\downarrow (2) \\ [S^W X, X(W) \wedge Y]^0_G & \xleftarrow[(3)]{} & [S^{U \oplus V} X, X(W) \wedge Y]^0_G . \end{array}$$

The map (1) is suspension with $\mathrm{id}(S^U)$, the map (2) is induced by $\varphi_{W,V}$ and the canonical homeomorphism $S^U S^V \cong S^{U \oplus V}$, and (3) by an isomorphism $U \oplus V \cong W$. One checks that $b_{W,V} b_{V,U} = b_{W,U}$. The direct limit of this system of maps is denoted by

$$\chi_0^G(X; Y) \quad \text{where} \quad \chi = (X(U), \varphi_{U,V})$$

stands for the spectrum at hand. Again, $\chi_0^G(X; Y)$ carries a natural structure of an abelian group, is contravariant in X and covariant in Y. We define

$$\begin{aligned} \chi_n^G(X; Y) &= \chi_0^G(S^n X; Y), \quad n \geq 0 \\ \chi_{-n}^G(X; Y) &= \chi_0^G(X; S^n Y), \quad n \geq 0 \\ \chi_n^G(X; Y) &= \chi_G^{-n}(X; Y). \end{aligned}$$

From elementary homotopy theory one obtains the following properties of these functors. Let $f\colon X \to Y$ be a pointed G-map. Let $C(f)$ denote the mapping cone of f. This is the space obtained from $X \times [0,1] + Y$ by identifying $X \times \{0\} \cup \{*\} \times I$ with a point and $(x, 1)$ with $f(x)$ for each $x \in X$; the image of $X \times \{0\}$ is the base point of $C(f)$. One has a canonical pointed G-map $P(f)\colon Y \to C(f)$, $y \mapsto$ class of y. For each pointed G-space A, the sequence

$$[X, A]^0_G \xrightarrow[f^*]{} [Y, A]^0_G \xrightarrow[Pf^*]{} [C(f), A]^0_G$$

is an exact sequence of pointed sets.

(6.3) Proposition. *The sequences*

$$\chi_n^G(A; X) \xrightarrow[f_*]{} \chi_n^G(A; Y) \xrightarrow[Pf_*]{} \chi_n^G(A; C(f))$$

and

$$\chi_G^n(X; A) \xleftarrow[f^*]{} \chi_G^n(Y; A) \xleftarrow[Pf^*]{} \chi_G^n(C(f); A)$$

are exact for each n, each pointed G-map $f\colon X \to Y$, and each pointed G-space A.

Proof. See Bröcker-tom Dieck [1970], p. 45. □

Let A be a real representation. The following maps

$$\begin{array}{ccc} [S^V X, X(V) \wedge Y]^0_G & \xrightarrow[(1)]{} & [S^A S^V X, S^A X(V) \wedge Y]^0_G \\ & \searrow^{\tau_A} & \downarrow (2) \\ & & [S^V S^A X, X(V) \wedge S^A Y]^0_G, \end{array}$$

where (1) is suspension with S^A and (2) is induced by interchanging factors, are compatible with the $b_{W,V}$ and therefore induce a natural transformation

$$\sigma^A \colon \chi^G_0(X; Y) \to \chi^G_0(S^A X; S^A Y).$$

(6.4) Proposition. *For each real representation, σ^A is an isomorphism.*

Proof. Let A be a complex representation. Then an inverse is defined by

$$\begin{aligned} [S^U S^A X, X(U) \wedge S^A \wedge Y]^0_G &\xrightarrow[(1)]{} [S^U S^A X, S^A \wedge X(U) \wedge Y]^0_G \\ &\xrightarrow[(2)]{} [S^{U \oplus A} X, X(U \oplus A) \wedge Y]^0_G. \end{aligned}$$

Here (1) interchanges factors and (2) uses the spectrum maps $\varphi_{U \oplus A, U}$. One verifies that this construction gives an inverse to σ^A. Thus σ^A is an isomorphism for complex A. For a real representation A one uses $\sigma^A \sigma^A \cong \sigma^{A \oplus A}$ and observes that $A \oplus A$ carries a complex structure. □

If one applies the preceding proposition to the trivial one-dimensional representation with compactification S^1, one defines natural suspension isomorphisms

(6.5) $\quad \sigma \colon \chi^G_n(X; Y) \cong \chi^G_{n+1}(X; S^1 Y)$

which are the identity for $n < 0$ and defined as σ^1 using (6.4) for $n \geq 0$. Similarly, one obtains

(6.6) $\quad \sigma \colon \chi^n_G(X; Y) \cong \chi^{n+1}_G(S^1 X; Y)$.

Remarks. (i) Occasionally, it is useful to define a restricted kind of spectra. One uses only maps $\varphi_{W,V}$ where V and W belong to a class of representations which is closed under direct sums and isomorphisms. Then (6.4) only holds for complex representations in this class.
(ii) It can happen that one has a canonical space $X[U]$ for each isomorphism class of complex representations. Then one can define $X(U) = X[U]$ and $h(W, V) = \mathrm{id}$.

Guided by the preceding considerations, we define two notions of equivariant (co-)homology theories. They are defined as functors on the category of pointed

G-complexes. Similar definitions can be given for other suitable categories. Let G-Com^0 be the category of pointed spaces which admit the structure of a pointed G-complex. Let Abel be the category of abelian groups.

(6.7) An **(unstable) equivariant homology theory** consists of a sequence $(h_n(?)\,|\,n \in \mathbb{Z})$ of covariant functors $h_n\colon G\text{-}\mathrm{Com}^0 \to \mathrm{Abel}$ and a sequence of natural transformations $\sigma_n\colon h_n(X) \to h_{n+1}(S^1 X)$, $n \in \mathbb{Z}$, such that the following holds:

(i) h_n is homotopy invariant, i.e. G-homotopic maps f_0 and f_1 yield the same induced homomorphism $h_n(f_0) = h_n(f_1)$.

(ii) σ_n is an isomorphism (suspension isomorphism).

(iii) For any map $f\colon X \to Y$ in G-Com^0, the sequence

$$h_n(X) \xrightarrow[h_n(f)]{} h_n(Y) \xrightarrow[h_n(Pf)]{} h_n(C_f)$$

is exact.

(6.8) A **(stable) equivariant homology theory** graded over the abelian group A consists of the following data.

(a) A family $(h_a(?)\,|\,a \in A)$ of covariant functors $h_a\colon G\text{-}\mathrm{Com}^0 \to \mathrm{Abel}$.

(b) A homomorphism $R(G) \to A$ from the (additive group of the) complex representation ring $R(G)$ into A; this homomorphism is not supposed to be injective; nevertheless, we usually denote a representation V and its image in A by the same symbol.

(c) A homomorphism $\mathbb{Z} \to A$; the same notational convention as in (b).

(d) For each complex representation V, a family of natural transformations $\sigma^V\colon h_a(X) \to h_{a+V}(S^V X)$, $a \in A$.

(e) Finally, a family of natural transformations $\sigma_a\colon h_a(X) \to h_{a+1}(S^1 \wedge X)$, $a \in A$. These data satisfy the following axioms.

(i) For each $a \in A$, $(h_{a+n}(?)\,|\,n \in \mathbb{Z})$ and $(\sigma_{a+n}\,|\,n \in \mathbb{Z})$ form an unstable homology theory.

(ii) σ^V is an isomorphism (suspension by V).

(iii) The diagram

$$\begin{array}{ccc} h_a(X) & \xrightarrow{\sigma^V} & h_{a+V}(S^V X) \\ \Big\downarrow{\scriptstyle \sigma^{U \oplus V}} & & \Big\downarrow{\scriptstyle \sigma^U} \\ h_{a+U+V}(S^{U \oplus V} X) & \xleftarrow[\cong]{} & h_{a+U+V}(S^U S^V X) \end{array}$$

is commutative.

(iv) The diagram

$$
\begin{array}{ccc}
h_a(X) & \xrightarrow{\ \sigma_a\ } & h_{a+1}(S^1 X) \\
\Big\downarrow \sigma^V & & \Big\downarrow \sigma^V \\
h_{a+V}(S^V X) & & h_{a+1+V}(S^V S^1 X) \\
& {}_{\sigma_{a+V}}\searrow \quad \nearrow_{T_*} & \\
& h_{a+V+1}(S^1 S^V X)\,, &
\end{array}
$$

with T_* induced by the twisting map $S^1 \wedge S^V \to S^1 \wedge S^V$, is commutative.
(v) If U and V are isomorphic, then the following diagram is commutative

$$
\begin{array}{ccc}
& & h_{a+U}(S^U X) \\
& \nearrow^{\sigma^U} & \\
h_a(X) & & \Big\downarrow \cong \\
& \searrow_{\sigma^V} & \\
& & h^{a+V}(S^V X)\,.
\end{array}
$$

The isomorphism is induced by any isomorphism $U \to V$ and the corresponding homeomorphism $S^U \to S^V$.

One may think of σ^V as defining a natural isomorphism from the unstable homology theory $(h_{a+n}(?) \mid n \in \mathbb{Z})$, $(\sigma_{a+n} \mid n \in \mathbb{Z})$ into the unstable theory $(h_{a+V+n}(?) \mid n \in \mathbb{Z})$, $(T_* \sigma_{a+V+n} \mid n \in \mathbb{Z})$.

A natural candidate for the index group A in (6.8) is the real representation ring $RO(G)$. In this case, the map $R(G) \to A$ is given by forgetting the complex structure of a representation. The map $\mathbb{Z} \to RO(G)$ sends 1 to the trivial one-dimensional representation.

Another candidate is $A = \mathbb{Z}$ and $R(G) \to \mathbb{Z}$ is the augmentation $V \mapsto \dim_{\mathbb{R}} V$.

Equivariant unstable or stable cohomology theories are defined in the obvious dual way: One has contravariant functors $h^a(?)$, suspension isomorphisms $h^a(X) \cong h^{a+V}(S^V X)$ etc.

There is a standard procedure for obtaining homology theories defined on pairs of spaces from homology theories defined on pointed spaces. In this context, the homology theory on pointed spaces is called a **reduced** homology theory and denoted by a tilde, $\tilde{h}_n(X)$. Suppose (X, A) carries the structure of a G-complex and a subcomplex. Then one defines $h_a(X, A) := \tilde{h}_a(X/A)$. One derives the usual exact sequences for homology and the excision from the axioms (6.8); see Switzer [1975], Ch. 7 for details.

A **natural transformation** between two stable homology theories $(h_a \mid a \in A)$ and $(k_a \mid a \in A)$ consists of a family of natural transformations $t_a(X)\colon h_a(X) \to k_a(X)$ which commute with the suspension maps σ_a and σ^V.

A homology theory is called **additive** if, for each family $(X_j \mid j \in J)$ of pointed G-complexes, the inclusions $i_j\colon X_j \to \bigvee_j X_j$ induce an isomorphism

$$(h_a(i_j) \mid j \in J)\colon \bigoplus_j h_a(X_j) \to h_a(\textstyle\bigvee_j X_j).$$

A cohomology theory is called additive if there is always an isomorphism

$$(h^a(i_j) \mid j \in J)\colon h^a(\textstyle\bigvee_j X_j) \to \prod_j h^a(X_j).$$

Important consequences of such additivity axioms are compatibility with direct limits (Milnor [1962], Switzer [1975], 7.53, 7.66).

Theories $\chi_0^G(X; Y)$ are additive in the variable Y but, in general, not in X. They should be restricted to finite-dimensional G-complexes X. If one wants to deal with arbitrary G-complexes, one has to construct more suitable categories of spectra.

For the purpose of induction over orbit bundles, it is necessary to reduce G-equivariant stable homotopy groups $\omega_0^G(G^+ \wedge_H X; Y)$ and $\omega_0^G(X; G^+ \wedge_H Y)$ to H-equivariant stable homotopy groups. We begin with the former. We recall I(4.9) that there are natural adjunction isomorphisms

$$[G^+ \wedge_H A, B]_G^0 \cong [A, \mathrm{res}_H B]_H^0$$

for pointed H-spaces A and G-spaces B. We apply these isomorphisms to the representing homotopy sets for $\omega_0^G(G^+ \wedge_H X; Y)$. Let $i\colon S^V X \to S^V(G^+ \wedge_H X)$ be the map given by $(v, x) \mapsto (v, e, x)$. Then the composition

$$[S^V(G^+ \wedge_H X), S^V Y]_G^0 \xrightarrow[\mathrm{res}]{} [S^V(G^+ \wedge_H X), S^V Y]_H^0 \xrightarrow[i^*]{} [S^V X, S^V Y]_H^0$$

is a bijection which is compatible with the suspension maps $b_{W,V}$ and therefore yields an isomorphism

(6.9) $$\omega_0^G(G^+ \wedge_H X; Y) \cong \omega_0^H(X; Y)$$

for pointed H-complexes X and G-complexes Y.

In dealing with $\omega_0^G(X; G^+ \wedge_H Y)$, one would rather expect the multiplicative induction (I, 4.11) to appear instead of $G^+ \wedge_H Y$. But since in the ideal world of stable homotopy finite sums and products coincide, the multiplicative induction can be replaced by the additive induction. Unfortunately, this program only works for finite groups G. So let us suppose for the moment that G is finite. Then the multiplicative induction provides us with the G-space $\mathrm{Top}_H(G, X)$ for any

H-space X. If X is a pointed space, then $\mathrm{Top}_H(G, X)$ is pointed by the constant map. In this case, the bijective correspondence (I, 4.11) reduces to

(6.10) $[\mathrm{res}_H A, B]_H^0 \cong [A, \mathrm{Top}_H(G, B)]_G^0$.

There is a canonical pointed G-map

(6.11) $j\colon G^+ \wedge_H B \to \mathrm{Top}_H(G, B)$

given by

$$j(g, b)(g_1) = \begin{cases} g_1 g b & \text{if } g_1 g \in H \\ * & \text{otherwise}. \end{cases}$$

We look at the induced map

$$j_*\colon [A, G^+ \wedge_H B]_G^0 \to [A, \mathrm{Top}_H(G, B)]_G^0.$$

This map j_* is a bijection if the dimension function of A is less than the connectivity of j; see (2.6). Thus we have to study the behaviour of j on fixed point sets. Given $K \subset G$, we decompose $\mathrm{res}_K j$ according to double cosets $KgH \in D = K \backslash G/H$.

$$j\colon \bigvee_{KgH \in D} (KgH^+ \wedge_H B) \to \prod_{KgH \in D} \mathrm{Top}_H(Hg^{-1}K, B).$$

If we pass to K-fixed points, we obtain

$$(KgH^+ \wedge_H B)^K \cong \begin{cases} \mathrm{Fix}(g^{-1}Kg, B) & \text{if } g^{-1}Kg \subset H \\ * & \text{otherwise} \end{cases}$$

$$\mathrm{Top}_H(Hg^{-1}K, B)^K \cong \mathrm{Fix}(g^{-1}Kg \cap H, B)$$

with j inducing the canonical inclusion between these spaces. In the case of $B = S^V Y$, the map j^K has the form

$$\bigvee_{KgH \in D(1)} \mathrm{Fix}(g^{-1}Kg, S^V Y) \to \prod_{KgH \in D(1)} \mathrm{Fix}(g^{-1}Kg, S^V Y) \times \prod_{KgH \in D(2)} \mathrm{Fix}(g^{-1}Kg \cap H, S^V Y)$$

with $D(1) = \{KgH \in D \mid g^{-1}Kg \subset H\}$ and $D(2) = D \backslash D(1)$. Thus the connectivity is at least the minimum of the numbers $\dim \mathrm{Fix}(g^{-1}Kg \cap H, S^V) - 1$, $KgH \in D(2)$, and $2 \dim \mathrm{Fix}(g^{-1}Kg, S^V) - 1$. Given an integer n, there exist representations V such that for $H \supset K, H \neq K$ the relations $\dim V^H + n < \dim V^K$ and $\dim V^G > n$ hold (e.g. take any V containing $n + 1$ copies of the regular representation as a direct summand). For such representations and G-complexes of dimension at most n, the map

$$j_*\colon [S^V X, G^+ \wedge_H S^V Y]_G^0 \to [S^V X, \mathrm{Top}_H(G, S^V Y)]_G^0$$

is bijective. If we compose with the canonical bijection (6.10), we obtain the following result in which $p\colon S^V(G^+ \wedge_H Y) \to S^V Y$ is the H-map given by $p(v, g, y) = (v, gy)$ for $g \in H$ and $= *$ otherwise.

(6.12) Proposition. *Let G be a finite group and X a finite-dimensional G-complex. Then, for all sufficiently large representations V, the composition*

$$[S^V X, S^V(G^+ \wedge_H Y)]^0_G \xrightarrow[\text{res}]{} [S^V X, S^V(G^+ \wedge_H Y)]^0_H \xrightarrow[p_*]{} [S^V X, S^V Y]^0_H$$

is a bijection. These maps induce a natural isomorphism

$$\omega_0^G(X; G^+ \wedge_H Y) \cong \omega_0^H(X; Y). \quad \square$$

We are now going to construct a map $\omega_0^H(X; Y) \to \omega_0^G(X; G^+ \wedge_H Y)$ which will turn out to be inverse to the isomorphism (6.12). Let G again be an arbitrary compact Lie group and H a closed subgroup. Let V be an orthogonal representation of G such that H is an isotropy subgroup at some point $v \in V$. We thus obtain a G-embedding $i\colon G/H \to V, gH \mapsto gv$. The tangent bundle T of G/H and the normal bundle N of this embedding have the form

$$T = G \times_H L \quad \text{and} \quad N = G \times_H W$$

where L and W are H-submodules of V and $V = W \oplus L$. Using a G-invariant tubular neighbourhood $G \times_H W \leftarrow U \subset V$ of $i(G/H)$, we obtain a pointed G-map (Pontrjagin-Thom construction)

$$\begin{array}{ccl} t\colon S^V & \to & G^+ \wedge_H S^W = (G \times_H W) \cup \{\infty\} \\ \cup & & \cup \\ U & \xrightarrow[\cong]{} & G \times_H W \end{array}$$

with $t(S^V \backslash U) = \{\infty\}$.

Using this map t, we define a natural transformation w as the following composition

$$\text{(6.13)} \quad \omega_0^H(X; S^L Y) \xrightarrow[\cong]{S^W} \omega_0^H(S^W X; S^W S^L Y)$$

$$\xrightarrow[\cong]{} \omega_0^H(S^W X; S^V Y) \xrightarrow{G^+ \wedge_H} \omega_0^G(G^+ \wedge_H (S^W X); G^+ \wedge_H S^V Y))$$

$$\xrightarrow[\cong]{} \omega_0^G((G^+ \wedge_H S^W) \wedge X; S^V(G^+ \wedge_H Y))$$

$$\xrightarrow{t^*} \omega_0^G(S^V X; S^V(G^+ \wedge_H Y)) \xrightarrow[\cong]{(\sigma^V)^{-1}} \omega_0^G(X; G^+ \wedge_H Y).$$

(6.14) Proposition. *The natural transformation w in (6.13) is an isomorphism.*

Proof. For the case of a compact Lie group, we refer to Wirthmüller [1974], Theorem 2.1. For finite G, we check that w is inverse to the isomorphism of Proposition (6.12). In this case, $W = V$ and it suffices to show that

$$[S^V X, S^V Y]_H^0 \to [(G^+ \wedge_H S^V) \wedge X, G^+ \wedge_H S^V Y]_G^0$$
$$\xrightarrow{t^*} [S^V X, G^+ \wedge_H S^V Y]_G^0 \to [S^V X, S^V Y]_H^0$$

is the identity. But this composition is obtained by mapping f to $(q \wedge \mathrm{id}(X))f$ with

$$q\colon S^V \xrightarrow[t]{} G^+ \wedge_H S^V \xrightarrow[p]{} S^V;$$

and q is H-homotopic to the identity.

For general G, an isomorphism in the opposite direction is constructed as a composition

$$\varrho\colon \omega_0^G(X; G^+ \wedge_H Y) \xrightarrow[\mathrm{res}_H]{} \omega_0^H(X; G^+ \wedge_H Y) \xrightarrow[p_*]{} \omega_0^H(X; S^L Y).$$

In this case, p has the form

$$p = G^+ \wedge_H p'\colon G^+ \wedge_H Y \to S^L H^+ \wedge_H Y = S^L Y$$

where $p'\colon G \to S^L H^+$ is the Pontrjagin-Thom map applied to the embedding $H \subset G$. The reader should check that

$$[S^W X, Y]_H^0 \to [(G^+ \wedge_H S^W) \wedge X, G^+ \wedge_H Y]_G^0 \xrightarrow{t^*} [S^V X, G^+ \wedge_H Y]_G^0$$
$$\xrightarrow[p_* \mathrm{res}_H]{} [S^V, S^L Y]_H^0$$

coincides with suspension by S^L, using $S^L S^W \cong S^V$. From this fact it follows directly that ϱw is an isomorphism. However, in general, it is not the identity because of the interchange of factors in $S^L S^L$. At least, this shows that w is injective in general. □

The isomorphism of the last proposition is related to Spanier-Whitehead duality in the category of G-complexes; compare Wirthmüller [1975], Dold-Puppe [1980].

Important examples of equivariant (co-)homology theories are: The Bredon-Illman theories, see section 9. The equivariant bordism theories Conner-Floyd [1964], Conner [1979] and their stable analogues tom Dieck [1970], [1970a], [1972a], [1972b], [1974a]. For the relation between unstable and stable bordism theories, see Bröcker-Hook [1972]. The equivariant K-theories Atiyah [1967], Segal [1968], Atiyah-Segal [1968], [1969]. The Borel-cohomology groups, see Chapter III. Stable equivariant (co-)homotopy.

Further references: Becker-Schultz [1974], [1976], Kosniowski [1974], Matumoto [1973], May [1982], Lewis-May-McClure [1981].

7. Homology with families.

The method of induction over orbit bundles has a particular formalisation in the context of homology theories. Let F_1 and F_2 be open isotropy families for the compact Lie group G, see (I, 6.). Assume $F_1 \supset F_2$. There is a unique G-homotopy class $EF_2 \to EF_1$ of maps between the classifying spaces. We can choose our models such that EF_i are G-complexes and $EF_2 \subset EF_1$ is a closed G-fibration (e.g. a subcomplex). Under these assumptions, the G-homotopy type of the pair (EF_1, EF_2) is uniquely determined up to unique G-homotopy equivalence (exercise).

Let (h_n) be an unstable G-homology theory defined on the category of pairs of G-complexes. We define a new homology theory by

$$h_n[F_1, F_2](X, A) := h_n((EF_1, EF_2) \times (X, A)).$$

The definition of the suspension isomorphism and the verification of the axioms is left as exercise. If F_2 is empty, we write $h_n[F_1]$ instead. The inclusions $EF_2 \to EF_1 \to (EF_1, EF_2)$ lead to exact homology sequences

$$\begin{aligned} \ldots &\to h_n[F_2](X, A) \to h_n[F_1](X, A) \to h_n[F_1, F_2](X, A) \\ &\to h_{n-1}[F_2](X, A) \to \ldots . \end{aligned}$$

The morphisms in this sequence are components of a natural transformation of homology theories (exercise). Note that if F is the family of all subgroups, then naturally $h_n[F] = h_n$ because EF is a contractible G-space. Thus, by enlarging the family F gradually, we obtain better and better approximations to our original homology theory.

Recall that families F_1, F_2 are called **adjacent** or H-adjacent if $F_1 \supset F_2$ and $F_1 \backslash F_2$ consists of the conjugacy class (H). The inductive approximation of the theory (h_n) is based on the analysis of $h_n(F_1, F_2]$ for adjacent families F_1, F_2.

In the setting of homology theories for pointed spaces, we define, of course,

$$\tilde{h}_n[F_1, F_2](X) := \tilde{h}(EF_1/EF_2 \wedge X).$$

Now let (h_n) and (k_n) be additive unstable homology theories and suppose $(t_n\colon h_n \to k_n)$ is a natural transformation of homology theories. Then we have induced natural transformations $t_n[F_1, F_2]\colon h_n[F_1, F_2] \to k_n[F_1, F_2]$ for each pair (F_1, F_2) of families.

(7.1) Proposition. *Suppose $(t_n[F_1, F_2])$ is a natural isomorphism for each pair of open adjacent families. Then (t_n) is a natural isomorphism.*

Proof. Let $\mathfrak{M}$ be the set of open families and $\mathfrak{M}'$ the set of families $F \in \mathfrak{M}$ such that $(t_n[F])$ is a natural isomorphism. Then $\mathfrak{M}'$ is not empty because it contains the family consisting of the trivial group alone. Let $A = \bigcup \{F \mid F \in \mathfrak{M}'\}$. Then A is an open family. We claim that $A \in \mathfrak{M}'$. In order to prove this, we show first that there is a sequence $F_1 \subset F_2 \subset \dots$ of families in $\mathfrak{M}'$ with union A. We know that A contains a countable number of conjugacy classes, see I, 5. Ex. 8. Let $F_1 \in \mathfrak{M}'$ be given and $H \in A \backslash F_1$. Let $F \in \mathfrak{M}'$ be a family containing H. We claim that $F_1 \cup F \in \mathfrak{M}'$. If $F \in \mathfrak{M}'$ and $F' \subset F$, then $F' \in \mathfrak{M}'$ because the projection $EF \times EF' \to EF'$ is a G-homotopy equivalence. Thus $F_1 \cap F \in \mathfrak{M}'$ and we can take (ex. 1)

$$E(F_1 \cup F) = EF_1 \cup_{E(F_1 \cap F)} EF.$$

We conclude that $F_1 \cup F \in \mathfrak{M}'$ by using the Mayer-Vietoris sequences

$$\dots \to h_n[F_1 \cap F] \to h_n[F_1] \oplus h_n[F] \to h_n[F_1 \cup F] \to \dots$$

and the five-lemma. Put $F_2 = F_1 \cup F$ and proceed inductively to obtain $F_3, F_4, \dots$. We now choose a sequence of closed G-cofibrations

$$EF_1 \subset EF_2 \subset \dots$$

and observe that $EA = \bigcup EF_i$ with the colimit topology (exercise 2). The additivity of the homology theories implies colim $h_n[F_i] \cong h_n[A]$; from this we see that $A \in \mathfrak{M}'$. If A were different from the set of all closed subgroups, we would choose a minimal subgroup $K \notin A$ and let $A' = A \cup (K)$. Then (A', A) is a K-adjacent pair. Since $(t_n[A', A])$ and $(t_n[A])$ are isomorphisms, we see that $(t_n[A'])$ is an isomorphism, contradicting the maximality of A. □

We want to use the preceding proposition in order to show that the homology groups $\omega_n^G(X)$ split naturally into a direct sum. Apart from that, we need properties of adjacent families in a homotopy setting. Let (F_1, F_2) be H-adjacent and abbreviate $N = NH$, $W = WH = NH/H$. We use $EF_2 \subset EF_1$ as an inclusion of G-complexes. We also note that

(7.2) $$EF_1 = (G \times_N EW) * EF_2$$

is a convenient model for EF_1 once EF_2 is given. From this model we see immediately that

$$EF_1^H = EW, \quad (EF_1/EF_2)^H = EW^+.$$

Passing to H-fixed points yields a natural map

(7.3) $$b\colon [X, EF_1/EF_2 \wedge Y]_G^0 \to [X^H, EW^+ \wedge Y^H]_W^0$$

for each pair X, Y of pointed G-spaces. Let

$$X_0 = \{x \in X \mid G_x \text{ not subconjugate to } H\}.$$

Then any map $f\colon X \to EF_1/EF_2 \wedge Y$ sends X_0 to the base-point. The projection $X \to X/X_0$ thus induces a commutative diagram

$$\begin{array}{ccc}
[X, EF_1/EF_2 \wedge Y]^0_G & \xrightarrow[b]{\qquad} & [X^H, EW^+ \wedge Y^H]^0_W \\
\uparrow \cong & & \uparrow \\
[X/X_0, EF_1/EF_2 \wedge Y]^0_G & \xrightarrow[b]{\qquad} & [(X/X_0)^H, EW^+ \wedge Y^H]^0_W .
\end{array}$$

The left vertical map is a bijection. We show that the map b at the bottom is essentially always a bijection.

(7.4) Proposition. *Suppose all isotropy groups of X, apart from the base-point, are in F_1. Suppose that open subsets of X and $X \times I$ are paracompact and that $X' \subset X$ and $X' \times I \cup X \times \partial I \subset X \times I$ are G-neighbourhood retracts for $X' = \{x \in X \mid (G_x) \geq (H)\}$. Let the inclusion of the base-point in Y be a G-cofibration. (E.g. X and Y pointed G-complexes.) Then b is a bijection.*

Proof. (i) Suppose the closed G-subset D of X contains X' and is a G-neighbourhood retract. Then each G-map $f: D \to EF_1/EF_2 \wedge Y$ has an extension to X.

Proof. Since D is a G-neighbourhood retract, there exists an extension of f to an open G-neighbourhood U of D in X, say $f_1: U \to EF_1/EF_2 \wedge Y$. Since $X \backslash D$ is open in X and paracompact and has only isotropy groups in F_2, there exists a G-map $f_0: X \backslash D \to EF_2$ by universality of EF_2. Let (t_0, t_1) be a numeration of the covering $(X \backslash D, U)$ by G-invariant functions t_i. Then

$$h: X \to EF_2 * (EF_1/EF_2 \wedge Y), x \mapsto (t_0(x) f_0(x), t_1(x) f_1(x))$$

extends $f: D \to EF_1/EF_2 \wedge Y \subset EF_2 * (EF_1/EF_2 \wedge Y)$. The next lemma, whose proof is deferred for a while, completes the proof of (i).

(7.5) Lemma. *There exists a G-map $EF_2 * (EF_1/EF_2 \wedge Y) \to EF_1/EF_2 \wedge Y$ which is the identity on $RF_1/EF_2 \wedge Y$.*

(ii) We come to the proof of (7.4). We show that b is surjective. Let $f_0: X^H \to EW^+ \wedge Y^H$ be given. Compose with $EW^+ \wedge Y^H = (EF_1/EF_2 \wedge Y)^H \subset EF_1/EF_2 \wedge Y$. The resulting map has a unique extension to a G-map $f: X' \to EF_1/EF_2 \wedge Y$. The hypotheses of the proposition show that (i) can be applied to f; the resulting map h satisfies $b[h] = [f]$.

In order to show injectivity, we start with two G-maps $f_0, f_1: X \to EF_1/EF_2 \wedge Y$ such that their restrictions to X^H are N-homotopic; hence their restrictions to X' are G-homotopic. A homotopy between them together with f_0 and f_1 yields a G-map $f: X' \times I \cup X \times \partial I \to EF_1/EF_2 \wedge Y$. We can apply (i) again to this map f and obtain an extension to $X \times I$, which is a G-homotopy between f_0 and f_1. □

Proof of (7.5). Consider the map

$$A \times I \times B \to A * B, (a, t, b) \mapsto (ta, (1-t)b)$$

and let $A *_1 B$ be the join with quotient topology induced by this map. The set theoretical identity $A *_1 B \to A * B$ is a G-homotopy equivalence. There exists a homotopy inverse which is the identity on the canonical subspaces A and B (exercise 4). Let $CA = I \times A/\{0\} \times A$ be the cone over A with subspace $A = \{1\} \times A$. There is a continuous map $A *_1 B \to (CA/A) \wedge B$ which is given on representatives by $(ta, (1-t)b) \mapsto (t, a, b)$. Altogether, we obtain $A * B \to (CA/A) \wedge B$ which is $i: b \mapsto (0, *, b)$ on the subspace B. We apply the preceding discussion to $A = EF_2$, $B = EF_1/EF_2 \wedge Y$. We show that, in this case, i is a G-homotopy equivalence. Since it is also a G-cofibration, it is a strong G-deformation retract. Composing with a retraction yields the desired map of the lemma. Since, by assumption, Y has a nice base point, we must only show that $i: EF_1/EF_2 \to CEF_2/EF_2 \wedge EF_1/EF_2$ is a G-homotopy equivalence. This map is induced by a map of pairs

$$j: (EF_1, EF_2) \to (CEF_2 \times EF_1, EF_2 \times EF_1 \cup CEF_2 \times EF_2).$$

It therefore suffices to show that each individual map of j is a G-homotopy equivalence. This is clear for $EF_1 \to CEF_2 \times EF_1$ because CEF_2 is contractible. For $EF_2 \to EF_2 \times EF_1 \cup CEF_2 \times EF_2$, we remark that this map is homotopic to the map induced by the diagonal $d: EF_2 \to EF_2 \times EF_2$. Since the partial maps $d: EF_2 \to EF_2 \times EF_2$, $d: EF_2 \to CEF_2 \times EF_2$, and $d: EF_2 \to (EF_2 \times EF_1) \cap (CEF_2 \times EF_2) = EF_2 \times EF_2$ are homotopy equivalences, d itself must be a homotopy equivalence (by the glueing theorem for homotopy equivalences). □

We are going to define a natural transformation of homology theories which will be used in the splitting of $\omega_n^G(X)$. Consider the following composition ζ_H of maps

$$\omega_n^W(EW^+ \wedge X^H) \xrightarrow[(1)]{}$$
$$\omega_n^N(S^L \wedge EW^+ \wedge X) \xrightarrow[(2)]{}$$
$$\omega_n^G((G \times_N EW)^+ \wedge X) \xrightarrow[(3)]{}$$
$$\omega_n^G(X).$$

Explanation: The map (1) is induced by the inclusions $X^H \subset X$ and $EW^+ \to S^L \wedge EW^+$, $z \mapsto (0, z)$; moreover, W-spaces are viewed as N-spaces via the quotient homomorphism $N \to W$. The map (2) is the Wirthmüller-isomorphism (6.14). The map (3) is induced by the projection onto X. It should be clear that ζ_H is a natural transformation of (unstable) homology theories (for the group G!). Let

$$(7.6)\qquad \zeta = (\zeta_H): \bigoplus_{(H)} \omega_n^{WH}(EWH^+ \wedge X^H) \to \omega_n^G(X)$$

be the map with components ζ_H where H runs through a complete set of representatives for the conjugacy classes of subgroups.

(7.7) Theorem. *The natural transformation ζ is an isomorphism.*

Proof. We use Proposition (7.1). Thus it suffices to show that ζ is an isomorphism for spaces $X = EF_1/EF_2 \wedge Y$ for arbitrary Y and adjacent families (F_1, F_2). Suppose (F_1, F_2) are K-adjacent and let $(H) \neq (K)$. We claim that $\omega_n^{WH}(EW^+ \wedge EF_1^H/EF_2^H \wedge Y^H) = 0$. If this group were non-zero, then EF_1^H would be non-empty and therefore $H \in F_1$. If $H \in F_2$, then EF_1 and EF_2 would be H-homotopy equivalent and thus the group above would be zero. So the group can only be non-zero if $H \in F_1 \backslash F_2$, i.e. if $(H) = (K)$.

Thus it remains to show that $\zeta_H[F_1, F_2]$ is an isomorphism for H-adjacent (F_1, F_2). We show that the three maps in the definition of ζ_H are isomorphisms in this case. To begin with, we prove that

$$\mathrm{pr}_*: \omega_n^G(G \times_N EW^+ \wedge EF_1/EF_2 \wedge Y) \to \omega_n^G(EF_1/EF_2 \wedge Y)$$

is an isomorphism. We write $Z = G \times_N EW$, suppress Y in the notation, and use homology of pairs. Consider the following diagram, to be explained in a moment.

$$\begin{array}{ccc}
\omega_n^G(Z \times EF_1, Z \times EF_2) & \xrightarrow{\;p\;} & \omega_n^G(Z \times CEF_2, Z \times EF_2) \\
\Big\downarrow \mathrm{pr}_* & & \Big\downarrow q \\
\omega_n^G(EF_1, EF_2) & \xrightarrow{\;r\;} & \omega_n^G(CZ \times EF_2 \cup Z \times CEF_2, CZ \times EF_2) .
\end{array}$$

Explanation: CA denotes the cone on A with A as a canonical subspace. A model for EF_1 is the space $CZ \times EF_2 \cup Z \times CEF_2$ and the inclusion $EF_2 \to EF_1$ is the map $EF_2 \to CZ \times EF_2$ with first component mapping to the cone-point and with second component the identity. Using these conventions, r is induced by the identity and q by the inclusion. The map p is induced by a map which is the identity on Z and any extension from $EF_2 \subset CEF_2$ to $EF_1 \to CEF_2$.

The map q is an isomorphism by excision. The map p is an isomorphism because the maps $Z \times EF_1 \to Z \times CEF_2$ and $Z \times EF_2 \to Z \times EF_2$ are homotopy equivalences. For the first map, this follows from the fact that CEF_2 is contractible and $\mathrm{pr}: Z \times EF_1 \to Z$ is a homotopy equivalence. Thus, if we can show that the diagram is commutative, then we know that pr_* is an isomorphism. We show that both paths in the diagram are induced by homotopic maps. The two maps $Z \times EF_2 \to CZ \times EF_2$ are obviously homotopic. The two maps $Z \times EF_1 \to (CZ \times EF_2) \cup (Z \times CEF_2) = EF_1$ are uniquely determined up to homotopy by the universal property of EF_1. The maps are actually homotopic as maps of pairs: This follows from the following observation. Let $A \subset B$ be a G-cofibration of F_1-numerable spaces. Then each G-map $A \to EF_1$ has an extension to B. This completes the proof that pr_* and hence the third map in the definition of ζ_H is an isomorphism.

Finally, we look at the first map in the definition of ζ_H. This is

$$\omega_n^W(EW^+ \wedge (EF_1/EF_2)^H \wedge Y^H) \to \omega_n^N(S^L \wedge EW^+ \wedge EF_1/EF_2 \wedge Y).$$

A left inverse of this map is given on representative homotopy classes by passing to the H-fixed point set, and this latter map is bijective by (7.4). In applying (7.4), one has to use the fact that $L^H = \{0\}$, see exercise 5. This completes the proof of Theorem (7.7). □

The splitting of equivariant homotopy groups for finite groups G is due to Segal [1971]. This section is based on tom Dieck [1975]. A geometric proof of the splitting, using ideas from cobordism theory, has been given by H. Hauschild [1975], [1977a]. Further references: Lewis-May-McClure [1982], Ulrich [1983], Becker [1986], Jackowski [1985], Schultz [1973], [1973a], [1977].

(7.8) Exercises.

1. Let F and F_1 be open families. Show that a model for $E(F \cup F_1)$ has the form $EF_1 \cup_{E(F_1 \cap F)} EF$.
2. Let $F_1 \subset F_2 \subset \ldots$ be a sequence of open families with union F. Let $EF_1 \subset EF_2 \subset \ldots$ be a corresponding sequence of closed G-cofibrations. Show that $\bigcup EF_i$ with colimit topology is a model for EF.
3. Let $F_1 \supset F_2$ be H-adjacent open families. Show that $(G \times_{NH} E(WH)) * EF_2$ is a model for EF_1.
4. Verify the statements about the map $A *_1 B \to A * B$ in the proof of (7.5).
5. Show, with notation of (7.7) and its proof: The tangent bundle of G/N is $G \times_N L$. The component of G/N^H containing eN is a point. Use this to show that $L^H = \{0\}$.

8. The Burnside ring and stable homotopy.

We want to compute $\omega_0^G = \omega_0^G(\text{Point})$. This is the direct limit over homotopy groups of the type $[S^V, S^V]_G^0$ for complex G-modules V. This stable homotopy set carries the structure of a commutative ring. The multiplication is induced by composition of representatives of the form $f: S^V \to S^V$ and $g: S^V \to S^V$, or by the smash-product $f \wedge g$ of representatives $f: S^V \to S^V$ and $g: S^W \to S^W$. The reader should check that these two assignments induce a well-defined ring structure and that the two multiplications are the same. Given any map $f: S^V \to S^V$, we can associate to it the degree function: This is a function $d_f: \phi(G) \to \mathbb{Z} \in C(G)$ sending (H) to degree f^H. Since degrees do not change under suspension, each element of ω_0^G has a well-defined degree function. By the equivariant Hopf theorem (4.11), we obtain an embedding

$$\delta: \omega_0^G \to C(G)$$

by associating to $x \in \omega_0^G$ its degree function, see exercise 5. This map δ is a ring homomorphism since degrees multiply under composition of maps.

In the previous section we obtained a direct sum decomposition

(8.1) $$\omega_0^G \cong \bigoplus_{(H)} \omega_0^{WH}(EWH^+).$$

This asks for a computation of $\omega_0^{WH}(EWH^+)$. We claim

(8.2) Proposition. *$\omega_0^{WH}(EWH^+) \cong 0$ if $\dim WH > 0$, and $\cong \mathbb{Z}$ if $\dim WH = 0$.*

Proof. We recall (see I (8.22., Ex. 15)) that EW can be obtained as a geometric realization of a simplicial space whose space of n-simplices is the $(n+1)$-fold product W^{n+1}. For each homology theory, the skeleton filtration of this geometric realization leads to a spectral sequence; the E^1-term for the theory $\omega_*^W(-)$ is

$$E_{p,q}^1 = \omega_q^W((W^{p+1})^+), \quad p \geq 0.$$

By (6.14), these groups are isomorphic to

(8.3) $$\omega_q(S^M \wedge (W^p)^+),$$

where ω_q denotes non-equivariant stable homotopy and M is the tangent space to W at the unit element. For $n = p + q = 0$ and $\dim W > 0$, the groups (8.3) are zero and therefore $\omega_0^W(EW^+)$ must be zero in this case. If W is finite, only the term $E_{0,0}^1 \cong \mathbb{Z}$ remains. □

Applying (8.2) to (8.1), we see that, additively, ω_0^G is the free abelian group with a basis $(x_H \mid (H) \in \phi(G))$, x_H the image of $1 \in \omega_0^{WH}(EWH^+)$ under ζ_H. We want to give a more explicit description of this basis.

We begin by describing a construction of elements in ω_0^G. Let X be a compact Euclidean G-neighbourhood retract (G-ENR); i.e. there exist an open G-subspace U of a (complex) G-representation V and G-maps $i: X \to U$, $r: U \to X$ such that $ri = \mathrm{id}(X)$. Given data (X, r, i), we construct an associated G-map $f: S^V \to S^V$ as follows. Let $\varrho: U \to [1, \infty[$ be a proper G-map which is equal to 1 on a neighbourhood of iX. Now form

$$f\colon S^V \to V/(V\backslash U) \cong U \cup \{\infty\} \xrightarrow[h]{} S^V$$

with $h(u) = \varrho(u)(u - iru)$. One can show by homotopy theory that the class of f in ω_0^G is independent of i, r, and ϱ (Dold [1974]). We shall prove this fact by using mapping degrees. By the Lefschetz fixed point theorem (Dold [1972], VII, 6.6), the degree of f equals the Euler characteristic $\chi(X)$ of X. The construction of f is compatible with passage to the fixed point sets; thus f^H has degree $\chi(X^H)$. By the Hopf theorem (4.11), we conclude that the class of f in ω_0^G only depends on the family of Euler characteristics $\chi(X^H)$. We denote $[f] \in \omega_0^G$ by $I(X)$ and call it the **Lefschetz-Dold index** of X.

These considerations suggest the definition of an additive group $A(G)$ as follows: The set $A(G)$ is the set of equivalence classes of compact G-ENR X; two such X, Y are called equivalent if for all $H \subset G$ the equality $\chi(X^H) = \chi(Y^H)$ holds. An additive group structure is induced on $A(G)$ by disjoint union. The assignment $X \mapsto I(X)$ induces a well-defined injective homomorphism

(8.4) $\quad I\colon A(G) \to \omega_0^G.$

Actually, cartesian product induces a multiplication on $A(G)$ and I becomes a homomorphism of commutative rings (exercise 1). We shall show now that I is an isomorphism. The ring $A(G)$ is called the Burnside ring. It will be studied in detail in Chapter IV.

A computational proof of the bijectivity of I is obtained by observing that $A(G) \subset C(G)$ and $\omega_0^G \subset C(G)$ are described by the same set of congruences; see IV. 5. Another description of ω_0^G is obtained by using the equivariant analogue of Dold [1974].

Now we show that $I(G/H)$ is the element x_H obtained from the splitting of ω_0^G, thus proving surjectivity of I. To this end we need a refinement of the Lefschetz-Dold index, called transfer (compare Dold [1976]). For simplicity, we consider compact differentiable G-manifolds M. Let $M \subset V$ be an embedding into a G-module V with normal bundle ν and tangent bundle τ of M. Let $M(\nu)$ be the Thom space of ν, $S^V \to M(\nu)$ the Pontrjagin-Thom map, and $M(\nu) \to M(\tau \oplus \nu) \cong M^+ \wedge S^V$ the map induced by the zero section of τ and $\tau \oplus \nu \cong M \times V$. The resulting composition $S^V \to M(\nu) \to M^+ \wedge S^V$ is called a **transfer map**. The composition with the projection $M^+ \wedge S^V \to S^V$ represents $I(M)$.

(8.5) Proposition. *The image of* $1_H \in \omega_0^{WH}(EWH^+) \cong \mathbb{Z}$ *in* ω_0^G *under* ζ_H *is* $I(G/H)$.

Proof. The element 1_H is the image of the generator [id] under

$$\omega_0 \to \omega_0^{WH}(WH^+) \to \omega_0^{WH}(EWH^+);$$

the first map is the extension homomorphism (6.13) and the second map is induced by the inclusion $WH \subset EWH$. There exists a G-representation V such that WH has an NH-embedding $j\colon WH \to \operatorname{res}_{NH} V$. Let $j'\colon S^V \to WH^+ \wedge S^V$ be an associated transfer map. Then we infer from construction (6.13) that 1_H is represented by

$$S^V \xrightarrow{j'} WH^+ \wedge S^V \subset EWH^+ \wedge S^V.$$

Next we have to apply the isomorphism $\omega_0^{WH}(EWH^+) \to \omega_0^G((G \times_N EWH)^+)$ in the definition of ζ_H. This is another instance of (6.13). Thus a representing map j_H for the image of 1_H is obtained as follows. Let $i\colon G/N \subset U$ be a G-embedding into a G-module and

$$i'\colon S^U \to G/N^+ \wedge S^U \cong G^+ \wedge_N S^U$$

be the corresponding transfer map. Then j_H is given as the composition

$$S^U S^V \xrightarrow{i' \wedge S^V} (G^+ \wedge_N S^U) \wedge S^V \cong G^+ \wedge_N (S^U S^V)$$
$$\xrightarrow{G^+ \wedge_N (S^U j')} G^+ \wedge_N (S^U \wedge W^+ \wedge S^V) \to (G^+ \wedge_N EW^+) \wedge S^U S^V.$$

Using the embedding $G/H \cong G \times_N W \to U \oplus V$, $(g, n) \mapsto (i(g), gj(n))$ and the corresponding transfer map

$$S^U S^V = S^{U \oplus V} \to (G \times_N W)^+ \wedge S^{U \oplus V} = (G \times_N W)^+ \wedge S^U S^V$$

together with $G \times_N W \subset G \times_N EW$, we obtain a map which is homotopic to j_H. □

The construction of the Lefschetz-Dold index uses the notion of G-ENR. For the convenience of the reader, we collect a few facts about G-ENR's.

(8.6) Proposition. *If X is a G-ENR and $i\colon X \to W$ a G-embedding into a G-module W, then iX is a G-retract of a neighbourhood.*

Proof. Dold [1972], p. 81, using I(3.11). □

(8.7) Proposition. A differentiable G-manifold with a finite number of orbit types is a G-ENR.

Proof. Such a G-manifold has a differentiable embedding into a G-module (Wasserman [1969]) and is a retract of an equivariant tubular neighbourhood. □

If there is no group acting, we simply talk about ENR's. Borsuk has shown that the property of being an ENR is a local property. A space X is called **locally contractible** if every neighbourhood V of a point $x \in X$ contains a neighbourhood W of x such that $W \subset V$ is nullhomotopic fixing x. An ENR is locally contractible (Dold [1972], p. 81). A space is **locally** ***n*-connected** if every neighbourhood V of a point x contains a neighbourhood W such that each map $S^j \to W, j \leq n$, is nullhomotopic in V.

(8.8) Proposition. *If* $X \subset \mathbb{R}^n$ *is locally* $(n-1)$*-connected and locally compact, then* X *is an* ENR.

Proof. Dold [1972], IV. 8.12, and 8.13, exercise 4. □

A basic theorem of point set topology states that a separable metric space of dimension $\leq n$ can be embedded into $\mathbb{R}^{2n+1}$; see Hurewicz-Wallman [1948]. Hence a space is an ENR if and only if it is locally compact, separable metric, finite-dimensional, and locally contractible.

(8.9) Proposition. *Let* X *be a* G-ENR. *Then the orbit space* X/G *is an* ENR.

Proof. Let $X \xrightarrow{i} U \xrightarrow{r} X$ be a presentation of X as a neighbourhood retract, i.e. U is an open G-subset of a G-module and $ri = \mathrm{id}$. We pass to orbit spaces. A retract of an ENR is an ENR. Hence it suffices to prove the proposition for differentiable manifolds X (and then apply it to the manifold U). Let $p: X \to X/G$ be the quotient map. Given $x \in V \subset X/G$ with V open, the preimage $p^{-1}V$ contains a G-invariant tubular neighbourhood W of the orbit $p^{-1}(x)$. Hence $p(W)$ is contractible. Therefore, X/G is locally contractible. The space X/G is locally compact (I, 3.6, iv) and separable metric (Palais [1960], 1.1.12); moreover, $\dim X/G \leq \dim X$ (Hurewicz-Wallman [1948]). □

(8.10) Proposition. *Let* X *be a* G*-space which is separable metric and finite-dimensional. Then* X *is a* G-ENR *if and only if* X *is locally compact, has a finite number of orbit types, and, for every isotropy group* $H \subset G$, *the fixed point set* X^H *is an* ENR.

Proof. Jaworowski [1976]. See also Ulrich [1983]. □

(8.11) Corollary. *If* X *is a* G-ENR, *then* $X_{(H)}$ *is a* G-ENR *for every* $H \subset G$. □

(8.12) Proposition. *A finite* G*-complex* X *is a* G-ENR.

Proof. Using (8.10), it essentially remains to be shown that fixed point sets X^H are locally contractible. This is left to the reader. □

(8.13) Exercises.

1. Show that (8.4) is compatible with multiplication.
2. If X is a compact ENR, then the Euler characteristic $\chi(X)$ is defined.
3. Let $E \to B$ be a fibre bundle with typical fibre F. If F and B are ENR's, then E is an ENR.
4. Let $F\colon (X, A) \to (Y, B)$ be a continuous map between compact ENR's such that $F(X \backslash A) = Y \backslash B$. Suppose $F\colon X \backslash A \to Y \backslash B$ is a fibration whose typical fibre Z is a compact ENR. Then: $\chi(X, A) = \chi(Z)\chi(Y, B)$. The Euler characteristic $\chi_c(X \backslash A)$ of $X \backslash A$, computed with Alexander-Spanier cohomology with compact support and coefficients in a field, exists and $\chi(X, A) = \chi_c(X \backslash A)$.
5. Show that $\delta\colon \omega_0^G \to C(G)$ is injective. Hint: If V is a complex representation such that $\dim V > 2$, then $\delta\colon [S^V, S^V]_G^0 \to C(G)$ is injective by the equivariant Hopf theorem. A direct limit of injective maps is injective.
6. Let W be infinite. Show that, for each W-complex X, $\omega_0^W(EW \times X) = 0$. If W is finite, show that $\omega_0^W(EW \times X)$ is free abelian; generators correspond to components of X/W.
7. Let X be a G-complex and let $f\colon G/H \to X$ be a G-map. The transfer map corresponding to G/H yields a homomorphism $\mathrm{tr}_H\colon \omega_0^G \to \omega_0^G(G/H^+)$. Let $x(f) \in \omega_0^G(X^+)$ be the element $f_* \circ \mathrm{tr}_H(1)$. Show that $\omega_0^G(X^+)$ is free abelian on elements $x(f)$, where f runs through the isomorphism classes of objects in $\pi_0(G, X)$; see I(10.3).
8. Use exercise 7 and $\omega_G^0(G/H^+; G/K^+) \cong \omega_0^H(G/K^+)$ to describe a $\mathbb{Z}$-basis for the stable G-equivariant maps $G/H \to G/K$.
9. Show that groups $\omega_0^G(X; Y)$ become modules over the Burnside ring ω_0^G by using smash-product with representing elements $f\colon SV \to SV$ for $x = [f] \in \omega_0^G$ to define multiplication by x. Show that the groups of any stable equivariant (co-)homology theory are naturally modules over the Burnside ring (suspend by SV and apply the morphism induced by $f\colon SV \to SV$).

9. Bredon-homology and Mackey functors.

Ordinary homology is characterized on the category of finite complexes by the seven axioms of Eilenberg and Steenrod [1952], Chapter I. There are analogous axioms for equivariant homology theories. In (6.7) we have given axioms for theories on pointed spaces. We now state briefly the axioms for pairs of spaces, assuming that the readers knows in principle how to deal with such axioms.

(9.1) An **equivariant homology theory** on the category of pairs of G-spaces consists of a sequence $(h_q \mid q \in \mathbb{Z})$ of covariant functors on such spaces into abelian groups together with natural transformations $\delta: h_q(X, A) \to h_{q-1}(A, \emptyset)$ such that each pair (X, A) leads to an exact sequence

$$\dots \to h_q(A, \emptyset) \to h_q(X, \emptyset) \to h_q(X, A) \xrightarrow{\delta} h_{q-1}(A, \emptyset)$$

in the usual way and such that G-homotopic maps $f_0, f_1: (X, A) \to (Y, B)$ induce the same homomorphisms $h_q(f_0) = h_q(f_1)$. Moreover, the **excision axiom** holds: Inclusion $(X \backslash A, U \backslash A) \to (X, U)$ induces an isomorphism

$$h_q(X \backslash A, U \backslash A) \xrightarrow{\cong} h_q(X, U)$$

whenever A is a G-subspace of U such that $\bar{A} \subset U$. The theory is called ordinary or classical homology if, moreover, the **dimension axiom** holds: For a homogeneous space G/H one has $h_q(G/H) = 0$ for $q \neq 0$.

Suppose an ordinary equivariant homology theory is given. The covariant functor h_0 restricted to the category $\mathrm{Or}(G)$ of homogeneous spaces, I (10.1), is called the **coefficient system** of the theory. It is a left $\mathbb{Z}\mathrm{Or}(G)$-module in the terminology of I.11. More or less by the same proof as in the non-equivariant case, an ordinary equivariant homology theory is determined on the category of finite G-complexes by specifying its coefficient system. We assume this fact and leave the proof to the reader.

Axioms for cohomology are the obvious dual ones. The coefficient system in this case is a contravariant functor $\mathrm{Or}(G) \to \mathbb{Z}$-Mod, alias right $\mathbb{Z}\mathrm{Or}(G)$-module.

Given a finite group G and a coefficient system, Bredon [1967] constructs (co-) homology theories satisfying the axioms (on a category of G-complexes). Bredon's theories have been described, using singular chains, by Bröcker [1971]. A construction of singular equivariant (co-)homology theories for spaces with compact Lie group action has been given by Illman [1972], [1972a], [1975].

A detailed investigation of such theories in the context of stable equivariant homotopy is contained in the work of Lewis, May, and McClure [1981]; in particular, these authors show that under suitable hypotheses on the coefficient system (Mackey functor), the theories are stable (as in (6.8)), a result which seems to be not obtainable by using singular chains.

Since, occasionally, we use Bredon-theories, we give a brief description of their construction for finite groups and refer to the references above for further details. We are mainly interested in the case that the coefficient system has some extra structure. We begin by describing this structure.

Let G be a finite group and $\hat{G}$ or G-Set be the category of finite G-sets and G-maps. A **bi-functor**

$$M = (M^*, M_*): G\text{-Set} \to \mathbb{Z}\text{-Mod}$$

consists of a contravariant functor $M^*: G\text{-Set} \to \mathbb{Z}\text{-Mod}$ and a covariant functor $M_*: G\text{-Set} \to \mathbb{Z}\text{-Mod}$; the functors are assumed to coincide on objects. Therefore, we write $M(S) = M_*(S) = M^*(S)$ for finite G-sets S. If $f: S \to T$ is a morphism, we often use the notation $M_*(f) = f_*$, $M^*(f) = f^*$. In this context, the $*$-index has nothing to do with a grading but only indicates the variance of the functor, in a notation which is analogous to the use in homology and cohomology. A bi-functor is called a **Mackey functor** if it has the following properties:

(9.2) For each pullback diagram

$$\begin{array}{ccc} U & \xrightarrow{F} & S \\ {\scriptstyle H}\downarrow & & \downarrow{\scriptstyle h} \\ T & \xrightarrow{f} & V \end{array}$$

in G-Set, we have $F_* H^* = h^* f_*$.

(9.3) The two embeddings $S \to S + T \leftarrow T$ into the disjoint union define an isomorphism $M^*(S + T) \cong M^*(S) \oplus M^*(T)$.

The notion of a Mackey functor was introduced by Dress [1973] in his axiomatic representation theory. Later we shall generalize this notion to compact Lie groups by using topological methods. The reader should also recall I (4.5) and (4.13) at this point.

Now let M be a Mackey functor on G-Set. We want to define homology and cohomology groups

$$H_*(X, A; M) \quad \text{and} \quad H^*(X, A; M)$$

with coefficients in this Mackey functor. Let C_* be the functor singular chain complex, i.e. $C_*(X, A)$ is the singular chain complex of the pair of spaces (X, A). If this is a pair of G-spaces, we can define a chain complex of right $\mathbb{Z}\hat{G}$-modules by setting

$$S \mapsto C_*(\mathrm{Hom}_G(S, X), \mathrm{Hom}_G(S, A)).$$

We take the tensor product I (11.6)

(9.4) $\quad C_*(\mathrm{Hom}_G(?, X), \mathrm{Hom}_G(?, A)) \otimes_{\mathbb{Z}\hat{G}} M_*(?)$,

which yields again a chain complex. Its homology groups are denoted by $H_*(X, A; M)$. Analogously, the cohomology groups $H^*(X, A; M)$ are defined via the cochain complex

(9.5) $\mathrm{Hom}_{\mathbb{Z}\hat{G}}(C_*(\mathrm{Hom}_G(?, X), \mathrm{Hom}_G(?, A)), M^*(?))$.

The verification that $H_*(X, A; M)$ is an ordinary equivariant homology theory with $H_0(?, M)\colon \hat{G} \to \mathbb{Z}$-Mod naturally equivalent to M_* is not difficult and follows more or less by naturality from the non-equivariant case (compare Bröcker [1971]). We leave the verification to the reader.

Of course, the homology groups only use the covariant part of the Mackey functor and are thus definable in the same manner for each covariant coefficient system.

The point of using a Mackey functor is that homology and cohomology become related in an expected way and, moreover, homology and cohomology themselves become Mackey functors as we shall explain in a moment.

Let $h_*(X, A)$ be an equivariant homology theory. Then the functor $S \mapsto h_*(S \times (X, A)) =: h_*[S](X, A)$ is a covariant functor from G-Set into the category of equivariant homology theories. We are interested in cases when this extends to a Mackey functor. A **Mackey structure** on h_* is a contravariant functor $h_*^![?]$ from G-Set into equivariant homology theories such that

$$h_n^![S](X, A) = h_n[S](X, A),$$

as functors in the variable (X, A), and such that for each $n \in \mathbb{Z}$ and each (X, A), the pair

$$(h_n[?](X, A), h_n^![?](X, A))$$

is a Mackey functor. Let $f^!$ denote the induced maps for the $h_*^!$-functors. A Mackey structure on a cohomology theory is defined dually in the obvious way.

(9.6) Proposition. *Let M be a Mackey functor. Then $H_*(-; M)$ resp. $H^*(-; M)$ carry canonical Mackey structures such that $H_0(-; M)$ and $H_0^!(-; M)$ resp. $H^0(-; M)$, $H_!^0(-; M)$ are, as functors on G-Set, naturally equivalent to M.*

Proof. Let $f\colon S \to T$ be a morphism in G-Set. We define the induced contravariant morphism $f^!$ on the chain level (9.4) and suppress A from the notation. A typical generator for the group of n-chains has the form $\alpha \otimes a$ where $\alpha \in C_n(\mathrm{Hom}_G(V, T \times X))$ is a singular n-simplex and $a \in M(V)$. The simplex α is given as a map $\alpha = (\alpha(1), \alpha(2))\colon V \times \Delta^n \to T \times X$. Since T is discrete, the map $\alpha(1)$ is a G-map of the form

$$V \times \Delta^n \xrightarrow[\mathrm{pr}]{} V \xrightarrow[\alpha_1]{} T.$$

Now form the pullback diagram

$$\begin{array}{ccc} U & \xrightarrow{\beta_1} & S \\ {\scriptstyle F}\downarrow & & \downarrow{\scriptstyle f} \\ V & \xrightarrow{\alpha_1} & T \end{array}$$

which is used to define the singular simplex

$$\beta = (\beta(1), \beta(2)) \colon U \times \Delta^n \to S \times X$$

with $\beta(1) = \beta_1 \circ \mathrm{pr}$ and $\beta(2) = \alpha(2) \circ (F \times \mathrm{id})$. The map $f^!$ is now defined on the chain level by sending $\alpha \otimes a$ to $\beta \otimes F^* a$. The fact that M is a Mackey functor is used to verify that this map $f^!$ is well-defined, i.e. compatible with the relation I (11.6) appearing in the definition of the tensor product. The verification that $f^!$ is a chain map and that the induced maps on homology have functorial properties as claimed is straightforward and left to the reader. The definition of $f_!$ for cohomology is analogous: Let c be an n-cochain for $S \times X$. Then $f_! c$ is the cochain defined by $f_! c(\alpha) = F_* c(\beta)$ where α and β have the same meaning as above. □

We now discuss a few examples and applications. Let (X, A) be a relatively free G-complex. We have the cellular cochain complex with group of cellular n-cochains $H^n(X_n, X_{n-1}; M)$. If $(\phi_j) \colon \coprod_j G \times (D^n, S^{n-1}) \to (X_n, X_{n-1})$ is a characteristic map for the n-cells, then, by excision and suspension as usual,

$$\begin{aligned} H^n(X_n, X_{n-1}; M) &\cong H^n\Big(\coprod_j G \times (D^n, S^{n-1}); M\Big) \\ &\cong \mathrm{Hom}_{\mathbb{Z}G}(H_n(X_n, X_{n-1}), M(G)). \end{aligned}$$

Looking at the boundary map in this chain complex, we see that only the effect of M on G-maps $G \to G$ is used, i.e. only the $\mathbb{Z}G$-module $M(G)$ is used. Therefore, the cellular cohomology groups of this complex are exactly those which where denoted by $\mathfrak{H}_G^*(X, A; M(G))$ in section 3; so here they appear as a special case of Bredon cohomology groups.

We use this occasion to mention a few more properties of these groups.

(9.7) Proposition. *Let (X, A) be a relatively free G-complex.*

(i) *If $M = K[G]$ is the permutation module of the group G over the abelian group K, then $\mathfrak{H}_G^*(X, A; M) \cong H^*(X, A; M)$.*

(ii) *If M is a G-module with trivial G-action, then $\mathfrak{H}_G^*(X, A; M) \cong H^*(X/G, A/G; M)$.*

Proof.
(i) follows by applying a natural isomorphism $\mathrm{Hom}_{\mathbb{Z}G}(F, K[G]) \cong \mathrm{Hom}_{\mathbb{Z}}(F, K)$ for $\mathbb{Z}G$-modules F.
(ii) follows by applying a natural isomorphism $\mathrm{Hom}_{\mathbb{Z}G}(F, M) \cong \mathrm{Hom}_{\mathbb{Z}}(F \otimes_{\mathbb{Z}G} \mathbb{Z}, M)$ for a free $\mathbb{Z}G$-module F and by observing $H_n(X_n, X_{n-1}) \otimes_{\mathbb{Z}G} \mathbb{Z} \cong H_n(X_n/G, X_{n-1}/G)$. □

(9.8) Example. We give an example of a Mackey functor M. Let K be an abelian group. For a finite G-set S, define $M(S) = \mathrm{Abb}(S/G, K)$ as the group of all maps $S/G \to K$. Let $f: S \to T$ be given. Define $f_*: M(S) \to M(T)$ by $(f_*\alpha)(D) = \sum\{\alpha(C) \mid C \in (f/G)^{-1}(D)\}$. A point $C \in S/G$ is an orbit of S. We let $|C|$ be its cardinality. The map $f^*: M(T) \to M(S)$ is defined as $f^*\alpha(C) = \dfrac{|C|}{|fC|}\alpha(fC)$. It is a straightforward exercise to verify (9.2).

If $H_*(X, A; K)$ is ordinary homology with coefficients in the abelian group K, then $(X, A) \mapsto H_*(X/G, A/G; K)$ defines an ordinary homology theory. The coefficient system is the functor M_* defined in **(9.8)**. Thus we have, for G-complexes (X, A), a natural isomorphism

(9.9) $\quad H_*(X/G, A/G; K) \cong H_*(X, A; M).$

By (9.6), this homology theory has a Mackey structure. For $S = G/H$, we have $(S \times X)/G \cong X/H$. If $H \subset L \subset G$, then the covariant map induced by the projection $p_{H,L}: S = G/H \to G/L = T$ corresponds to the map induced by the projection $X/H \to X/L$. But, for G-complexes (X, A), there is also a homomorphism

$$p^!_{H,L}: H_*(X/L, A/L; K) \to H_*(X/H, A/H; K)$$

called **transfer**, which comes from the Mackey structure given by (9.6) and (9.8).

Let $p: G \to G/G$. Then p_* and $p^!$ are the basic natural transformations. The composition $p_* p^!$ is multiplication by $|G|$ on the coefficient system, hence, in general,

(9.10) $\quad |G| = p_* p^!: H_*(X/G, A/G) \to H_*(X/G, A/G).$

In order to evaluate $p^! p_*$, we apply (9.2) to the pullback

$$\begin{array}{ccc} G \times G & \xrightarrow{P_2} & G \\ \Big\downarrow{\scriptstyle P_1} & & \Big\downarrow{\scriptstyle p} \\ G & \xrightarrow{p} & G/G \end{array}$$

and obtain

(9.11) $p^! p_* = \sum_{g \in G} l_{g*} : H_*(X, A) \to H_*(X, A)$.

Here, l_{g*} is the map induced by left translation with g. In order to derive (9.11) from (9.2), one has to decompose $G \times G$ into G-orbits, $G \times G \cong \coprod_{g \in G} G \times \{g\}$. Then $P_1 | G \times \{g\} : (h, g) \mapsto h$ and $P_2 | G \times \{g\} : (h, g) \mapsto hg$. Using the homeomorphisms $(G \times X)/G \to X$, $(h, x) \mapsto h^{-1}x$ and $(G \times \{g\} \times X)/G \to X$, $(h, g, x) \mapsto h^{-1}x$, we see that $P_1 | G \times \{g\}$ reduces to the identity and $P_2 | G \times \{g\}$ to left translation by g^{-1}.

For any Mackey functor M, the following holds: Let $f: S \to S$ be an isomorphism. Then

$$\begin{array}{ccc} S & \xrightarrow{f} & S \\ {\scriptstyle f}\downarrow & & \downarrow{\scriptstyle \mathrm{id}} \\ S & \xrightarrow[\mathrm{id}]{} & S \end{array}$$

is a pullback and therefore $f_* f^* = \mathrm{id}^* \mathrm{id}_* = \mathrm{id}$. We see that $l_g^!$ is inverse to l_{g*}. Since $p = p \circ l_g$, we conclude that the image of $p^!$ is G-invariant

(9.12) $\operatorname{Im} p^! \subset H_*(X, A)^G$.

Since $p^! p_*$ is multiplication by $|G|$ on $H_*(X, A)^G$, we obtain

(9.13) Proposition. *Suppose multiplication by $|G|$ is an isomorphism on the coefficient group K. Then, for G-complexes (X, A), we have inverse isomorphisms*

$$p_* : H_*(X, A)^G \to H_*(X/G, A/G)$$
$$p^! : H_*(X/G, A/G) \to H_*(X, A)^G. \quad \square$$

We advice the reader to study the elementary derivation of such results in Bredon [1972], Ch. III. The itention here was to indicate another viewpoint and to relate the material to general machinery. Of course, analogous dual results hold for cohomology.

Let V be a complex representation of G with real dimension $2n$. Then $H^{2n}(DV, SV; \mathbb{Z}) \cong \mathbb{Z}$ and multiplication by a generator induces a natural suspension isomorphism

$$H^k(X, A)^G \cong H^{2n+k}((DV, SV) \times (X, A))^G.$$

Thus if multiplication by $|G|$ is an isomorphism on the coefficient group, then, using the cohomology version of (9.12), we obtain a natural isomorphism

(9.14) $H^k(X/G, A/G) \cong H^{2n+k}((DV, SV) \times_G (X, A))$.

This is induced by lifting elements along the projection

$(DV, SV) \times_G (X, A) \xrightarrow{q} (X/G, A/G)$ and then multiplying with a suitable element $t(X, A) \in H^{2n}((DV, SV) \times_G (X, A))$. The exact cohomology sequence of the pair $DV \times_G X$, $SV \times_G X$ then leads to an exact Gysin sequence

(9.15) $$H^k(X/G) \xrightarrow[e]{} H^{k+2n}(X/G) \xrightarrow[q]{} H^{k+2n}(SV \times_G X) \to H^{k+1}(X/G)$$

where e is multiplication by a suitable $e = e(X) \in H^{2n}(X/G)$, namely the restriction of $t(X)$ to $H^{2n}(DV \times_G X) \cong H^{2n}(X/G)$.

Let X be an S^1-complex of finite orbit type. Then there exists a finite group $G \subset S^1$ such that the induced S^1/G-action on X/G is **semi-free**, i.e. has only fixed points and free orbits. If we use the rational numbers as coefficients, we have a natural isomorphism $H^*(X, A) \cong H^*(X/G, A/G)$; so we can reduce the (co-)homological study to the semi-free case.

Transfer maps as in (9.13) for compact Lie groups are constructed in Oliver [1982], and Lewis-May-Mc Clure [1981].

Stable equivariant homology and cohomology theories always have a Mackey structure. Geometric equivariant bordism theories are unstable theories with Mackey structure. The use of a Mackey structure for the splitting of homology theories is explained in tom Dieck [1973].

(9.16) Exercises.

1. Verify the axioms for equivariant (co-)homology for the theories $H_*(X, A; M)$ and $H^*(X, A; M)$ with coefficients in a Mackey functor.
2. Complete the proof of (9.6).
3. Verify (9.2) for the example (9.8).
4. Show (9.9) for G-complexes. Is this an isomorphism for general G-spaces?
5. Define a canonical Mackey structure on stable equivariant homology and cohomology theories.

10. Homotopy representations.

We have already considered unit spheres of representations as transformation groups. A large part of the theory of transformation groups may be unified under the heading of geometric representation theory. This theory, which will be the subject of another monograph, has the following aspects:

(i) Geometrical investigation of representations as transformation groups.
(ii) Investigation of transformation groups which, in some respect, resemble representations.
(iii) Geometric methods in (axiomatic) representation theory.

(iv) Investigation of the rôle of classical representation theory for the analysis of transformation groups.
(v) Representation theoretical methods in the theory of transformation groups.

We give an introduction to this theory by explaining the concept of a homotopy representation.

(10.1) Definition. A **homotopy representation** for the compact Lie group G is a G-complex X with the following properties:
(i) X is finite-dimensional.
(ii) X has finite orbit type.
(iii) For each $H \subset G$, the fixed point set X^H has the homotopy type of a sphere $S^{n(H)-1}$. (If X^H is empty, then $n(H) = 0$.)
(iv) For each $H \subset G$, the topological dimension of X^H equals $n(H) - 1$.
(v) If $H \in \operatorname{Iso}(X)$ and $L \supset H$, $L \neq H$, then $n(L) < n(H)$.
(vi) The set $\operatorname{Iso}(X)$ is closed with respect to intersections.

A G-complex X satisfying (i)–(iii) above is called a **generalized homotopy representation**. Occasionally, one does not want to require (v) and (vi). The (generalized) homotopy representation is called **finite** if X is a finite complex.

One aim is the classification of homotopy representations up to G-homotopy type. The actual cellular structure of the complexes is irrelevant. For this reason we also call such spaces homotopy representations which have the G-homotopy type of „actual“ homotopy representations.

(10.2) Example. Let V be an orthogonal G-representation. Then the unit sphere $S(V)$ is a finite homotopy representation, see exercise 4. Such homotopy representations (or their homotopy types) are called **linear**. Another aim of the theory is to understand the homotopy types of G-spaces $S(V)$ in terms of homotopical and representation theoretical data.

One of the basic invariants of a homotopy representation is its dimension function. Let X be a homotopy representation and assume $X^H \simeq S^{n(H)-1}$. We define

(10.3) $\operatorname{Dim}(X)(H) = n(H)$.

Then $\operatorname{Dim}(X)$ is an integer-valued function on conjugacy classes of closed subgroups. Let $\psi(G)$ be the set of conjugacy classes of closed subgroups with the Hausdorff topology and let $C'(G) = C(\psi(G), \mathbb{Z})$ be the ring of continuous functions $\psi(G) \to \mathbb{Z}$ into the discrete space $\mathbb{Z}$; see IV.3 for further details. Then $\operatorname{Dim}(X) \in C'(G)$ and this function is called the **dimension function** of X.

Another invariant of a generalized homotopy representation is its orientation behaviour. Let CX denote the cone on X (a point, if X is empty). For each $H \subset G$, we have $H^{n(H)}(CX^H, X^H; \mathbb{Z}) \cong \mathbb{Z}$. The action of WH on (CX^H, X^H)

induces a homomorphism

(10.4) $e_{X,H}: WH \to \mathrm{Aut}(\mathbb{Z}) = \{1, -1\} = \mathbb{Z}^*$.

The homomorphism $e_{X,H}$ is called the **orientation behaviour of X at H**. The family of all $e_{X,H}$ is called the **orientation behaviour of X**. An **orientation** of a generalized homotopy representation X is a choice of a generator $z(H) \in H^{n(H)}(CX^H, X^H; \mathbb{Z})$ for each subgroup H. The definition of an orientation is slightly artificial: Even if $X^H = X^K$, one may choose different generators. One also has to be careful in order to make this notion well-defined for conjugacy classes. If $gHg^{-1} = K$, then left translation $l_g: X^H \to X^K$, $x \mapsto gx$ induces an isomorphism

$$l_g^*: H^{n(H)}(CX^K, X^K) \to H^{n(H)}(CX^H, X^H)$$

which is independent of the choice of $g \in G$ with $gHg^{-1} = K$ only if $e_{X,H}$ is trivial. In general, $e_{X,K}\, c_g = e_{X,H}$ for

$$c_g: WH \to WK, nH \mapsto gng^{-1}K.$$

If V and W are orthogonal representations of G, then there is a canonical G-homeomorphism $S(V \oplus W) \cong SV * SW$. The join operation $*$ is a purely topological construction and can be applied to any pair X, Y of G-spaces: $X * Y = X \times [0,1] \times Y/\sim$, with $(x, 0, y) \sim (x', 0, y)$ and $(x, 1, y) \sim (x, 1, y')$ and G-action $g(x, t, y) = (gx, t, gy)$. The G-homotopy type of $X * Y$ depends only on the G-homotopy types of X and Y. Moreover, $(X * Y)^H = X^H * Y^H$, at least as sets. The isotropy group of $(x, t, y) \in X * Y$ is $G_x \cap G_y$ if $t \neq 0, 1$, is G_x if $t = 1$, and is G_y if $t = 0$. Therefore, if X and Y have finite orbit type, so has $X * Y$ (exercise 1). Of course, if we work with G-complexes, we have the usual difficulty that $X * Y$ itself may not be a G-complex in a natural way. Assume that it has at least the G-homotopy type of a G-complex. If X resp. Y are G-complexes and have topological dimension m resp. n, then assume that $X * Y$ has topological dimension $m + n + 1$. These technicalities granted, if X and Y are homotopy representations, then $X * Y$ is a homotopy representation. We have

(10.5) $\mathrm{Dim}(X * Y) = \mathrm{Dim}\, X + \mathrm{Dim}\, Y$

for the dimension functions. If X and Y have been oriented, there is a canonical orientation on $X * Y$ which is associative. Note also that

(10.6) $e_{X,H} \cdot e_{Y,H} = e_{X*Y,H}$

(pointwise multiplication of functions $WH \to \mathbb{Z}^*$).

Direct sum of representations induces the additive structure of the representation ring. We define the homotopical analogue. Let $\mathfrak{A}$ be a category of G-spaces with join operation; assume that the join operation is associative and commutative up to equivalence in that category. The set of equivalence classes together with the join as composition law is then a commutative semi-group $J^+(\mathfrak{A})$. We let $J(\mathfrak{A})$ be the associated Grothendieck group. We also write $J^+(G, \mathfrak{A})$ and $J(G, \mathfrak{A})$ if the dependence on G is to be emphasized. The letter J stands for join and also indicates that classical J groups, which come from homotopy equivalences of vector bundles, are special cases of this construction.

We next describe the categories $\mathfrak{A}$ which we have in mind.

Let $\mathfrak{H}^\infty$ be the category of homotopy representations for G. Equivalence is G-homotopy equivalence. We obtain $J^+(G, \mathfrak{H}^\infty)$ and the group $J(G, \mathfrak{H}^\infty)$. We call this group the **homotopy representation group** (similarly for analogous categories). If X is a homotopy representation, then $[X]$ shall denote its image in $J(G, \mathfrak{H}^\infty)$. This group is written additively so that elements in this group have the form $[X]-[Y]$. Moreover, by definition, $[X * Z]-[Y * Z]=[X]-[Y]$. Other relevant categories in this context are:

(10.7)

$\mathfrak{H}$	finite homotopy representations
$\tilde{\mathfrak{H}}^\infty$	generalized homotopy representations
$\tilde{\mathfrak{H}}$	finite generalized homotopy representations
$\mathfrak{L}$	linear homotopy representations

The group $J(G, \mathfrak{L})$ is a quotient of the additive group of the real representation ring $RO(G)$ and is denoted by $JO(G)$ or JO_G at other places in the literature. The groups $J(G, \mathfrak{A})$ are called $V(G, \mathfrak{A})$ in tom Dieck-Petrie [1982]. The following notations are also in use

$$\begin{aligned} \tilde{V}^\infty(G) &= J(G, \tilde{\mathfrak{H}}^\infty) \\ V^\infty(G) &= J(G, \mathfrak{H}^\infty) \\ V(G) &= J(G, \mathfrak{H}). \end{aligned}$$

Inclusion of categories induces the homomorphisms in the following commutative diagram

(10.8)
$$\begin{array}{ccccc} J(G, \mathfrak{L}) & \longrightarrow & J(G, \mathfrak{H}) & \longrightarrow & J(G, \mathfrak{H}^\infty) \\ & & \downarrow & & \downarrow \\ & & J(G, \tilde{\mathfrak{H}}) & \longrightarrow & J(G, \tilde{\mathfrak{H}}^\infty). \end{array}$$

It is shown in tom Dieck-Petrie [1982] that, for finite G, the horizontal maps are injective and the vertical maps are bijective.

From (10.5), we obtain a homomorphism

(10.9) $\quad \mathrm{Dim}\colon V^\infty(G) \to C'(G)$

and similarly for the other categories. The kernel of (10.9) is denoted by

(10.10) $\quad v^\infty(G) = \text{kernel of Dim}.$

From (10.6), we obtain a homomorphism

(10.11) $e_H: V^\infty(G) \to \mathrm{Hom}(WH, \mathbb{Z}^*)$.

The equivariant Hopf theorem can be used to give a partial computation of $v^\infty(G)$. To this end, another invariant is needed, the degree function. Suppose X and Y are oriented homotopy representations with $\mathrm{Dim}\, X = \mathrm{Dim}\, Y$. Given a G-map $f: X \to Y$, we can consider the mapping degrees

$$\operatorname{degree} f^H = d(f^H) \in \mathbb{Z}.$$

The function $(H) \mapsto d(f^H)$ is contained in $C'(G)$ and is called the **degree function** of f.

At this point one has to be more specific about orientations. For the rest of this section we only require conditions (10.1) (i)–(iv). We also change notation for the dimension function and write $X^H \simeq S^{n(H)}$ in the sequel. Also we use several results from Chapters III and IV.

The next two propositions are due to Laitinen [1986]. Let X be a homotopy representation for G.

(10.12) Proposition. *Suppose $H \subset K$ and $n(H) = n(K) = n$. Then the inclusion $X^K \subset X^H$ is a homotopy equivalence.*

Proof. For $n = -1$ both spaces are empty and for $n = 0$ both spaces are homeomorphic to S^0. Therefore, suppose that $n > 0$. It suffices to show that $i: X^K \subset X^H$ induces an isomorphism in (integral) cohomology. The exact sequence

$$0 \leftarrow H^{n+1}(X^H, X^K) \leftarrow H^n(X^K) \xleftarrow{i^*} H^n(X^H)$$

implies that i^* is surjective and hence bijective because $H^{n+1}(X^H, X^K) = 0$ as (X^H, X^K) has the homotopy type of a relative n-dimensional complex. □

(10.13) Proposition. *Suppose $K \supset H$, $L \supset H$ and $n = n(H) = n(K) = n(L)$. Let $M = \langle K, L\rangle$ be the group generated by K and L. Then $n = n(M)$.*

Proof. Let $n > 0$ and denote the inclusions by

$$i: X^K \to X^K \cup X^L, \quad j: X^K \cup X^L \to X^H.$$

The composition

$$H^n(X^K) \xleftarrow{i^*} H^n(X^K \cup X^L) \xleftarrow{j^*} H^n(X^H)$$

is an isomorphism by (10.12); hence j^* is injective. From the exact sequence

$$H^{n+1}(X^H, X^K \cup X^L) \leftarrow H^n(X^K \cup X^L) \xleftarrow{j^*} H^n(X^H)$$

we conclude that j^* is surjective, too. The Mayer-Vietoris-sequence

$$H^{k+1}(X^K \cup X^L) \leftarrow H^k(X^M) \leftarrow H^k(X^K) \oplus H^k(X^L) \leftarrow H^k(X^K \cup X^L)$$

shows, that $H^k(X^M) = 0$ for $k > n$ and $H^n(X^M) \neq 0$; hence $n(M) = n$. □

(10.14) Proposition. *Let $H \subset G$. The set*

$$I(H) := \{K \subset G \mid K \supset H, n(K) = n(H)\}$$

contains a unique maximal element $\hat{H}$. If $W\hat{H}$ is finite, then $\hat{H} \in \mathrm{Iso}(X)$.

Proof. Let L be an element of maximal dimension in $I(H)$ and let K be any other element. Then $M = \langle L, K \rangle \in I(H)$ by (10.13). Hence each element is contained in an element of maximal dimension. If $L, K \in I(H)$ have maximal dimension, then $M = \langle L, K \rangle$ has the same dimension. Therefore, for the components of e, $L_0 = M_0 = K_0$ holds. Suppose there is a proper ascending chain $K_1 \subset K_2 \subset K_3 \subset \dots$ of elements of maximal dimension in $I(H)$. Let $\bar{K}$ be the closure of $\bigcup K_i$. Since X has finite orbit type,

$$X^{\bar{K}} = X^{K_i}$$

for some i, hence $n(\bar{K}) = n(K_i)$ and therefore $\bar{K} \in I(H)$. But $\bar{K}$ must have larger dimension, a contradiction. Thus there exists a maximal element. The second assertion follows from the fact that there is only a finite number of isotropy groups larger than H. □

We only consider orientations $z(H, X) \in \tilde{H}^{n(H)}(X^H)$ for H such that $X^H \neq \emptyset$ and assume that for $i: X^{\hat{H}} \subset X^H$ equality $i^* z(\hat{H}, X) = z(H, X)$ holds. Suppose $K = gHg^{-1}$ and let $l_g: X^H \to X^K$ be the left translation by g. Then

$$l_g^* z(K, X) = e(g, H, X) z(H, X), \quad e(g, H, X) \in \mathbb{Z}^*.$$

(10.15) Definition. Let X and Y be homotopy representations with the same dimension functions. Orientations of X and Y are called **coherent** if for all (H, g) with $n(H) > -1$

$$e(g, H, X) = e(g, H, Y).$$

(10.16) Lemma. *Suppose X and Y are homotopy representations with the same dimension function. Then:*

(i) *X and Y have the same orientation behaviour.*

(ii) *X and Y possess coherent orientations.*

Proof. (i) $e_{X,H}: WH \to \mathbb{Z}^*$ is determined by its values on elements $w \in WH$ of order 2^k, $k > 0$. Suppose K/H is generated by such a w and let L/H be the

subgroup of index 2. Then $e_{X,H}(w) = 1$ if and only if $n(K) - n(L)$ is even (see III (4.45), Ex. 7).
(ii) The groups H such that $H = \hat{H}$ are determined by the dimension function and are therefore the same for X and Y. One chooses the orientations $z(H, X)$ and $z(H, Y)$ for a fixed group in the conjugacy class (H) with $H = \hat{H}$. If $K \sim H$, $K \neq H$, choose g such that $K = gHg^{-1}$ and define $z(K, X)$ by requiring $e(g, H, X) = 1$; similarly for Y. Using (i), one sees that there are unique orientations having these values and that they are coherent. □

Let $f\colon X \to Y$ be a G-map between coherently oriented homotopy representations with the same dimension function. Then we have a well-defined degree function $H \mapsto d(f^H)$. (We put $d(f^H) = 1$ if $n(H) = -1$.) In this case

(10.17) (i) $d(f^H) = d(f^K)$ if $H \sim K$.
(ii) $d(f^H) = d(f^K)$ if $n(H) = n(K)$.
(iii) $d(f^H) = 1$ if $n(H) = -1$.
(iv) $d(f^H) \in \{1, 0, -1\}$ if $n(H) = 0$.

(10.18) Proposition. *Suppose X and Y are homotopy representations with the same dimension function for the torus T. Then $d(f) = \pm d(f^T)$.*

Proof. III (5.16) and II (5.15). □

Let $o(G)$ be the least common multiple of the orders $|WH|$ for $(H) \in \phi(G)$; see IV (6.15). Let $c = ko(G)$ for some $k > 0$ in $\mathbb{Z}$.

(10.19) Lemma. *Let* $\operatorname{Dim} X = \operatorname{Dim} Y$. *A G-map $f\colon X \to Y$ has a degree function with values prime to c if and only if $(d(f^H), c) = 1$ for all $H \in \operatorname{Iso}(X)$, $|WH| < \infty$.*

Proof. Let $K \subset G$. There exists $L \subset G$, $L \triangleright K$, $L/K = T$ a torus, $(L) \in \phi(G)$; see IV.4. By (10.18), $(d(f^L), c) = 1$ implies $(d(f^K), c) = 1$. For L, we can apply (10.12)–(10.14): There exists $P \in \operatorname{Iso}(X)$, $P \supset L$, such that $d(f^L) = d(f^P)$. □

(10.20) Theorem. *Suppose X and Y are homotopy representations with the same dimension functions. There exists a G-map $f\colon X \to Y$ with degrees prime to c.*

Proof. Let $(H_1), \ldots, (H_n)$ be the conjugacy classes of $\operatorname{Iso}(X)$. Let $(H_i) < (H_j)$ imply $j < i$. Let $X(t) = \{x \in X \mid (G_x) = (H_i),\ i \leq t\}$. We construct inductively maps $f(t)\colon X(t) \to Y$ such that

$$f(t+1) \mid X(t) = f(t)$$
$$(d(f(t)^L), c) = 1 \quad \text{for} \quad (L) \in \phi(G, X),\ (L) = (X_i),\ i \leq t.$$

For $t = 1$, the group $K := H_1$ is a maximal isotropy group of X. Choose a WK-map $h: X^K \to Y^K$ such that $(d(h), c) = 1$. Such a map exists: Certainly, there exist WK-maps. If $|WK| < \infty$, let p be a prime divisor of $|WK|$ and choose $NK \supset L \rhd K$ with $|L/K| = p$. If $n(K) = 0$, then X^K and Y^K consist of two points and we can choose h of degree one. So assume $n(K) > 0$. Then $X^L = \emptyset$. By III (4.29), $d(h) \not\equiv 0 \bmod p$ and hence $(d(h), |WK|) = 1$ for each WK-map h. By (4.11), we can find h such that $d(h)$ is prime to c. The map h has a unique G-extension $f(1)$, see I.7, and $f(1)$ has the required properties. If $|WK| = \infty$, we choose any G-map $f(1)$.

Suppose $f(t)$ is given and let $K = H_{t+1}$. The G-extensions $f(t+1)$ of $f(t)$ correspond to the WK-extensions $h: X^K \to Y^K$ of $f(t)^K$. If $n(K) = 0$, then $X(t+1)^{>K} = \emptyset$ and we can choose h as a homeomorphism. Let $n(K) > 0$, $|WK| < \infty$, p a prime divisor of $|WK|$, and $L/K \subset WK$ a subgroup of order p. Then again by III (4.29),

$$d(h^K) \not\equiv 0 \bmod p \Leftrightarrow d(h^L) \not\equiv 0 \bmod p.$$

If $X^L = \emptyset$, then $d(h^L) = 1$ and we are done. If $X^L \neq \emptyset$, then $P = L \in \mathrm{Iso}(X)$, $(P) = (H_i)$ for $i \leq t$. By induction hypothesis, $(d(h^P), c) = 1$, hence $(d(h^L), p) = 1$ and $(d(h^K), p) = 1$. It follows again that $(d(h^K), |WK|) = 1$ for each WK-map; by (4.11), we can find h with degree prime to c. If $|WK| = \infty$, we choose any extension $f(t+1)$. The map $f = f(n)$ has the required properties. □

(10.21) Proposition. *Let $f: X \to Y$ be a G-map between two coherently oriented homotopy representations with the same dimension functions. Suppose f has degrees prime to c. Then there exists a G-map $h: Y \to X$ such that for all $H \subset G$*

(10.22) $d(f^H)d(h^H) \equiv 1 \bmod c.$

Proof. As in (10.19), one shows that it suffices to have (10.22) for $(H) \in \phi(G, Y)$. Note that in (10.18) one has $d(f^T) = d(f)$ in case $X = Y$. As above, let $\mathrm{Iso}(Y) = (H_1) \cup \ldots \cup (H_n)$; we construct $h(t): Y(t) \to X$ inductively.

Suppose $K = H_1$. If $n(K) = 0$, then f^K has degree ± 1, and we find a WK-map $h: Y^K \to X^K$ as homeomorphism of the same degree. Suppose $n(K) > 0$, $|WK| < \infty$ and let $h: Y^K \to X^K$ be any WK-map. For $u = f^K h: Y^K \to Y^K$, it follows from (4.19) that $d(u) \equiv 1 \bmod |WK|$. By (4.11), we can thus find h such that $d(u) \equiv 1 \bmod c$. Let $h(1)$ be the unique G-extension of h. In case $|WK| = \infty$ we let $h(1)$ be any G-map.

Now suppose $h(t)$ is given for $1 \leq t < n$ such that

$$d(f^L)d(h(t)^L) \equiv 1 \bmod c$$

for $(L) \in \phi(G, Y)$, $(L) = (H_i)$, $i \leq t$.

Let $K = H_{t+1}$, $|WK| < \infty$, $n(K) > 0$. Let $h(t+1)$ be any G-extension of $h(t)$. By (4.19), for $u = fh(t+1)$ there is a congruence of the type

(10.23) $d(u^K) \equiv \sum a(H,K)d(u^H) \bmod |WK|$,

with $a(H,K) \in \mathbb{Z}$ independent of u and the sum ranging over $(H) \in \phi(G, Y)$, $(H) > (K)$. By induction hypothesis, $d(u^H) \equiv 1 \bmod |WK|$. But since (10.23) is also satisfied by the identity of Y^K, we conclude that

$$d(u^K) \equiv \sum a(H,K) \equiv d(\mathrm{id}) \equiv 1 \bmod |WK|.$$

Again by (4.11), we can find $h(t+1)$ such that $d(u^K) \equiv 1 \bmod c$. In case $|WK| = \infty$ we let $h(t+1)$ be any G-extension of $h(t)$. The map $h = h(n)$ has the required properties. □

Since, by IV (6.15), $o(G)C(G) \subset A(G)$ we can form the rings

$$\bar{C} = \bar{C}(G) = C(G)/o(G)C(G) \supset \bar{A} = \bar{A}(G) = A(G)/o(G)C(G).$$

Let S^* denote the group of units of the ring S. Consider the multiplicative group

(10.24) $\mathrm{Pic}(G) = \bar{C}^*/C^*\bar{A}^*$.

If $x = [X] - [Y] \in v^\infty(G)$ and $f: X \to Y$ is a G-map with degrees prime to $o(G)$, then $d(f)$ represents an element $[d(f)]$ in $\mathrm{Pic}(G)$. The degree function $d(f)$ depends on the choice of orientations but the image $[d(f)]$ does not since we have factored out $C(G)^*$.

(10.25) Lemma. *Suppose $x = [X] - [Y] = [X'] - [Y']$. Let $f: X \to Y$ and $f': X' \to Y'$ be maps with degrees prime to c. Then $[d(f)] = [d(f')]$.*

Assuming this lemma (see exercise 7), we can define a map

(10.26) $D: v^\infty(G) \to \mathrm{Pic}(G)$

by $x \mapsto [d(f)]$. Obviously, this is a homomorphism. The first main result about homotopy representations is

(10.27) Theorem. *The homomorphism D is an isomorphism.*

We do not prove this theorem in this book. For finite groups, see tom Dieck-Petrie [1982]. One can show that $\mathrm{Pic}(G)$ is isomorphic to the Picard group of the Burnside ring $A(G)$.

We show that D is injective. Suppose $D(x) = 1$, $x = [X] - [Y]$, and $f: X \to Y$ is a G-map with degrees prime to $o(G)$. Since $d(f) \in A(G)C^*$, we can find, by II.8, a complex representation V and a G-map $h: S^V \to S^V$ such that, for $H \subset G$,

$$d(f^H)d(h^H) \equiv \varepsilon(H) \bmod o(G), \quad \varepsilon(H) \in \mathbb{Z}^*.$$

It is not an essential restriction to assume that f is a pointed map. Now consider $u = h \wedge f: S^V \wedge X \to S^V \wedge Y$ and write $d(u^H) = d(f^H)d(h^H) = \varepsilon(H) + o(G)k(H)$. Then $(H) \mapsto k(H)o(G)$ is contained in $o(G)C(G)$. We can find some V such that $S^V \wedge X$ and $S^V \wedge Y$ satisfy the hypotheses of (4.19). Therefore, there exists a map $v: S^V \wedge X \to S^V \wedge Y$ such that $\varepsilon(H) = d(v^H)$ for $(H) \in \phi(G, S^V \wedge X)$. But then $d(v^K) \in \mathbb{Z}^*$ for all $K \subset G$ and we see, by II.2, that v is a G-homotopy equivalence. Hence $x = [X] - [Y] = [S^V \wedge X] - [S^V \wedge Y] = 0$ in $v^\infty(G)$. □

(10.28) Exercises.

1. Let X and Y be G-spaces of finite orbit type. Show that $X \times Y$ with diagonal action has finite orbit type. (G compact Lie).
2. Construct the canonical orientation of $X * Y$, show it to be associative, and verify (10.6).
3. Let V be an orthogonal representation of G. If G is finite, show that $S(V)$ satisfies (10.1), (v) and (vi). For non-finite G, these properties do not hold in general.
4. Let X be a generalized homotopy representation. Show that there exists a representation V such that $X * SV$ has properties (10.1), (v) and (vi).
5. Show that the join of two homotopy representations has the homotopy type of a homotopy representation.
6. Let Z be a homotopy representation for G. Show that the degree function of any G-map $f: Z \to Z$ is contained in $A(G)$. (Use (4.19) and results from II.5 and IV.5)
7. Use exercise 6 and (10.21) in order to prove (10.25).

Chapter III: Localization

This chapter comprises the following sections.
1. Introduction of the bundle (Borel-) cohomology. Elementary methods of its computation.
2. The cohomology of classifying spaces of cyclic and quaternion groups.
3. Various forms of the localization theorem for equivariant cohomology.
4. Explanation of localization and its application to cyclic groups. The basic theorems of P. A. Smith and Borel.
5. Applications of the Borel-Smith theorems of section 4: The resemblance of actions of nilpotent groups on spheres with representation spheres.
6. Relations between Euler characteristics of a space and its fixed point set. Contractible orbit spaces.

1. Equivariant bundle cohomology.

Let $p: EG \to BG$ denote the universal principal G-bundle for the compact Lie group G (see I.8). If X is a G-space, then the associated fibre bundle with fibre X,

(1.1) $\quad p_X: X_G = EG \times_G X \to BG,$

can be considered as an invariant of the transformation group X. (This notation should not be confused with I (1.7).) The advantage of looking at X_G is that we no longer have a G-action but simply a topological space. We can use methods from algebraic topology to analyse the space X_G or rather the bundle p_X. But some information is lost if we pass from X to p_X. Let $f: X \to Y$ be a G-map which is a homotopy equivalence if we forget the G-action. Then the induced map $f_G = \mathrm{id} \times_G f: X_G \to Y_G$ is a fibre homotopy equivalence over BG; see I (8.18).

We apply ordinary cohomology to the bundle p_X and use the notation

(1.2) $\quad H_G^n(X) = H^n(X_G) = H^n(EG \times_G X).$

Here $H^n(-)$ denotes a suitable n-th cohomology group. It is important that the cohomology is „suitable": For general considerations, a type of Čech-cohomology is useful. If one is only interested in G-complexes, then singular cohomology will work too. It also matters which coefficient group is chosen. Our procedure will be partly axiomatic in that we point out which properties of the theory $H^*(-)$ we actually need.

We now collect some properties of the functors $H^n_G(-)$. Any such property will be a consequence of properties of ordinary cohomology (with coefficients unspecified for the moment).

(1.3) A G-map $f: X \to Y$ induces $f_G: X_G \to Y_G$ and $(f_G)^* = f^*: H^n_G(Y) \to H^n_G X)$. If f and g are G-homotopic, then f_G and g_G are homotopic over BG and therefore $f^*_G = g^*_G$. Hence $H^n_G(-)$ is a G-**homotopy-invariant** contravariant **functor**. This uses homotopy-invariance of $H^n(-)$.

If A is a G-subspace of X, then A_G is a subspace of X_G; see I (3.25, Ex. 12). We set $H^n_G(X, A) = H^n(X_G, A_G)$ and obtain the exact cohomology sequence

(1.4) $$\ldots \to H^n_G(X, A) \to H^n_G(X) \to H^n_G(A) \to H^{n+1}_G(X, A) \to \ldots.$$

For some theories $H^n(-)$, one has to assume that A_G is closed in X_G; see e.g. Massey [1978], p. 230.

We now assume that the theory $H^*(-)$ is multiplicative. This means that we are given product pairings (cup-product)

$$H^m(Y, Y_1) \otimes H^n(Y, Y_2) \to H^{m+n}(Y, Y_1 \cup Y_2)$$

with the usual properties: natural, associative, commutative, compatible with the boundary operator. In general, one has to add certain assumptions on (Y, Y_1, Y_2) in order to ensure that such pairings are defined (excisive triads; see Massey [1978], p. 253; Dold [1972], p. 214). The usual properties of products are listed in Dold [1972], p. 220–221, Switzer [1975], Ch. 13.

If (X, A, B) is a suitable triad of G-spaces, the products in $H^*(-)$ yield a product pairing

(1.5) $$H^m_G(X, A) \otimes H^n_G(X, B) \to H^{m+n}_G(X, A \cup B).$$

There is no problem in having these products if X is a G-complex and A and B are subcomplexes. In particular, $H^*_G(X, A)$ becomes a graded algebra (with unit if A is empty).

Most important for us is the fact that $H^*_G(X, A)$ is a graded module over $H^*(BG)$ in a canonical way. The module structure is defined as follows: For $x \in H^*(BG)$ and $y \in H^*_G(X, A)$ consider $p^*_X(x) \in H^*_G(X)$ and form the product $p^*_X(x)\, y$ in the sense explained above. This defines xy. The significance of this module structure stems from the fact that the ring $H^*(BG)$ is usually quite large and therefore has a rich module structure, so that methods from commutative algebra and algebraic geometry can be applied to analyse these module structures. The idea is that certain algebraic properties of the modules reflect geometric properties of the G-action. In the sequel, we shall see some elementary examples of this phenomenon.

The first use of the $H^*(BG)$-modules $H^*_G(X, A)$ was made in Borel [1960], Ch. IV. For this reason, the functors $H^*_G(X, A)$ are sometimes called Borel-cohomology-groups. Many workers have since contributed to the theory and I suggest to call it **equivariant bundle cohomology**.

For expositions of various aspects of the theory, see Swan [1960b], Bredon [1972], Ch. VII, Quillen [1971], W. Y. Hsiang [1975], Atiyah-Bott [1984], K. S. Brown [1982], Ch. VII. We only give an elementary introduction to the subject.

The reader may have noticed that instead of ordinary cohomology any generalized cohomology theory h^* can be used to define functors $X \mapsto h^*(X_G)$. Most of the methods and results to follow apply to this more general situation, cum grano salis. In this respect, useful theories are K-theories or cobordism theories.

We list further properties of the cohomology theory. Verifications are left as exercise.

Let K be a subgroup of G. Then there is a restriction homomorphism

(1.6) $\quad \operatorname{res}_K^G \colon H_G^*(X, A) \to H_K^*(X, A)$

defined as follows. We have a commutative diagram

$$\begin{array}{ccc} EG \times_G X & \xleftarrow{\quad f \quad} & EG \times_K X \\ \downarrow & & \downarrow \\ BG = EG/G & \xleftarrow[\quad Bi \quad]{} & EG/K = BK. \end{array}$$

The space $\operatorname{res}_K^G EG$ is a model for EK and therefore $(EG)/K$ is a model for BK and Bi is a model for the classifying map induced by the inclusion $i \colon K \subset G$; see I.8. We define res_K^G in (1.6) as $H^*(f)$. We also have $(Bi)^* = \operatorname{res}_K^G \colon H^*(BG) \to H^*(BK)$ and the module structures are related by

(1.7) $\quad \operatorname{res}_K^G(xy) = \operatorname{res}_K^G(x) \operatorname{res}_K^G(y)$

for $x \in H^*(BG)$ and $y \in H_G^*(X, A)$. For $L \subset K \subset G$, we have

(1.8) $\quad \operatorname{res}_L^G = \operatorname{res}_L^K \operatorname{res}_K^G$.

If X is a K-space, the canonical homeomorphism $EG \times_G (G \times_K X) \cong EG \times_K X$ yields an isomorphism

(1.9) $\quad r_K^G \colon H_G^*(G \times_K X) \xrightarrow{\cong} H_K^*(X)$

which satisfies $r_K^G(xy) = \operatorname{res}_K^G x \, r_K^G y$ for $x \in H^*(BG)$ and $y \in H_G^*(G \times_K X)$. For a G-space X, using the canonical G-homeomorphism $G \times_K X \cong G/K \times X$, $(g, x) \mapsto (g, gx)$, we have the composition

$$H_G^*(X) \xrightarrow[\mathrm{pr}^*]{} H_G^*(G/K \times X) \cong H_G^*(G \times_K X) \xrightarrow[r_K^G]{} H_K^*(X)$$

which coincides with res_K^G.

Let V be a unitary representation of G of complex dimension n with unit ball (resp. unit sphere) DV (resp. SV). Then there is a suspension isomorphism

(1.10) $H_G^i(X) \xrightarrow{\cong} H_G^{i+2n}(DV \times X, SV \times X)$

which, by definition, is the Thom isomorphism for the vector bundle $EG \times_G (V \times X) \to EG \times_G X$. (The complex structure on V induces a canonical orientation and Thom class for this vector bundle.)

If X is a numerable free G-space, the homotopy equivalence $EG \times_G X \to X/G$ induces a natural isomorphism

(1.11) $H_G^*(X) \cong H^*(X/G)$.

The $H^*(BG)$-module structure on the left corresponds to the $H^*(BG)$-module structure on $H^*(X/G)$ induced by the classifying map $X/G \to BG$ of $X \to X/G$.

Note that we have verified that H_G^* is a stable equivariant cohomology theory, graded over the integers.

As we have already said, our aim is to study pairs of G-spaces (X, A) using the $H^*(BG)$-module structure of the cohomology $H_G^*(X, A)$. Therefore, we need methods for the computation of this cohomology. The starting point is the fibration $X \to EG \times_G X \to BG$. Algebraic topology provides general methods for the computation of cohomology for fibrations, e.g. spectral sequences. For simplicity, let us assume that H^* is singular cohomology and (X, A) a pair of G-complexes.

We begin with a simple case. If G acts trivially on X, then $X_G = BG \times X$ and we can compute $H^*(X_G)$ using the Künneth theorem for cohomology. If we use a field F as coefficient group, we obtain in this case

$$H_G^* X) \cong H^*(BG) \otimes_F H^*(X),$$

an isomorphism of $H^*(BG)$-algebras.

Let us now look at cohomology with coefficients in a principal ideal domain R. For $b \in EG$, the map $j_b \colon X \to EG \times_G X$, $x \mapsto (b, x)$, is the inclusion of a typical fibre. Since EG is path-connected, any two of these j_b are homotopic. Therefore, we use the notation $j^* \colon H_G^*(X, A) \to H^*(X, A)$ for the induced map.

A **cohomology extension of the fibre** is an R-module homomorphism of degree zero

(1.12) $t \colon H^*(X, A) \to H_G^*(X, A)$

such that the composition $j^* t$ is the identity.

Given a cohomology extension t, we define a homomorphism

(1.13) $\phi_t \colon H^*(BG) \otimes_R H^*(X, A) \to H_G^*(X, A), u \otimes v \mapsto p^*(u) \cup t(v)$.

The following result is known as the Leray-Hirsch theorem (for more details, compare Spanier [1966], p. 258, and Dold [1970]).

(1.14) Proposition. *Suppose $H^*(X, A)$ is a finitely generated free R-module. Then ϕ_t is an isomorphism of graded R-modules.*

Proof. Let $U \subset EG$ be any G-subspace. We compose t with the map induced by the inclusion $H^*(EG \times_G (X, A)) \to H^*(U \times_G (X, A))$ and call the result t again. A definition like (1.13) gives a homomorphism

$$\phi_t\colon H^*(U/G) \otimes_R H^*(X, A) \to H^*(U \times_G (X, A))$$

which is natural in U. Suppose U is a finite G-complex such that the principal bundle $U \to U/G$ is trivial. We want to show that ϕ_t is an isomorphism in this case. By naturality, it suffices to consider path-connected U/G.

Use a trivialisation $U \cong U/G \times G$ in order to write $H^*(U \times_G (X, A)) \cong H^*(U/G \times (X, A))$ and use the Künneth isomorphism $H^*(U/G) \otimes H^*(X, A) \cong H^*(U/G \times (X, A))$. These isomorphisms transform t into a homomorphism $t'\colon H^*(X, A) \to H^*(U/G) \otimes H^*(X, A)$ and j into the map $u \otimes x \mapsto \varepsilon(u)x$ with $\varepsilon\colon H^*(U/G) \to R$ the agumentation. Let $(x_j | j \in J)$ be an R-basis of $H^*(X, A)$ and suppose that $t'(x_i) = \sum a_{ij} \otimes x_j$. Since $j^* t = \mathrm{id}$, we must have $\varepsilon(a_{ij}) = \delta_{ij}$, the Kronecker-$\delta$. The reader should convince himself that these facts imply that the $t'(x_i)$ form an $H^*(U/G)$-basis of $H^*(U/G) \otimes H^*(X/A)$. The map ϕ_t therefore sends a $H^*(U/G)$-basis into another such basis.

One can choose a model for EG which is a G-complex with finite n-skeleta EG_n. Since $H^k(E \times_G (X, A)) \cong H^k(EG_n \times_G (X, A))$ for large n, it suffices to show that ϕ_t is an isomorphism for $U = EG_n$. In this case, the bundle $U \to U/G$ is trivial over sets of a finite covering $V_1, \ldots, V_r$ of U/G by subcomplexes and one shows ϕ_t to be an isomorphism over $V_1 \cup \ldots \cup V_i$ by induction on i using Mayer-Vietoris sequences. Observe that $\otimes_R H^*(X, A)$ preserves exact sequences and therefore ϕ_t induces a morphism between Mayer-Vietoris sequences. □

We call (X, A) **totally nonhomologous to zero** in (X_G, A_G) with respect to $H^*(-)$ if

$$j^*\colon H_G^*(X, A) \to H^*(X, A)$$

is surjective. Since a surjective map onto a free module splits, we have

(1.15) If (X, A) is totally nonhomologous to zero and $H^*(X, A)$ is a free R-module, then a cohomology extension of the fibre exists.

If $l_g\colon X \to X$ denotes left translation by $g \in G$, then $j_b l_g = j_{bg}$. Therefore, the image of j^* is contained in the subgroup $H^*(X, A)^G$ of elements invariant under the G-action. Hence we conclude:

(1.16) If (X, A) is totally nonhomologous to zero in (X_G, A_G), then G acts trivially on $H^*(X, A)$.

We want to find conditions which ensure that (X, A) is totally nonhomologous to zero in (X_G, A_G). For this purpose, we look at the Leray-Serre spectral sequence of the fibration $p: X_G \to BG$. The spectral sequence has E_2-term

$$E_2^{p,q} = H^p(BG; H^q(X, A)).$$

Of course, it is necessary to use local coefficients; the G-action on (X, A) induces a G-action on $H^q(X, A)$ and hence a $\pi_0 G$-action on this cohomology group. Via the isomorphism $\pi_0 G \cong \pi_1 BG$ this corresponds to a local coefficient system on BG (see G.W. Whitehead [1978], VI. 1.12). For finite groups G, one can view this as the algebraic cohomology of G with coefficients in the G-module $H^q(X, A)$.

(1.17) Proposition. *(X, A) is totally nonhomologous to zero in (X_G, A_G) if and only if G acts trivially on $H^*(X, A)$ and $E_2^{0,*}$ consists of permanent cocycles (the latter meaning $E_\infty^{0,q} = E_2^{0,q}$).*

Proof. We know from (1.16) that the local system $H^q(X, A)$ is trivial if (X, A) is totally nonhomologous to zero in (X_G, A_G). The spectral sequence has associated to it the edge homomorphism

$$H^q(X_G, A_G) \to E_\infty^{0,q} \to E_2^{0,q} \to H^q(X, A).$$

Using $E_2^{0,q} = H^0(BG; H^q(X, A)) \cong H^q(X, A)^G$, the map $E_2^{0,q} \to H^q(X, A)$ becomes the inclusion $H^q(X, A)^G \subset H^q(X, A)$. The edge homomorphism is the map j^* defined above. (See G.W. Whitehead [1978], XIII. 7.6, p. 650). From these remarks the proof is finished easily. □

If G acts trivially on $H^*(X, A)$ and $H^*(X, A)$ is a finitely-generated, free R-module, then, by the universal coefficient theorem,

$$E_2^{p,q} \cong E_2^{p,0} \otimes E_2^{0,q};$$

and the multiplicative properties of spectral sequences imply that the spectral sequence degenerates (i.e. $E_2 = E_\infty$) if and only if $E_2^{0,*} = E_\infty^{0,*}$. Taking into account (1.15), we have

(1.18) Proposition. *Suppose $H^*(X, A)$ is a finitely-generated, free R-module. Then the following assertions* (i) *and* (ii) *are equivalent:*

(i) *(X, A) is totally nonhomologous to zero in (X_G, A_G).*
(ii) *G acts trivially on $H^*(X, A)$ and the Leray-Serre spectral sequence associated to $(X_G, X_G) \to BG$ degenerates.*
(iii) *If* (i) *and* (ii) *hold, then $H_G^*(X, A)$ is a free $H^*(BG)$-module. Any set*

$(x_\nu \mid \nu \in J)$, $x_\nu \in H_G^*(X, A)$, such that $(j^* x_\nu \mid \nu \in J)$ is an R-basis of $H^*(X, A)$, can be taken as $H^*(BG)$-basis. □

We look at an example. Suppose $H^*(X, A)$ is zero in odd degrees; then the spectral sequence degenerates for formal reasons. Suppose, moreover, that G is connected; then G acts trivially on $H^*(X, A)$. Finally, if $H^*(X, A)$ is a free R-module of finite type, then $H_G^*(X, A)$ is a free $H^*(BG)$-module.

Suppose X has the cohomology of complex projective space, say $H^*(X; \mathbb{Z}) \cong H^*(\mathbb{C}P^n; \mathbb{Z})$, and let G be connected. Then the previous considerations can be applied: Choose $y \in H_G^2(X)$ such that $j^* y$ is a generator of $H^2(X)$. Then $1, y, y^2, \ldots, y^n$ form an $H^*(BG)$-basis of $H_G^*(X)$. Of course, in general the multiplicative structure of $H_G^*(X)$ is different from $H^*(BG) \otimes_R H^*(X)$ as a tensor product of algebras. In fact, this difference of multiplicative structures reflects the G-action on X.

(1.19) Exercises.

1. Obtain (1.10) as special case of (1.14).
2. Let G be a compact Lie group. Show that there exists a model for EG which is a G-complex with finite skeleta.
3. Give a detailed proof of (1.18).

2. Cohomology of some classifying spaces.

Since our aim is the investigation of $H_G^*(X, A)$ as a module over $H^*(BG)$, we need some information about the latter groups. For finite G, the ring $H^*(BG)$ coincides with the algebraic cohomology of G; see Mac Lane [1963], IV. 11.

We begin with the circle group $G = S^1$ and cohomology with integer coefficients.

(2.1) Proposition. *$H^*(BS^1) \cong \mathbb{Z}[c]$ with generator $c \in H^2(BS^1)$.*

Proof. This is one of the first calculations of a cohomology ring in textbooks of algebraic topology. One uses the fact that BS^1 is the infinite complex projective space $P(\mathbb{C}^\infty)$. For a computation, see Milnor-Stasheff [1947], § 14; Spanier [1966], p. 265; Dold [1972], p. 222. Note that $H^*(P(\mathbb{C}^\infty)) \cong \operatorname{inv\,lim}_n H^*(P(\mathbb{C}^n))$. The generator c (suitably normalized) is called the **first Chern class** of the universal complex line bundle over $P(\mathbb{C}^\infty)$. □

One explanation for (2.1) is given by using the **Gysin sequence** of a vector bundle. We recall it because later we have to use it several times.

Let $p: E \to B$ be an orientable n-dimensional vector bundle. We assume that E

carries a Riemannian metric, so we can talk about the unit disk bundle $D(E)$ and the unit sphere bundle $S(E)$. We consider the exact cohomology sequence of the pair (DE, SE):

$$\ldots \to H^*(DE, SE) \to H^*(DE) \to H^*(SE) \to \ldots .$$

Since the projection $DE \to B$ is a homotopy equivalence (contractible fibre), we have

$$H^*(DE) \cong H^*(B).$$

We orient E by choosing a generator $t(E) \in H^n(DE, SE) \cong \mathbb{Z}$, a so called **Thom class** for E. The **Thom isomorphism theorem** for vector bundles says that multiplication with $t(E)$ induces an isomorphism of degree n

$$H^*(B) \cong H^{*+n}(DE, SE).$$

(This is a generalized suspension isomorphism. See Bröcker-tom Dieck [1970], p. 88 for a proof of the Thom isomorphism theorem from this point of view.) Using these isomorphisms in the exact sequence above, we obtain the **Gysin sequence** for E

$$\ldots \to H^*(B) \xrightarrow[e(E)]{} H^{*+n}(B) \xrightarrow[p^*]{} H^{*+n}(SE) \to \ldots .$$

Here, $p\colon SE \to B$ is the projection and $e(E)$ denotes multiplication (from the left, say) with the **Euler class** $e(E) \in H^n(B)$ which is, by definition, the image of the Thom class $t(E)$ under the mapping

$$H^n(DE, SE) \to H^n(DE) \cong H^n(B).$$

We return to the bundle $p\colon ES^1 \to BS^1$ and note that ES^1 can be identified with the sphere bundle in the universal line bundle $ES^1 \times_{S^1} \mathbb{C} \to BS^1$. We obtain a Gysin sequence

$$\ldots \to H^i(BS^1) \to H^{i+2}(BS^1) \to H^{i+2}(ES^1) \to \ldots .$$

In this case the Euler class $e \in H^2(BS^1)$ can be taken as the first Chern class $c \in H^2(BS^1)$ of the universal bundle (or vice versa).

Since ES^1 is contractible, multiplication by c induces an isomorphism

$$H^i(BS^1) \to H^{i+2}(BS^1), \quad i \geqq 0;$$

together with $H^0(BS^1) \cong \mathbb{Z}$ we obtain (2.1).

The n-fold product $T(n) = S^1 \times \ldots \times S^1$ is called the ***n*-dimensional torus**. The projections $\mathrm{pr}_i\colon T(n) \to S^1$ onto the i-th factor induce

$$B_i = B(\mathrm{pr}_i)\colon BT(n) \to BS^1$$

and these induce

$$\begin{aligned} H^*(BS^1) \otimes \ldots \otimes H^*(BS^1) &\to H^*(BT(n)) \\ x_1 \otimes \ldots \otimes x_n &\mapsto (B_1^* x_1) \cdot \ldots \cdot (B_n^* x_n). \end{aligned}$$

The Künneth theorem for products shows that this map is an isomorphism. Hence we obtain

(2.2) Proposition. *$H^*(BT(n)) \cong \mathbb{Z}[t_1, \ldots, t_n]$ with $t_i = B_i^* c \in H^2(BT(n))$.* □

We now consider cyclic groups $G = \mathbb{Z}/m$ of order m. We identify G with the subgroup of S^1 consisting of all m-th roots of unity.

Starting with the universal bundle $p\colon ES^1 \to BS^1$, we can use the space ES^1/G as model for BG because $ES^1 \to ES^1/G$ is a numerable principal G-bundle with contractible total space. Hence from p we obtain a map

$$q\colon BG \to BS^1$$

which has fibres isomorphic to $S^1/G \cong S^1$. Therefore, q is again a sphere bundle. We now want to use the Gysin sequence. For this purpose we identify BG with the unit sphere bundle of a suitable vector bundle.

Let γ be the standard bundle $ES^1 \times_{S^1} \mathbb{C} \to BS^1$ and γ^m the m-fold tensor product $\gamma \otimes \ldots \otimes \gamma$.

(2.3) Lemma. *$S(\gamma^m) \to BS^1$ is canonically isomorphic to $q\colon BG \to BS^1$.*

Proof. The total space of γ^m is isomorphic to $ES^1 \times_{S^1} V^m$ where V^m is the representation $S^1 \times \mathbb{C} \to \mathbb{C}$, $(\lambda, z) \mapsto \lambda^m z$. Hence $S(\gamma^m) = ES^1 \times_{S^1} S(V^m)$ and the latter space is $(ES^1 \times S^1)/\sim$ with equivalence relation $(e\lambda, z) \sim (e, \lambda^m z)$. The mapping $S^1 \to S(V^m)$, $z \mapsto z^m$ induces an S^1-homeomorphism $S^1/G \to S(V^m)$. Therefore, we have a homeomorphism $ES^1 \times_{S^1} S^1/G \to ES^1 \times_{S^1} S(V^m)$ over BS^1. Moreover,

$$ES^1 \times_{S^1} S^1/G \to ES^1/G,\ (e, z) \mapsto ez^{-1}$$

is a homeomorphism over BS^1. This completes the proof of the lemma. □

We now apply the Gysin-sequence to the bundle $S(\gamma^m) \to BS^1$. With suitable generator, the Euler class of γ is $e(\gamma) = c \in H^2(BS^1; \mathbb{Z})$. Therefore, $e(\gamma^m) = mc$. (This can be interpreted as follows: BS^1 is an Eilenberg-Mac Lane space $K(\mathbb{Z}, 2)$. Hence $H^2(X; \mathbb{Z}) = [X, BS^1]$ and complex line bundles are classified by their Euler class. By means of the tensor product of line bundles, the set of isomorphism classes of line bundles can be made into an abelian group. This group structure coincides with the cohomological group structure in $H^2(X; \mathbb{Z})$.)

The Gysin-sequence has the following form

$$\ldots \to H^{i-1}(S\gamma^m) \to H^{i-2}(BS^1) \xrightarrow{mc} H^i(BS^1) \to \ldots.$$

We can now read off the cohomology groups

$$H^i(B\mathbb{Z}/m;\mathbb{Z}) = H^i(S\gamma^m;\mathbb{Z}) = \begin{cases} 0 & i \equiv 1 \bmod 2 \\ \mathbb{Z}/m & i \equiv 0 \bmod 2,\ i \neq 0 \\ \mathbb{Z} & i = 0 \end{cases}$$

The mapping

$$H^{2i}(BS^1;\mathbb{Z}) \to H^{2i}(S\gamma^m;\mathbb{Z})$$

induced by the projection $q\colon S\gamma^m \to BS^1$ is surjective. The image of c^i will be denoted by t^i. This describes also the multiplicative structure of $H^*(B\mathbb{Z}/m;\mathbb{Z})$.

We now consider more closely the case $m = p$, p prime, and look at cohomology

$$H^*(B\mathbb{Z}/p;\mathbb{Z}/p)$$

with coefficients in $\mathbb{Z}/p$. Again we have a Gysin sequence. But in this case, the Euler class pc is zero and therefore the Gysin sequence splits into exact sequences

$$0 \to H^i(BS^1;\mathbb{Z}/p) \to H^i(S\gamma^m) \to H^{i-1}(BS^1;\mathbb{Z}/p) \to 0.$$

Using

$$H^i(BS^1;\mathbb{Z}/p) = \begin{cases} 0 & i \text{ odd} \\ \mathbb{Z}/p & i \text{ even} \end{cases},$$

we read off

$$H^i(B\mathbb{Z}/p;\mathbb{Z}/p) \cong \mathbb{Z}/p.$$

We want to obtain the ring structure. In even degrees we have an isomorphism

$$q^*\colon H^{2i}(BS^1;\mathbb{Z}/p) \to H^{2i}(B\mathbb{Z}/p;\mathbb{Z}/p)$$

so that $H^{2*}(B\mathbb{Z}/p;\mathbb{Z}/p)$ is seen to be a polynomial ring with generator $t \in H^2(B\mathbb{Z}/p;\mathbb{Z}/p)$. Note that t is the image of the mod p reduction of $c \in H^2(BS^1)$.

If $p \neq 2$, an element $s \in H^1$ satisfies $s^2 = (-1)^1 s^2$ so that s^2 must be zero.

(2.4) Lemma. *Multiplication by t induces an isomorphism*

$$H^i(B\mathbb{Z}/p;\mathbb{Z}/p) \to H^{i+2}(B\mathbb{Z}/p;\mathbb{Z}/p), \quad a \mapsto a \cup t.$$

Proof. We know this if i is even. For i odd, we use the definition of the relevant elements and consider the following diagram with $S = S\gamma^m$, $B = B\mathbb{Z}/p$.

$$\begin{array}{ccc}
H^2(S)\otimes H^{2i+1}(S) & \xrightarrow[(1)]{} & H^{2i+3}(S) \\
\cong \downarrow \mathrm{id}\otimes\delta & & \cong \downarrow \delta \\
H^2(S)\otimes H^{2i+2}(D,S) & \xrightarrow[(2)]{} & H^{2i+4}(D,S) \\
\cong \uparrow q^*\otimes\phi & & \cong \uparrow \phi \\
H^2(B)\otimes H^{2i}(B) & \xrightarrow[(3)]{} & H^{2i+2}(B).
\end{array}$$

The horizontal maps are cup-products and ϕ is the Thom isomorphism. The upper square is commutative up to sign (product rule for the boundary operator). It suffices to verify that map (2) is an isomorphism. The two paths in the lower square are

$$t\otimes x \mapsto q^*t\otimes(q^*x\cup t(\xi)) \mapsto q^*t\cup(q^*x\cup t(\xi))$$
$$t\otimes x \mapsto t\cup x \mapsto q^*(t\cup x)\cup t(\xi),$$

where $t(\xi)$ is the Thom class of $\xi=\gamma^m$. Therefore, this square is commutative. We know that (3) is an isomorphism. □

We collect some of the results obtained so far in

(2.5) Theorem. *For $p \neq 2$,*

$$H^*(B\mathbb{Z}/p;\mathbb{Z}/p)\cong \mathbb{Z}/p[t]\otimes_{\mathbb{Z}/p}\Lambda[s],$$

where $t\in H^2$, $s\in H^1$, Λ exterior algebra over $\mathbb{Z}/p$; hence H^{2i} is generated by t^i and H^{2i+1} is generated by $t^i s$, where $s^2=0$. For $p=2$,

$$H^*(B\mathbb{Z}/2;\mathbb{Z}/2)\cong\mathbb{Z}/2[s]$$

where $s\in H^1$.

Proof. Lemma (2.4) together with the remarks preceding it and the computation of the additive structure from the Gysin sequence give the result for $p\neq 2$.

For the case $p=2$, it is better to use the following proof. The universal covering $E\mathbb{Z}/2\to B\mathbb{Z}/2$ is the sphere bundle of the universal real line bundle. This line bundle has a Thom class for $\mathbb{Z}/2$-coefficients (this is true for any vector bundle) and one can again write down a Gysin sequence. Since $E\mathbb{Z}/2$ is contractible, this sequence degenerates to the isomorphisms

$$H^i(B\mathbb{Z}/2;\mathbb{Z}/2)\to H^{i+1}(B\mathbb{Z}/2;\mathbb{Z}/2)$$

given by multiplication with the Euler class $s \in H^1$ of the universal line bundle. The result follows. □

The element s in (2.5) is not yet uniquely determined. In order to achieve this, we use the Bockstein homomorphism. Comparing the Gysin sequences for $S\gamma^p$ with coefficients $\mathbb{Z}/p^2$ and $\mathbb{Z}/p$, one sees that the reduction modulo p, $\mathbb{Z}/p^2 \to \mathbb{Z}/p$, induces an isomorphism

$$H^{2i}(B\mathbb{Z}/p; \mathbb{Z}/p^2) \to H^{2i}(B\mathbb{Z}/p; \mathbb{Z}/p), \quad i > 0.$$

Therefore the Bockstein homomorphism β associated to $0 \to \mathbb{Z}/p \to \mathbb{Z}/p^2 \to \mathbb{Z}/p \to 0$ (see Spanier [1966], p. 240) is an isomorphism

$$\beta\colon H^{2i-1}(B\mathbb{Z}/p; \mathbb{Z}/p) \to H^{2i}(B\mathbb{Z}/p; \mathbb{Z}/p), \quad i > 0.$$

Hence $s \in H^1$ may be specified uniquely by requiring $\beta(s) = t$.

We need to know the cohomology of the quaternion groups. Let $Q = Q(4k)$ be the quaternion group of order $4k$ given by generators and relations as follows

(2.6) $\quad Q = \langle x, y \mid x^k = y^2,\ xyx = y \rangle.$

In this group, $x^{2k} = e = y^4$ and each element can be uniquely written in the form $x^m y^n$, $0 \leqq m < 2k$, $0 \leqq n < 2$. The element x generates a normal cyclic subgroup of order $2k$. The group Q may be regarded as a subgroup of the group S^3 of quaternions of norm 1 by setting

$$x \mapsto e^{2\pi i/k}, \quad y \mapsto j.$$

(This inclusion $Q \subset S^3 \subset \mathbb{H}$ defines a faithful irreducible representation of Q.)

The left translation action of Q on S^3 is free. Therefore, the projection $X := EQ \times_Q S^3 \to S^3/Q$ is a homotopy equivalence. In the Gysin sequence of the orientable S^3-bundle $X \to BQ$,

$$\dots \to H^i(BQ) \xrightarrow{\ e\ } H^{i+4}(BQ) \to H^{i+4}(X) \to \dots,$$

one has $H^{i+4}(X) = 0$ for $i \geqq 0$ because X is homotopy equivalent to a 3-dimensional space. Therefore multiplication by the Euler class e is an isomorphism

$$H^i(BG) \to H^{i+4}(BG), \quad i > 0$$

(for any coefficient group with trivial Q-action).

In order to determine $H^4(BQ; \mathbb{Z})$, we compare with the Gysin sequence of the trivial sphere bundle $EQ \times S^3 \to EQ$ and obtain a commutative diagram

$$\begin{array}{ccccccc}
H^3(S^3/Q) \cong H^3(EQ \times_Q S^3) & \longrightarrow & H^0(BQ) & \longrightarrow & H^4(BQ) & \longrightarrow & 0 \\
\downarrow p^* \qquad\qquad \downarrow & & \downarrow \cong & & & & \\
H^3(S^3) \cong H^3(EQ \times S^3) & \xrightarrow[(1)]{\cong} & H^0(EG). & & & &
\end{array}$$

The map (1) is an isomorphism because the Euler class of a trivial bundle is zero. The image of p^* consists of the multiples of $|Q|$ because p is a map of degree $|Q|$ between orientable 3-manifolds. Thus we obtain from the diagram that $H^4(BQ; \mathbb{Z}) \cong \mathbb{Z}/|Q|$.

An algebraic computation of $H^*(BQ; A)$ is given in Cartan-Eilenberg [1956], p. 254.

(2.7) Exercises.

1. Let M be an abelian group. Let $t \in H^2(B\mathbb{Z}/m; \mathbb{Z})$ be the reduction of the first Chern class $c \in H^2(BS^1; \mathbb{Z})$. Show that multiplication by t induces an isomorphism $H^k(B\mathbb{Z}/m; M) \cong H^{k+2}(B\mathbb{Z}/m; M)$ for $k > 0$.
2. Let $S(\mathbb{C}^n) = S^{2n-1}$ be the unit sphere with action of $G = \mathbb{Z}/m \subset S^1 \subset \mathbb{C}$ given by scalar multiplication. Compute $H^*(S(\mathbb{C}^n)/G)$. Show that the direct limit of $S(\mathbb{C}^n) \subset S(\mathbb{C}^{n+1}) \subset \ldots$ with this action is EG. Show that for $\mathbb{Z}/p$-coefficients and $G = \mathbb{Z}/p$, the inclusion $S(\mathbb{C}^n) \to EG$ induces a surjection in cohomology.
3. Let S^3 be the group of quaternions of norm 1. The bundle $ES^3 \to BS^3$ is the unit sphere bundle of the universal one-dimensional quaternionic vector bundle $ES^3 \times_{S^3} \mathbb{H} \to BS^3$. Let $d \in H^4(BS^3; \mathbb{Z})$ be its Euler class. Use the Gysin sequence to obtain an isomorphism $H^*(BS^3; \mathbb{Z}) \cong \mathbb{Z}[d]$.
4. Let $\mathrm{pr}_i\colon B((\mathbb{Z}/p)^n) = (B\mathbb{Z}/p)^n \to B\mathbb{Z}/p$ be the projection onto the i-th factor. Let $\mathrm{pr}_i^* t = t_i$ and $\mathrm{pr}_i^* s = s_i$. Use the Künneth formula to show that, for $p \neq 2$, $H^*((B\mathbb{Z}/p)^n; \mathbb{Z}/p) \cong \mathbb{Z}/p[t_1, \ldots, t_n] \otimes_{\mathbb{Z}/p} \Lambda[s_1, \ldots, s_n]$ where $\Lambda[s_1, \ldots, s_n]$ denotes the (graded) exterior algebra over $\mathbb{Z}/p$ with generators $s_1, \ldots, s_n$ in degree 1. Similarly, $H^*((B\mathbb{Z}/2)^n; \mathbb{Z}/2) \cong \mathbb{Z}/2[s_1, \ldots, s_n]$.
5. Use the model of exercise 2 in order to obtain $E\mathbb{Z}/m$ as a CW-complex whose cellular chain complex is the standard periodic complex

$$\mathbb{Z}G \xleftarrow[T]{} \mathbb{Z}G \xleftarrow[N]{} \mathbb{Z}G \xleftarrow[T]{} \ldots;$$

   here, $G = \mathbb{Z}/m$ is generated by g and T resp. N is multiplication by $1-g$ resp. $1+g+g^2+\ldots+g^{m-1}$. Use this to show that, for a $\mathbb{Z}G$-module L, $H^i(BG; L)$ is $\ker T/\operatorname{im} N$ for $i > 0$ even and $\ker N/\operatorname{im} T$ for $i > 0$ odd.

3. Localization.

We consider a multiplicative cohomology theory $H^*(-)$. Let $S \subset H^*(BG)$ be a **multiplicatively closed subset** of homogeneous elements, i.e. $1 \in S$ and $a, b \in S \Rightarrow ab \in S$. For simplicity, we assume that S is contained in the center of $H^*(BG)$, i.e.

$$s \in S,\ b \in H^*(BG) \Rightarrow sb = bs.$$

This commutation property is satisfied if S contains only elements of even dimension or if $H^*(-)$ is a theory with $\mathbb{Z}/2$-coefficients.

We look at $\mathbb{Z}$-graded $H^*(BG)$-modules M and consider the localization of M with respect to S, denoted by $S^{-1}M$. This is obtained from M by formally inverting the left multiplication by elements of S. The module $S^{-1}M$ is constructed as follows: Consider the product $S \times M$ and on it the equivalence relation

$$(s_1, m_1) \sim (s_2, m_2) \Leftrightarrow \text{there exists } t \in S \text{ such that } ts_2 m_1 = ts_1 m_2.$$

The equivalence class of (s, m) is denoted by $\dfrac{m}{s}$ or m/s. Addition is given by the usual rules for calculus of fractions. We regard $S^{-1}M$ as $S^{-1}H^*(BG)$-module which can also be obtained as

$$S^{-1}M \cong S^{-1}H^*(BG) \otimes_{H^*(BG)} M.$$

Fundamental properties of localization are

(3.1) Proposition.

(i) *If $M \to N \to P$ is an exact sequence of $H^*(BG)$-modules, then $S^{-1}M \to S^{-1}N \to S^{-1}P$ is an exact sequence of $S^{-1}H^*(BG)$-modules.*
(ii) *The kernel of the canonical map $M \to S^{-1}M$, $m \mapsto m/1$ consists of those $m \in M$ which are annihilated by some element of S.*
(iii) *Localization commutes with colimits.* □

For these and other properties of localization, we refer the reader to Bourbaki [1961 b]; Atiyah-Mac Donald [1969], Ch. III; Lang [1965], II. 3. We point out that in forming m/s we only use homogeneous elements m and s but this does not effect the arguments in an essential way.

What happens to the grading in $S^{-1}M$? To $(s, m) \in S \times M$ we assign the degree $|m| - |s|$, $|m| =$ degree m. This is compatible with the equivalence relation and induces a grading on $S^{-1}M$. But if S contains elements s of positive degree $d = |s|$, then multiplication by s induces an isomorphism of $(S^{-1}M)^i$ with $(S^{-1}M)^{i+d}$ so that, essentially, we have only to look at degrees modulo d.

The following family $\mathfrak{F}(S)$ of closed subgroups of G will be important:

$$\mathfrak{F}(S) = \{H \mid S \cap \operatorname{kernel}(H_G^*(G/G) \to H_G^*(G/H)) \neq \emptyset\}.$$

Here, of course, $H_G^*(G/G) = H^*(BG)$ and $G/H \to G/G$ induces a homomorphism $H_G^*(G/G) \to H_G^*(G/H)$. The family $\mathfrak{F}(S)$ is open in the sense of I.6.

For the purpose of the localization theorems to be proved shortly, we use the following

(3.2) Definition. A G-space X is of **finite S-type** if the following holds: There exists a numerable, finite-dimensional G-covering $(U_\alpha | \alpha \in A)$ of X, a finite number of subgroups $H_1, \ldots, H_r$ in $\mathfrak{F}(S)$, and G-maps $f_\alpha: U_\alpha \to G/H_{n(\alpha)}$, $n(\alpha) \in \{1, \ldots, r\}$.

If X is of finite S-type and $Y \subset X$, then Y is of finite S-type, too.

The significance of these notions will become clear through the following theorem and its proof.

(3.3) Theorem. *Let X be of finite S-type. Then*

$$S^{-1} H_G^*(X) = 0.$$

In the proof of this theorem we use an elementary lemma which, for instance, holds for singular cohomology.

(3.4) Lemma. *Suppose $A_1, \ldots, A_r$ is an open covering of X by G-sets. Suppose we are given elements $x_i \in H_G^*(X)$ whose restriction to A_i is zero. Then the product $x_1 x_2 \ldots x_r$ is zero.*

Proof. There exist elements $y_i \in H_G^*(X, A_i)$ which map to x_i (exactness). Since the A_i are open, the product of the y_i is defined and contained in $H_G^*(X, \cup A_i)$. But the latter group is zero. □

Proof of (3.3). We take

$$A_i = \cup \{U_\alpha | n(\alpha) = i\}.$$

Then $A_1, \ldots, A_r$ is an open G-covering of X. We show that for each i there exists $s_i \in S$ with image in $H_G^*(A_i)$ being zero. Then $s = s_1 s_2 \ldots s_r$ is zero in $H_G^*(X)$. Hence each element in $H_G^*(X)$ is annihilated by s, which implies, by (3.1), that $S^{-1} H_G^*(X) = 0$. The covering $(U_\alpha | \alpha \in n^{-1}(i))$ of A_i is finite-dimensional and also numerable. Hence it suffices to consider the case $r = 1$. We set $H = H_1$ in this case. There exists a covering $V_0, \ldots, V_n$ of X with the following property: Eeach V_i is a disjoint union of open G-sets which are contained in at least one of the U_α; compare I (6.12). In particular, each V_i has a G-map $h_i: V_i \to G/H$. Since $H \in \mathfrak{F}(S)$, we can choose an element $s \in S$ in the kernel of $H_G^*(G/G) \to H_G^*(G/H)$. This element s is then contained in the kernel of

$$H_G^*(G/G) \to H_G^*(G/H) \to H_G^*(V_i)$$

where the second map is h_i^*. Hence $H_G^*(V_i)$ is annihilated by s. Again using (3.1), we see that $H_G^*(X)$ is annihilated by s^{n+1}. □

We now derive a version of Theorem (3.3) for relative $\mathfrak{F}(S)$-spaces. We have to assume a certain continuity property of the theory $H_G^*(-)$ for the pair (X, A) which we put down in the next definition.

(3.5) Definition. The G-subspace A of X is **taut in X with respect to $H_G^*(-)$** if the canonical map

$$\operatorname{colim}_V H_G^*(X, V) \to H_G^*(X, A)$$

is an isomorphism. The colimit (alias direct limit) is taken over the open G-neigbourhoods V of A in X.

(3.6) Localization Theorem. *Let A be taut in X and closed. Let $X \setminus A$ be of finite S-type. Then the inclusion $A \to X$ induces an isomorphism*

$$S^{-1} H_G^*(X) \cong S^{-1} H_G^*(A).$$

Proof. Since localization preserves exactness, it suffices to show that $S^{-1} H_G^*(X, A)$ is zero; and since localization commutes with colimits, it suffices to show that $S^{-1} H_G^*(X, V)$ is zero for open G-neighbourhoods V of A. We assume the following excision isomorphism (3.7) which holds for singular cohomology if A is closed in X.

(3.7) $$S^{-1} H_G^*(X, V) \cong S^{-1} H_G^*(X \setminus A, V \setminus A).$$

But $X \setminus A$ and hence $V \setminus A$ satisfies the hypothesis of Theorem (3.3). Using this theorem and the exact cohomology sequence, we see that the right hand side of (3.7) is zero. This completes the proof. □

We are mainly interested in G-complexes and the algebraic aspects of the localization theorem. Therefore, we prove a version of the localization theorem for G-complexes.

(3.8) Localization Theorem. *Let (X, A) be a finite-dimensional relative G-complex. Suppose $X \setminus A$ has finite orbit type with orbits isomorphic to G/H, $H \in \mathfrak{F}(S)$. Then the inclusion $A \subset X$ induces an isomorphism*

$$S^{-1} H_G^*(X) \cong S^{-1} H_G^*(A).$$

Proof. By induction over the relative skeleta: Assume that X is obtained from A by attaching n-cells with characteristic map

$$\phi: \coprod G/H_j \times (D^n, S^{n-1}) \to (X, A).$$

Then

$$S^{-1}H_G^*(X, A) \cong S^{-1}H_G^*(\coprod G/H_j \times (D^n, S^{n-1})).$$

By hypothesis, there are only finitely many distinct H_j (up to conjugation) and $H_j \in \mathfrak{F}(S)$. So it sufficies to show that $S^{-1}H_G^*(G/L \times B) = 0$ for $L \in \mathfrak{F}(S)$. But if $s \in$ kernel $(H_G^*(G/G) \to H_G^*(G/L))$, then s annihilates $H_G^*(G/L \times B)$. □

There are, of course, localization theorems for pairs of spaces. For G-complexes we have to consider

$$(A_1, A_2) \to (X_1, X_2)$$

where (X_1, A_1) and (X_2, A_2) are relative finite-dimensional G-complexes with $X_i \backslash A_i$ of finite orbit type with orbits in $\mathfrak{F}(S)$. Then inclusion induces an ismorphism

$$S^{-1}H_G^*(X_1, X_2) \cong S^{-1}H_G^*(A_1, A_2).$$

Some finiteness assumptions in (3.6), (3.8) are essential. This is seen by looking at $X = EG$; then X_G is homotopy equivalent to BG.

As the reader may have noticed, the localization theorem is completely formal. Thus it holds for any equivariant cohomology theory as long as the appropriate axioms are satisfied.

Let k_G^* be an equivariant multiplicative cohomology theory. "Multiplicative" means that products

$$k_G^*(X, A) \otimes k_G^*(X, B) \to k_G^*(X, A \cup B)$$

are given. Let h_G^* be another equivariant cohomology theory which is a module over k_G^*, i.e. a pairing

$$k_G^*(X, A) \otimes h_G^*(X, A) \to h_G^*(X, A)$$

is given. These products and pairings should have the usual properties: associative, compatible with exact sequences etc.; in particular, $h_G^*(X, A)$ is a graded module over $k_G^*(X, A)$. (The grading group is not assumed to be the integers, but it should be the same for k_G^* and h_G^*; even this is not strictly necessary – if one uses pairings of grading groups.)

One localizes with respect to a multiplicatively closed subset $S \subset k_G^*$ (Point), contained in the center, and obtains such objects as $S^{-1}h_G^*(X, A)$. The family $\mathfrak{F}(S)$ is now, of course, defined using k_G^* instead of h_G^*. Then (3.6) and (3.8) still hold provided one assumes the appropriate tautness conditions. (See exercise 4.)

We describe a particular instance of the localization process in greater detail. Let $\tilde{h}_G^*$ be a stable equivariant cohomology theory in the sense of II.6. Then, among other things, we have suspension isomorphisms

$$\sigma^V : \tilde{h}_G^*(X) \to \tilde{h}_G^{*+V}(S^V X)$$

for complex representations V. The inclusion $\{0\} \subset V$ induces s^V: $\tilde{h}_G^{*+V}(S^V X) \to \tilde{h}_G^{*+V}(X)$ and the composition $s^V \sigma^V = e(V)$ is called **multiplication with the Euler class of** V. (This is a special case of the previously discussed module structure, coming from an action of stable equivariant cohomotopy on $\tilde{h}_G^*$.) We have $e(V)\,e(W) = e(V \oplus W)$ and $e(V)$ only depends on the isomorphism class of V. Given any class of representations containing $\{0\}$ and the direct sum of any two members, the set of its Euler classes may be viewed as a multiplicatively closed subset S which allows us to form $S^{-1}\tilde{h}_G^*(X)$. In this case, the localization theorem says that we can excise the set $X(S)$ of all orbits which can be mapped equivariantly into $V\backslash\{0\}$ where V is a representation in the given class. Thus if $X^S = X\backslash X(S)$, then the inclusion $X^S \to X$ induces an isomorphism

(3.9) $\quad S^{-1}\tilde{h}_G^*(X) \to S^{-1}\tilde{h}_G^*(X^S).$

In order to prove this, we reduce to a situation as in (3.3) and show

(3.10) Proposition. *Suppose X is a G-space admitting a map $u: X \to V\backslash\{0\}$. Then $e(V)$ annihilates $\tilde{h}_G^* X^+)$.*

Proof. This follows from the fact that $X^+ \to S^V X$, $x \mapsto (0, x)$ is homotopic to $x \mapsto (u(x), x)$ by linear connection in S^V and thus homotopic to the constant map $x \mapsto (\infty, x)$. □

If we use the set S of Euler classes $e(V)$ of representations without trivial direct summand, then $X^S = X^G$.

A similar localization procedure can be applied to homology theories with suspension isomorphisms $\sigma^V: \tilde{h}_*^G(X) \to \tilde{h}_{*+V}^G(S^V X)$. Then multiplication with the Euler class $e(V)$ is a map

$$\tilde{h}_*^G(X) \to \tilde{h}_*^G(S^V X) \xrightarrow{\cong} \tilde{h}_{*-V}^G(X).$$

In this case, the localization can be described as homology with families. Let $\mathfrak{F}_\infty$ be the family of all closed subgroups and $\mathfrak{F}_S$ the family of isotropy groups appearing on unit spheres $S(V)$ for V in the given class of representations (which is still assumed to be closed under direct sums). Suppose that h_*^G is additive.

(3.11) Proposition. *There exists a natural isomorphism of equivariant homology theories*

$$(S^{-1}h^G(X, A))_a \cong h_a^G[\mathfrak{F}_\infty, \mathfrak{F}_S](X, A).$$

Proof. We first show that $S^{-1}h_*^G(X, A)$ is a suitable direct limit over groups $h_*^G((DV, SV) \times (X, A))$. Choose a sequence $V_1 \subset V_2 \subset V_3 \subset \ldots$ of representations in the system such that each other representation V in the system is isomorphic to a subrepresentation of some V_i (a „cofinal sequence"). Then $(S^{-1}h^G(X))_a$, the localization in degree a, is the direct limit of the system

$$\ldots \to h^G_{a-V_i}(X) \xrightarrow[e_i]{} h^G_{a-V_{i+1}}(X) \to \ldots,$$

where e_i is multiplication with the Euler class of W_i, $V_i \oplus W_i \cong V_{i+1}$. Using the suspension isomorphism, this is transformed into the direct limit of

$$\ldots \to h^G_a((DV_i, SV_i) \times X) \to h^G_a((DV_{i+1}, SV_{i+1}) \times X) \to \ldots$$

of maps induced by the inclusions $DV_i \subset DV_{i+1}$. If we let $V_\infty = \cup V_i$ with colimit topology, we obtain, by additivity of the homology theory, that this direct limit is isomorphic to $h^G_a((DV_\infty, SV_\infty) \times X)$. Moreover, the canonical map $h^G_a(X) \to (S^{-1}h^G_*(X))_a$ corresponds to $h^G_a(X) \cong h^G_a(DV_\infty \times X) \to h^G_a((DV_\infty, SV_\infty) \times X)$. It remains to be shown that SV_∞ is a model for $E\mathfrak{F}_S$; and this is accomplished by I (6., Ex. 7) since SV^H, $H \in \mathfrak{F}_S$, are infinite-dimensional spheres and therefore contractible. (By the method of proof in I (6.14), one can also show directly that any two G-maps $SV_\infty \to SV_\infty$ are G-homotopic; moreover, one can directly construct a G-map $E\mathfrak{F}_S \to SV_\infty$.) □

(3.12) Remark. By the preceding proposition, we have embedded the natural transformation of homology theories $h^G_*(X) \to S^{-1}h^G_*(X)$ into an exact triangle, the third term being $h^G_*[\mathfrak{F}_S](X)$. An analogous embedding into an exact triangle can be constructed for the cohomology map $h^*_G(X) \to S^{-1}h^*_G(X)$. The third term is a direct limit over Gysin homomorphisms $h^*_G(S(V) \times X) \to h^{*+W}_G(S(V \oplus W) \times X)$. This is carried out in tom Dieck [1971 a].

The localization excision (3.6) and (3.8) is transformed by (3.11) into an excision isomorphism for homology with families (see exercise 5.).

Let $(p) \in \mathbb{Z}$ be a prime ideal. A **p-torus of rank** n shall be $(\mathbb{Z}/p)^n$ for $p \neq 0$ and $(S^1)^n$ for $p = 0$. We let F_p denote the prime field of characteristic p; hence $F_0 = \mathbb{Q}$.

We investigate the equivariant cohomology H^*_G for p-tori G and use F_p as coefficients. We want to apply the localization theorem and use the set $S \subset H^*(BG; F_p)$ which consists of 1 and of Euler classes of representations without trivial summand. We make this more explicit.

Suppose V is a complex representation without trivial direct summand, i.e. $V^G = \{0\}$, (for $p \neq 2$) or a real representation of the same type (for $p = 2$). Consider the associated bundle $EG \times_G V \to BG$. This is orientable for the theory $H^*(-; F_p)$ and therefore has an Euler class $e(V)$. Since

$$e(V \oplus W) = e(V)\,e(W),$$

it suffices to consider irreducible representations.

Let G be the 0-torus of rank n and suppose V is given by

$$S^1 \times \ldots \times S^1 \to S^1,\ (z_1, \ldots, z_n) \mapsto z_1^{a_1} \cdot \ldots \cdot z_n^{a_n}.$$

Then $H^*(BG) \cong F_0[t_1, \dots, t_n]$ and $e(V)$ is the element $a_1 t_1 + \dots + a_n t_n$. In general, the Euler classes are products of linear polynomials. The situation is completely analogous for $p \neq 0$; for $p = 2$, one obtains linear polynomials in dimension one; for $p \neq 2$, linear polynomials in dimension two.

(3.13) Theorem. *Let G be a p-torus. Let X be a finite-dimensional G-complex of finite orbit type. The inclusion $X^G \to X$ induces an isomorphism*

$$S^{-1} H_G^*(X) \cong S^{-1} H_G^*(X^G).$$

Proof. We apply (3.8). We know that (X, X^G) is a relative G-complex of finite orbit type. We have to show that orbits of $X \setminus X^G$ are contained in $\mathfrak{F}(S)$. This means: Let H be a subgroup different from G; then there exists $s \in S$ in the kernel of $H_G^*(G/G) \to H_G^*(G/H)$. For $p \neq 0$, we can assume without loss of generality that $H = (\mathbb{Z}/p)^k \subset (\mathbb{Z}/p)^n = G$ consist of the first k factors. Then

$$H^*(BG) = H_G^*(G/G) \to H_G^*(G/H) = H^*(BH)$$

maps $t_1, \dots, t_n$ to $t_1, \dots, t_k, 0, \dots, 0$, respectively. Obviously, there are linear polynomials in the kernel of this map. For $p = 0$, let H_0 be the component of e in H. Then, the coefficients being the rationals, the natural map $H^*(BH) \to H^*(BH_0)$ is an isomorphism (why?). One can now choose coordinates in G such that H_0 consist of the first k factors and argue as for $p \neq 0$. (One can also give a proof using (3.9).) □

A simple application of (3.13) is

(3.14) Proposition. *Let X be a finite-dimensional G-complex of finite orbit type. Then the following are equivalent:*

(i) *X has a fixed point.*
(ii) *$q: EG \times_G X \to BG$ has a section.*
(iii) *$q^*: H^*(BG) \to H_G^*(X)$ is injective.*

Proof. If $x \in X$ is a fixed point, then $EG \to EG \times X$, $e \mapsto (e, x)$ induces a section of q. If q has a section, then q^* is obviously injective. If q^* is injective, then $S^{-1} q^*$ is injective. For a p-torus G, the group $S^{-1} H^*(BG)$ is non-zero. Hence $S^{-1} H_G^*(X)$ is nonzero so that, by (3.13), there exists a fixed point. □

(3.15) Exercises.

1. A G-space is called **G-finitistic** if each covering by open G-sets has a finite-dimensional refinement by open G-sets. Show that X is finitistic if one of the following conditions is satisfied: (i) X is compact. (ii) X has finite covering dimension and G is finite.

2. Let G be a compact Lie group. Let X be a G-finitistic paracompact Hausdorff space. Assume that A is closed and taut in X and that $X \setminus A$ has finite orbit type with isotropy groups in $\mathfrak{F}(S)$. Then (3.6) holds.
3. Let G be a compact Lie group and let $A \subset X$ be an inclusion of G-spaces. Then A is taut in X with respect to H_G^* (defined via Alexander-Spanier cohomology) if one of the following conditions is satisfied:
   (i) A compact, X Hausdorff.
   (ii) A closed, X paracompact Hausdorff.
   (iii) A arbitrary, X metrizable.
   (iv) A G-retract of a G-neighbourhood of A in X.
   (See Massey [1978].)
4. With respect to the remarks after (3.8), show that the family $\mathfrak{F}(S)$ has to be defined as follows: $\mathfrak{F}(S) \ni H$ if and only if multiplication by some $s \in S$, $s \in k_G^*(\text{Point})$, annihilates $h_G^*(BH)$.
   With this definition of $\mathfrak{F}(S)$, theorems (3.6) and (3.8) hold for k_G^* and h_G^*.
5. Let X be a G-complex, $\mathfrak{F}$ an open family, and consider $X^{\mathfrak{F}} = \{x \in X \mid G_x \notin \mathfrak{F}\}$. Show that the inclusion $X^{\mathfrak{F}} \to X$ induces an isomorphism

$$h_*[\mathfrak{F}_\infty, \mathfrak{F}](X^{\mathfrak{F}}) \to h_*[\mathfrak{F}_\infty, \mathfrak{F}](X)$$

provided the canonical map

$$h_*[\mathfrak{F}_\infty, \mathfrak{F}](X^{\mathfrak{F}}) \to \operatorname{inv\,lim} h_*[\mathfrak{F}_\infty, \mathfrak{F}](U)$$

into the inverse limit over the open G-neighbourhoods U of $X^{\mathfrak{F}}$ in X is an isomorphism. Hint: Use I (6.17) and exact homology sequences as in the proof of the localization theorem. (See tom Dieck [1972], Satz 6.)

4. Applications of localization.

We consider the localization theorem for cyclic groups and p-tori in greater detail. As an application we prove important theorems of P. A. Smith and Borel about actions on disks and spheres.

Let $G = \mathbb{Z}/m$ or S^1. Unless specified otherwise, we use cohomology H^* and H_G^* with coefficients in an abelian group M. For simplicity, pairs of G-spaces (X, A) will consists of finite-dimensional G-complexes X of finite orbit type and subcomplexes A. Note that some finiteness condition is necessary for the following developments.

Let $t \in H^2(BS^1; \mathbb{Z})$ be a generator. By reduction we obtain an element $t \in H^2(B\mathbb{Z}/m; \mathbb{Z})$. Let S denote the multiplicative subset $\{1, t, t^2, \ldots\}$. Let $\mathfrak{F} = \{H \subset G \mid \tilde{H}^*(BH; M) = 0\}$ and use the notation $FX = X \setminus X(\mathfrak{F})$. If $M \neq 0$, then $S^1 \notin \mathfrak{F}$. If $H = \mathbb{Z}/k$, then

(4.1) $\quad \tilde{H}^*(BH; M) = 0 \Leftrightarrow$ multiplication by k is an isomorphism on M.

If $m = p$ is a prime and $M = F_p$ the prime field of characteristic p or if $M = F_0 = \mathbb{Q}$ and $G = S^1$, then $FX = X^G$ is the fixed point set. If $M = \mathbb{Z}$, then FX is the singular set of X.

In this case, the localization theorem reads

(4.2) Theorem. *The inclusion* $i: (FX, FA) \to (X, A)$ *induces an isomorphism*

$$S^{-1} i^* : S^{-1} H_G^*(X, A) \to S^{-1} H_G^*(FX, FA).$$

Proof. We have to show that $\mathfrak{F} = \mathfrak{F}(S)$. By (3.8) and (3.15, Ex. 4), the set $\mathfrak{F}(S)$ consists of those groups $H \subset G$ for which multiplication by some t^i annihilates $H^*(BH; M)$. But multiplication with t^i is an isomorphism $H^k(BH; M) \cong H^{k+2i}(BH; M)$ for $k > 0$ (see (2.7, Ex. 1) and also Lemma (4.7)). □

The following considerations should give the reader a better feeling for the localization theorem and its proof.

Multiplication by $t \in H^2(BG; \mathbb{Z})$ yields a map

(4.3) $$t \cup : H_G^r(X) \to H_G^{r+2}(X).$$

This map has the following interpretation. The element t is the Euler class of the standard vector bundle $EG \times_G V \to BG$ where $V = \mathbb{C}$ with $G \subset S^1$ acting by scalar multiplication. Therefore, we have a Gysin sequence

(4.4) $$H_G^{r+1}(X \times SV) \to H_G^r(X) \xrightarrow{t \cup} H_G^{r+2}(X) \to H_G^{r+2}(X \times SV).$$

The space SV is a free G-space and, consequently,

(4.5) $$H_G^i(X \times SV) \cong H^i((X \times SV)/G).$$

(4.6) Lemma. *Suppose* $H^i(X) = 0$ *for* $i > n$. *Then* $H^i((X \times SV)/G) = 0$ *for* $i > n - \dim G + 1$.

Proof. If $G = S^1$, then $(X \times SV)/G \cong X$ and there is nothing to prove. If G is finite, we use the Serre spectral sequence of the fibration $X \to (X \times SV)/G \to SV/G$. □

From (4.4)–(4.6) we obtain

(4.7) Lemma. *Suppose* $H^i(X) = 0$ *for* $i > n$. *Then* (4.3) *is an isomorphism for* $r > n - \dim G$ *and an epimorphism for* $r > n - \dim G - 1$. □

In general, localization is a colimit over multiplications. In our case, in degree r it is the colimit of

$$H^r_G(X) \xrightarrow{t\cup} H^{r+2}_G(X) \xrightarrow{t\cup} H^{r+4}_G(X) \xrightarrow{t\cup} \dots .$$

Hence we obtain from (4.7)

(4.8) Proposition. *Suppose $H^i(X) = 0$ for $i > n$. Then the canonical map*

$$H^r_G(X) \to (S^{-1}H^*_G(X))^r$$

is an isomorphism for $r > n - \dim G$ and an epimorphism for $r > n - \dim G - 1$. □

We use this information in the localization theorem and consider the diagram with $F = FX$

$$\begin{array}{ccc} H^r_G(X) & \xrightarrow{(1)} & H^r_G(F) \\ \downarrow & & \downarrow (2) \\ (S^{-1}H^*_G(X))^r & \xrightarrow{\cong} & (S^{-1}H^*_G(F))^r . \end{array}$$

It shows: (2) is surjective. Since $H^*_G(F)$ is a free $H^*(BG)$-module, (2) is always injective. Hence

(4.9) Proposition. *Suppose $H^i(X) = 0$ for $i > n$. Then the inclusion $F \to X$ induces an isomorphism $H^r_G(X) \to H^r_G(F)$ for $r > n - \dim G$ and an epimorphism for $r > n - \dim G - 1$.* □

From (4.9) we conclude that, with F instead of X, (4.3) is an isomorphism (resp. epimorphism) for $r > n - \dim G$ (resp. $r > n - \dim G - 1$). Using (4.4) for F, we then see that $H^{r+1}(F \times SV/G) = 0$ for $r > n - \dim G$ and this in turn implies $H^r(F) = 0$ for $r > n$. Thus

(4.10) Proposition. *Suppose $H^i(X) = 0$ for $i > n$. Then $H^i(F) = 0$ for $i > n$.* □

We can also look at the exact sequence of the pair (X, F) and obtain from (4.9) that, under the hypothesis of (4.9),

$$H^r_G(X, F) \cong H^r(X/G, F) = 0 \quad \text{for} \quad r > n - \dim G.$$

The preceding results still hold if we replace X by a pair (X, A). We now specialize to the following situation

(4.11) We fix a prime ideal $(p) \subset \mathbb{Z}$ and set:

$$G = \mathbb{Z}/p \quad \text{for} \quad p \neq 0$$
$$G = S^1 \quad \text{for} \quad p = 0$$
$$K = F_p \quad \text{prime field of characteristic } p.$$

We use cohomology with coefficients in $M = K$. We have seen in section 2 that

(4.12) $$H^*(BG) \cong \begin{cases} K[t] & p = 0 \\ K[s] & p = 2 \\ K[t] \otimes \Lambda[s] & p \neq 0,2 \end{cases}$$

with $t \in H^d(BG)$ and $s \in H^1(BG)$. In case $p = 2$, we have $t = s^2$.

For $p \neq 2$, we let $H^{(*)}(X, A)$ be the $\mathbb{Z}/2$-graded module (algebra) with $\bigoplus_k H^{2k}(X, A)$ resp. $\bigoplus_k H^{2k+1}(X, A)$ as the groups in degree 0 resp. 1 For $p = 2$, we let $H^{(*)}(X, A) = \bigoplus_k H^k(X, A)$ be the ungraded algebra. Of course, the same notation applies to H_G^*. Thus, for $p \neq 2$, $H_G^{(*)}(X, A)$ is a $\mathbb{Z}/2$-graded $H^{(*)}(BG)$-module. We have a homomorphism

(4.13) $$\eta: H^{(*)}(BG) \to K$$

of ($\mathbb{Z}/2$-graded) K-algebras defined by mapping $\eta(t) = 1$, $\eta(s) = 0$ for $p \neq 2$ and $\eta(s) = 1$ for $p = 2$. Here, K is concentrated in degree zero.

We use (4.13) in order to reformulate the localization theorem. Note that in all cases $\eta(t)$ is invertible so that η extends to a homomorphism

(4.14) $$\eta: S^{-1} H^{(*)}(BG) \to K.$$

Note also that $S^{-1} H^{(*)}(BG)$ is still $\mathbb{Z}/2$-graded for $p \neq 2$ and (4.14) preserves the grading. We consider K via (4.13) or (4.14) as $H^{(*)}(BG)$-resp. $S^{-1} H^{(*)}(BG)$-module and use ${}_\eta K$ as notation for this module. The localization theorem is used in the following chain of natural isomorphisms of graded algebras.

(4.15)
$$\begin{aligned} & H_G^{(*)}(X) \otimes_{H^{(*)}(BG)} {}_\eta K \cong \\ & S^{-1} H_G^{(*)}(X) \otimes_{S^{-1}H^{(*)}(BG)} {}_\eta K \cong \\ & S^{-1} H_G^{(*)}(X^G) \otimes_{S^{-1}H^{(*)}(BG)} {}_\eta K \cong \\ & H^{(*)}(X^G) \otimes_K S^{-1} H^{(*)}(BG) \otimes_{S^{-1}H^{(*)}(BG)} {}_\eta K \cong \\ & H^{(*)}(X^G) \otimes_K {}_\eta K \cong H^{(*)}(X^G). \end{aligned}$$

Similarly for (X, A) in place of X. In particular, (4.15) says: The K-algebra $H^{(*)}(X^G)$ is obtainable by a purely algebraic process from the $H^{(*)}(BG)$-algebra $H_G^{(*)}(X)$.

(4.16) Proposition. *Suppose* $\dim_K \bigoplus_k H^k(X, A)$ *is finite and let* $H^r(X, A) = 0$ *for* $r > n$. *Then:*

(i) $$\dim_K \bigoplus_k H^k(X^G, A^G) \leq \dim_K \bigoplus_k H^k(X, A).$$

Moreover, the following assertions are equivalent:

(ii) *Equality of dimensions holds in* (i).

(iii) (X, A) *is totally nonhomologous to zero in* (X_G, A_G) *with respect to* $H^*(-)$.

(iv) $$\dim_K H_G^r(X, A) = \dim_K \bigoplus_k H^k(X, A) \quad \textit{for } r > n.$$

(v) *G acts trivially on* $H^*(X, A)$ *and the Serre spectral sequence of* $(X, A) \to (X_G, A_G) \to BG$ *degenerates.*

Proof. The equivalence (iii) $\Leftrightarrow$ (v) was shown in (1.18).
Consider the Serre spectral sequence. The groups $E_\infty^{k-i,i}$ are quotients of successive terms in a filtration of $H_G^k(X, A)$, hence

$$\dim H_G^k(X, A) = \sum_i \dim E_\infty^{k-i,i}.$$

The group $E_\infty^{k,l}$ is a subquotient of $E_2^{k,l}$. Therefore,

$$\dim H_G^k(X, A) \leqq \sum_i \dim E_2^{k-i,i} = \sum_i \dim H^{k-i}(BG; H^i(X, A)).$$

For $p \neq 0$, one has, for any G-module L

$$H^i(BG; L) = \begin{cases} \ker T & i = 0 \\ \ker T/\operatorname{im} N & i > 0 \text{ even} \\ \ker N/\operatorname{im} T & i > 0 \text{ odd} \end{cases}$$

where $T: L \to L, m \mapsto (1-g)m$ and $N: L \to L, m \mapsto (1 + g + \ldots + g^{p-1})m$ with g a generator of G. If L is a K-vector space, then

$$\dim \ker N/\operatorname{im} T \leqq \dim L/\operatorname{im} T = \dim \ker T;$$

therefore, in general,

$$\dim H^i(BG; L) \leqq \dim L^G.$$

Consequently, for $p \neq 0$,

(4.17) $$\dim H_G^k(X, A) \leqq \dim \bigoplus_i H^i(X, A)^G \leqq \dim \bigoplus_i H^i(X, A).$$

For $p = 0$, one obtains

(4.18) $$\dim H_G^k(X, A) \leqq \dim \bigoplus_{k \equiv i(2)} H^i(X, A).$$

By (4.9), we have $\dim H_G^k(X, A) = \dim H_G^k(X^G, A^G)$ for $k > n$. By the Künneth theorem,

(4.19) $$\begin{aligned} \dim H_G^k(X^G, A^G) &= \dim \bigoplus_i H^i(X^G, A^G), \quad p \neq 0 \\ &= \dim \bigoplus_{k \equiv i(2)} H^i(X^G, A^G), \quad p = 0. \end{aligned}$$

Thus from (4.17)–(4.19) we obtain (i).

Equality holds in (4.17) if and only if G acts trivially on $H^*(X, A)$ and $E_2^{k-i,i} = E_\infty^{k-i,i}$.

This proves (iv) $\Leftrightarrow$ (v) for $p \neq 0$; similarly for $p = 0$. Using the isomorphism $H_G^r(X, A) \cong H_G^r(X^G, A^G)$ for $r > n$, we finally see the equivalence (ii) $\Leftrightarrow$ (iv). □

(4.20) Proposition. *Assume the hypothesis of* (4.16) *and assume that* (4.16) (ii)–(v) *hold. Then:*

(i) $H_G^*(X, A)$ *is a free* $H^*(BG)$*-module of rank* $\dim_K \bigoplus_k H^k(X, A)$.

(ii) *The graded K-modules* $H^{(*)}(X, A)$ *and* $H^{(*)}(X^G, A^G)$ *are isomorphic.*

Proof. Use (1.18, iii) to conclude that $H_G^*(X, A)$ is a free $H^*(BG)$-module isomorphic to $H^*(X, A) \otimes_K H^*(BG)$. The isomorphism (4.15) then proves the remaining statements. □

(4.21) Definition. Let $n \geq 0$ be an integer. A pair (X, A) is called a **K-cohomology n-disk** if $H^i(X, A; K) = 0$ for $i \neq n$ and $H^n(X, A; K) \cong K$. A space X is called a **K-cohomology n-sphere** if $H^*(X; K) \cong H^*(S^n; K)$.

We now come to the theorems of P.A. Smith. (4.11) and the general hypotheses of this section are still in force.

(4.22) Theorem. *Suppose* (X, A) *is a K-cohomology n-disk. Then* (X^G, A^G) *is a K-cohomology r-disk for some integer* $r \in [0, n]$. *If* $p \neq 2$, *then* $n - r$ *is even.*

Proof. G can only act trivially on $H^n(X, A)$ and the Serre spectral sequence degenerates because of dimensional reasons. (4.10) and (4.20, ii) give the desired conclusion. □

(4.23) Theorem. *Suppose X is a K-cohomology n-sphere. Then* X^G *is a K-cohomology r-sphere for any integer* $r \in [-1, n]$. ($X^G = \emptyset \Leftrightarrow r = -1$). *If* $p \neq 2$, *then* $n - r$ *is even.*

Proof. Let $CX = X \times [0,1]/X \times \{0\}$ denote the cone on X with the induced G-action (trivial on $[0,1]$). Then $C(X^G) = (CX)^G$. Moreover, X is a cohomology n-sphere if and only if (CX, X) is a cohomology $(n+1)$-disk. Now apply (4.22). □

In (4.23), $r = n$ is possible. We shall see in exercise 4 that for each finite group G and $n \geq 2$ there exists a finite-dimensional G-complex X such that $X \simeq S^n \simeq X^G$ where $X \setminus X^G$ carries a free action and the degree of $X^G \to X$ is a given integer prime to the order of G. This integer has to be prime to $|G|$ because of the next result.

(4.24) Proposition. *Under the hypothesis of* (4.23) *assume $r = n$. Then $i\colon X^G \subset X$ induces an isomorphism $i^*\colon H^*(X) \to H^*(X^G)$.*

Proof. It suffices to show that $i\colon (CX^G, X^G) \to (CX, X)$ induces an isomorphism. Suppose this is not the case. Then i^* is the zero map. From the commutative diagram

$$\begin{array}{ccc} H^{n+1}(CX_G, X_G) & \xrightarrow{\;i_G^*\;} & H^{n+1}(CX_G^G, X_G^G) \\ \downarrow \cong\, j^* & & \downarrow \cong\, j^* \\ H^{n+1}(CX, X) & \xrightarrow{\;i^*\;} & H^{n+1}(CX^G, X^G) \end{array}$$

we see that i_G^* is the zero map. Both $H_G^*(CX, X)$ and $H_G^*(CX^G, X^G)$ are free $H^*(BG)$-modules with a single generator in dimension $n+1$.

Therefore i_G^* is the zero map in all degrees, in contradiction to the localization theorem. □

We still retain the hypothesis and notation of (4.23). Then $H_G^*(CX, X)$ is a free $H^*(BG)$-module of rank one by the Thom isomorphism theorem or by (1.18). A basis element $t(X) \in H_G^{n+1}(CX, X)$ is called a **Thom class**. The choice of a Thom class is called an **orientation** of X. We have a relation of the type

(4.25) $i^* t(X) = e(X, X^G)\, t(X^G)$

with suitable $e(X, X^G) \in H^{n-r}(BG)$, called **Euler class** of the (oriented) pair (X, X^G).

(4.26) Lemma. $e(X, X^G) \neq 0$.

Proof. If $e(X, X^G)$ were zero, then $S^{-1} t^*$ would be the zero map, in contradiction to the localization theorem. □

Now suppose we have two cohomology n-spheres X and Y such that X^G and Y^G are both cohomology r-spheres. Let $f\colon X \to Y$ be a G-map. We have two mapping degrees

$$d(f), \quad d(f^G) \in K$$

defined by

(4.27) $f^* t(Y) = d(f)\, t(X), \quad f^* t(Y^G) = d(f^G)\, t(Y)$.

From (4.25) and (4.27), we obtain

(4.28) $d(f^G)\, e(Y, Y^G) = d(f)\, e(X, X^G)$.

In particular, this implies that

(4.29) $d(f^G) \neq 0 \Leftrightarrow d(f) \neq 0.$
$\qquad r = -1 \Rightarrow d(f) \neq 0.$

(4.30) Remark. The preceding results can be generalized. Let P be a finite p-group. Then P has a normal series

$$1 = P_0 \lhd P_1 \lhd P_2 \lhd \ldots \lhd P_k = P$$

such that $P_{i+1}/P_i \cong \mathbb{Z}/p$. If (X, A) is a pair of finite-dimensional P-complexes and (X, A) is an F_p-cohomology n-disk, then, by induction on k using (4.22) and the fact that $X^{P_{i+1}}$ is the fixed point set of a P_{i+1}/P_i-action on X^{P_i}, we see that (X^P, A^P) is an F_p-cohomology r-disk for some $r \leq n$; moreover, $n \equiv r \bmod 2$ if p is odd. Similarly for (4.23). For $p = 0$, one generalizes in the same manner from S^1 to a k-dimensional torus.

(4.31) Proposition. *Let $G = \mathbb{Z}/4$ act on the F_2-cohomology n-sphere X. Suppose $X^{\mathbb{Z}/2}$ is an F_2-cohomology r-sphere. Then $n - r$ is even.*

Proof. We use cohomology with F_2-coefficients. By (4.2), we have

$$S^{-1} H^*_{\mathbb{Z}/4}(CX, X) \cong S^{-1} H^*_{\mathbb{Z}/4}(CX^{\mathbb{Z}/2}, X^{\mathbb{Z}/2}).$$

The left hand side is a free $S^{-1} H^*(B\mathbb{Z}/4)$-module with a single generator in dimension $n+1$, the right hand side is such a module with generator in dimension $r+1$. Both generators differ by a unit of $S^{-1} H^*(B\mathbb{Z}/4) \cong F_2[t, t^{-1}] \otimes \Lambda[s]$. Units only occur in even degrees. (Or use (4.15)) □

Let $Q(2^k)$ be the quaternion group of order $2^k \geq 8$ (see section 2). It contains a single subgroup C of order two. This subgroup is the center and is contained in each non-trivial subgroup.

(4.32) Proposition. *Let $G = Q(2^k)$ act on the F_2-cohomology n-sphere X with X^C an F_2-cohomology r-sphere. Then $n - r$ is divisible by 4.*

Proof. The proof is analogous to the proof of (4.31). One needs the following information about $H^*(BG)$ with F_2-coefficients: $H^{4*}(BG) \cong F_2[u]$, $u \in H^4(BG)$. Elements in $H^r(BG)$ are nilpotent if $r \not\equiv 0 \bmod 4$. Therefore, one can define a homomorphism of $\mathbb{Z}/4$-graded algebras $\eta\colon H^{(*)}(BG) \to F_2$ by mapping u to 1 and, if $r \not\equiv 0 \bmod 4$, $x \in H^r(BG)$ to zero. If $S = \{1, u, u^2, \ldots\}$, then, in this case, the analogue of (4.2) is

(4.33) $S^{-1} H^*_G(X, A) \cong S^{-1} H^*_G(X^C, A^C).$

Note that X^C is the singular set of X. As in (4.15), one obtains a natural isomorphism of $\mathbb{Z}/4$-graded algebras

(4.34) $H_G^{(*)}(X, A) \otimes_{H^*(BG)} {}_\eta F_2 \cong H^{(*)}(X^C, A^C)$.

Using (4.34), one can now continue as for (4.31). □

We now turn our attention to actions of p-tori and make the following assumptions.

(4.35) We fix a prime ideal $(p) \subset \mathbb{Z}$.
G is a p-torus of rank $k \geq 2$.
$K = F_p$ is the prime field of characteristic p.
X is a finite-dimensional G-complex of finite orbit type which is a K-cohomology n-sphere.

For each subtorus H of G, we know that X^H is a K-cohomology $n(H)$-sphere for some integer $n(H) \leq n$. We let $S \subset H^*(BG)$ be the set of Euler classes of representations without trivial summand; see (3.13). We set

$$S_H = \{s \in S \mid s \notin \ker(H^*(BG) \to H^*(BH))\}.$$

In this case, the localization theorem (3.8) says, slightly generalizing (3.13),

(4.36) $S_H^{-1} H_G^*(Y, B) \cong S_H^{-1} H_G^*(Y^H, B^H)$.

Let CX denote the cone on X. We consider the cohomology sequence of the triple (CX, X, X^G)

$$\ldots \to H_G^*(CX, X) \xrightarrow{i^*} H_G^*(CX, X^G) \xrightarrow{j^*} H_G^*(X, X^G) \to \ldots.$$

We know that $H_G^*(CX, X)$ is a free $H^*(BG)$-module with a single generator $t(X)$ of degree n; similarly, $H_G^*(CX, X^G)$ is free on $t(X^G)$ in degree r. We set $i^* t(X) = e t(X^G)$, $e \in H^{n-r}$ the Euler class of (X, X^G). The map i^* is injective because a free $H^*(BG)$-module maps injectively into its localization (since $H^*(BG) \to S^{-1} H^*(BG)$ is injective in this case) and $S^{-1} i^*$ is an isomorphism. Therefore, j^* is surjective and we obtain an isomorphism of $H^*(BG)$-modules

$$H_G^*(X, X^G) \cong H^*(BG)/J$$

where J is the annihilator ideal of $j^* t(X^G)$. From the exact cohomology sequence, we see that J is the principal ideal (e) generated by the Euler class $e = e(X, X^G)$.

Recall that $H^*(BG)$ contains a polynomial ring $P^*(G) = K[t_1, \ldots, t_k]$ with $t_i \in H^2$ for $p \neq 2$ and $t_i \in H^1$ for $p = 2$. The elements in S are products of linear polynomials in the t_i. Elements not in $P^*(G)$ are nilpotent. Each element $u \in H^*(BG)$ can be written uniquely in the form $u = z + n$, with $z \in P^*(G)$ and n nilpotent. Let

(4.37) $e = z + n$

be such a decomposition of the Euler class.

(4.38) Lemma. $z \in S$.

Proof. $S^{-1}H^*(BG)/(e) \cong S^{-1}H^*_G(X, X^G) = 0$ by (4.36). Therefore: Given $x \in H^*(BG)$, there exists $s \in S$ such that $sx \in (e)$. Take $x = 1$; there exist $s \in S$, $y \in H^*(BG)$ such that $s = ye$. Write $y = z' + n'$ as in (4.37). Then $s = zz'$. Since $P^*(G)$ is a unique factorization domain, z is a product of linear factors. □

(4.39) Lemma. $S_H^{-1}H^*(BG)/(e) = 0 \Leftrightarrow z \in S_H$.

Proof. The implication $\Rightarrow$ is proved like (4.38). Conversely: Suppose $z \in S_H$. Then $z^2 = (z+n)(z-n)$; hence $z^2 \in S_H$ annihilates $1 \in H^*(BG)/J$ and therefore the whole module. □

Now let H be a subtorus of rank $k-1$. The inclusion $H \subset G$ induces $P^*(G) \to P^*(H)$. The kernel is a principal ideal (s_H), s_H linear. Let

$$z = e_1^{k_1} \cdot \ldots \cdot e_s^{k_s}$$

be a factorization of z into relatively prime linear factors $e_1, \ldots, e_s$.

The following theorem is due to Borel [1960]. The notation (4.35) is still valid.

(4.40) Theorem. (i) *For a subtorus H of G of rank $k-1$ we have $n(H) > r$ if and only if $(s_H) = (e_i)$ for some $i \in \{1, \ldots, s\}$. The equation $(s_H) = (e_i)$ determines H uniquely and we write $H = H_i$ in this case.*

(ii) $$n - r = \sum_{i=1}^{s} (n(H_i) - r).$$

Proof. Let H of rank $k-1$ be given. We write $G = H \times L$ and obtain

$$\begin{aligned} H^*_G(X^H, X^G) &\cong H^*(BH \times EL \times_L (X^H, X^G)) \\ &\cong H^*(BH) \otimes_K H^*_L(X^H, X^G) \\ &\cong H^*(BH) \otimes_K H^*(BL)/(e_H) \\ &\cong H^*(BG)/(e_H) \end{aligned}$$

where e_H is a certain power of the polynomial generator s_H of $P^*(L) \subset H^*(BL)$ and proportional to the Euler class $e(X^H, X^G)$. From (4.39), we obtain the following equivalences: $S_H^{-1}H^*_G(X, X^G) \neq 0 \Leftrightarrow z \in S_H \Leftrightarrow z \in \ker(H^*(BG) \to H^*(BH)) \Leftrightarrow s_H$ divides z. From (4.36), we obtain the equivalences

$0 \neq S_H^{-1} H_G^*(X, X^G) \cong S_H^{-1}(X^H, X^G) \cong S_H^{-1} H^*(BG)/(e_H) \Leftrightarrow (e_H) \neq (1) \Leftrightarrow n(H) > r$. This proves (i).

Now let us write $z = s_H^a u$, $e_H = s_H^b$ with s_H prime to u. From $S_H^{-1} H^*(BG)/(e) \cong S_H^{-1} H^*(BG)/(e_H)$, we obtain equality of ideals $(e) = (e_H)$ in $S_H^{-1} H^*(BG)$; hence $e = \varepsilon e_H$ with a unit $\varepsilon \in S_H^{-1}(BG)$. This is only possible if $a = b$. If $H = H_i$, then, of course, $a = k_i$. On the other hand, $n(H) - r =$ degree (e_H) and $n - r =$ degree (z). This shows (ii). □

Let X be a finite-dimensional G-complex on which the finite group G acts freely. Assume that $H^*(X; \mathbb{Z}) \cong H^*(S^n; \mathbb{Z})$. Let p be a prime number and $P = G(p)$ a p-Sylow group of G. If P has a subgroup $H \cong \mathbb{Z}/p \times \mathbb{Z}/p$, then H acts freely on X. Let $H_0, \ldots, H_p$ denote the subgroups of order p in H. The Borel-formula (4.40) yields $n = -1$ because $n(H_i) = -1$ by assumption; this is absurd. Therefore, we have shown that $\mathbb{Z}/p \times \mathbb{Z}/p$ cannot act freely on a finite-dimensional $\mathbb{Z}$-cohomology n-sphere. Slightly more general is

(4.41) Proposition. *$H = \mathbb{Z}/p \times \mathbb{Z}/p$ cannot act semi-freely (i.e. with orbits only of type H and H/H) on a finite-dimensional F_p-cohomology n-sphere such that $n > n(H)$.*

Proof. Otherwise, $X^{H_i} = X^H$ and $n = n(H)$ by the Borel-formula (4.40). □

(4.42) Remark. In case $n = n(H)$, there exist non-trivial semi-free actions with the property that $X^H \subset X$ has a degree $\not\equiv 0 \bmod p$.

The group G acting freely on X must have all abelian subgroup cyclic. Such groups can be classified. We only mention

(4.43) Proposition. *Let G be a p-group. All abelian subgroups of G are cyclic if and only if:*
(i) *(for $p \neq 2$) G is cyclic.*
(ii) *(for $p = 2$) G is cyclic or a generalized quaternion group.*

Proof. Wolf [1967], Theorem 5.3.2. □

A finite group G is said to have **periodic cohomology** if there exists $e \in H^n(BG; \mathbb{Z})$ such that for $i > 0$ multiplication with e is an isomorphism $H^i(BG; \mathbb{Z}) \to H^{i+n}(BG; \mathbb{Z})$. Any such e is called a **periodicity generator**. For basic facts about periodic cohomology, see Cartan-Eilenberg [1956], XII. 11.

Suppose G acts freely on the finite-dimensional G-complex X with $H^*(X; \mathbb{Z}) \cong H^*(S^{d-1}; \mathbb{Z})$. We look at the sphere bundle $EG \times_G X \to BG$. Unless $G \cong \mathbb{Z}/2$, we must have $d \equiv 0 \bmod 2$ and the action of G on $H^{d-1}(X; \mathbb{Z})$ is trivial.

Therefore, we assume that $d \equiv 0 \bmod 2$ and that $H^{d-1}(X; \mathbb{Z})$ has trivial G-

action. We call such actions orientable. The sphere bundle $EG \times_G X \to BG$ is then orientable, has an Euler class $e(X) \in H^d(BG)$, and a Gysin sequence. Moreover, under these assumptions we have

(4.44) Proposition. *G has periodic cohomology and the Euler class $e(X)$ is a periodicity generator.*

Proof. Since G acts freely on X, we have a homotopy-equivalence $EG \times_G X \simeq X/G$. Since $H^j(X/G) = 0$ for j large enough, we see from the Gysin sequence that multiplication by $e(X)$ is an isomorphism $H^k(BG) \to H^{k+n}(BG)$ for large k. By algebra (Cartan-Eilenberg [1956], XII. Prop. 11.1), this implies the assertion. If X/G is assumed to be of dimension $n-1$, then we conclude directly that $H^{k+n}(X/G) = 0$ for $k \geq 0$. □

Swan [1960] has proved a converse of (4.44): Given a periodicity generator $e \in H^n(BG)$, there exists a finite-dimensional G-complex $X \simeq S^{n-1}$ with free G-action, unique up to oriented G-homotopy type, such that $e = e(X)$.

Results of the type (4.22) and (4.23) were first proved by P. A. Smith [1938a]. P. A. Smith initiated the use of algebraic topology for transformation groups to a large extend; the references P. A. Smith [1938–1967] collect some of his publications.

For other proofs of P. A. Smith-type theorems, see Borel [1955], Floyd [1952], [1960], Bredon [1972], V. Puppe [1984]. For the methods of this section, see Bredon [1968], [1972]. The problem of the converse of the P. A. Smith-theorems began with work of Jones [1971], [1972] and has since attracted considerable attention, see e.g. Löffler [1981], [1981a].

The study of free group actions on topological or smooth spheres is sometimes called the spherical space form problem. It has been successfully dealt with by using surgery theory. Here is a small collection of references concerned with this problem: Browder-Livesay [1967], Madsen [1977], [1983], Madsen-Thomas-Wall [1976], [1983], Milnor [1957], R. Lee [1973], Lopez de Medrano [1971], Petrie [1971], Wall [1970].

(4.45) Exercises.

1. Let X be an $(n-1)$-dimensional G-complex with free G-action, homotopy equivalent to S^{n-1}. The cellular chain complex of X yields an exact sequence of $\mathbb{Z}G$-modules $0 \to \mathbb{Z} \to C_{n-1}(X) \to \ldots \to C_0(X) \to \mathbb{Z} \to 0$ if we assume that G acts trivially on $\mathbb{Z} = H_{n-1}(X; \mathbb{Z})$. Any such exact sequence, being an n-extension, defines an element in $\mathrm{Ext}^n_{\mathbb{Z}G} \cong H^n(BG, \mathbb{Z})$ (Mac Lane [1963], Ch. IV). Identify this element with the Euler class $e(X)$.
2. Verify (4.33) and (4.34).
3. Let p and q be two odd prime numbers, let q divide $p-1$ and let G be the

semi-direct product $\mathbb{Z}/p \to G \to \mathbb{Z}/q$ where Z/q acts on Z/p by an injective homomorphism $\mathbb{Z}/q \to \mathrm{Aut}(\mathbb{Z}/p) = \mathbb{Z}/p^*$. Show that $H^*(BG; F_p) = F_p[u]$, $u \in H^{2q}(BG; F_p)$. Use this to define a homomorphism $H^*(BG; F_p) \to F_p$ of $Z/2q$-graded algebras which maps u to 1. Formulate and prove an analogue of (4.34).

4. Let G be a finite group. For each $n > 1$, there exists a finite-dimensional G-complex X such that $X \simeq S^n \simeq X^G$, $X \backslash X^G$ carries a free action, and the degree of $X^G \subset X$ is a given integer k prime to $|G|$. The construction is as follows: There exist free $\mathbb{Z}G$-modules F_1 and F_2 and an isomorphism

$$\varphi: \mathbb{Z} \oplus F_2 \to \mathbb{Z} \oplus F_1$$

such that

$$\mathbb{Z} \xrightarrow[\subset]{} \mathbb{Z} \oplus F_1 \xrightarrow[\varphi^{-1}]{} \mathbb{Z} \oplus F_2 \xrightarrow[\mathrm{pr}]{} \mathbb{Z}$$

is multiplication by k (see Swan [1960]). Now consider the exact sequence

$$0 \to F_2 \to \mathbb{Z} \oplus F_1 \to \mathbb{Z} \to 0$$

and realize $F_2 \to \mathbb{Z} \oplus F_1$ geometrically as the cellular chain complex of a space X as follows: Start with S^n and trivial G-action. Attach cells of type $G \times D^n$ to S^n by trivial attaching maps, one for each element of a $\mathbb{Z}G$-basis of F_1. Let Y be the resulting G-complex. Then $\pi_n Y \cong H_n Y \cong \mathbb{Z} \oplus F_1$. For each basis element u of F_2, attach a cell of type $G \times D^{n+1}$ with attaching map $\{e\} \times S^n \to Y$ representing $\varphi(0, u) \in \mathbb{Z} \oplus F_1 \cong \pi_n Y$. The resulting space X has the desired properties.

5. Let T be a torus and let $f: X \to Y$ be a T-map between generalized homotopy representations. Assume that for p-groups $H \subset T$ we have $\mathrm{Dim}\, X(H) = \mathrm{Dim}\, Y(H)$. Let degree $f^T = 1$ if $X^T \neq \emptyset$. Then f has degree ± 1. Hint: (4.27)–(4.30).

6. Let $G = \mathbb{Z}/2$. Let $S \subset H^*(BG; \mathbb{Z})$ be the set of Euler classes of complex representations without trivial direct summand. Show that $S^{-1} H^{(*)}(BG; \mathbb{Z})$ ist still $\mathbb{Z}/2$-graded, concentrated in even degrees, and isomorphic to $F_2[t, t^{-1}]$ with $t \in H^2(BG; \mathbb{Z})$. Therefore, we have a homomorphism of $\mathbb{Z}/2$-graded algebras

$$\eta: S^{-1} H^{(*)}(BG; \mathbb{Z}) \to F_2,$$

$\eta(t) = 1$. As in (4.15), the localization theorem yields an isomorphism of $\mathbb{Z}/2$-graded algebras

$$H_G^{(*)}(X, A; \mathbb{Z}) \otimes_{H^{(*)}(BG;\mathbb{Z})} F_2 \cong H^{(*)}(X^G, A^G; F_2).$$

A similar statement holds for cohomology with $\mathbb{Z}_{(2)}$-coefficients (the integers localized at (2)).

7. Let (X, A) be a finite-dimensional G-complex with finitely generated ho-

mology $H_*(X, A; \mathbb{Z})$ where $G = \mathbb{Z}/2$. Assume that (X, A) is an F_2-cohomology n-disk ($n > 0$). Show that G acts trivially on $H^*(X, A; \mathbb{Z}_{(2)})$ if and only if G acts trivially on $H^n(X, A; \mathbb{Z})/\text{Torsion}$. Suppose G acts trivially on $\mathbb{Z}$-cohomology. Show that the Serre-spectral sequence associated to $(X_G, A_G) \to BG$ for $H^*(-; \mathbb{Z}_{(2)})$ degenerates. Show that $H_G^*(X, A; \mathbb{Z}_{(2)})$ is a free $H^*(BG; \mathbb{Z}_{(2)})$-module with a generator in degree n, (1.18). Use the previous exercise to show that $n - r$ is even if (X^G, A^G) is an F_2-r-disk, (4.22). Suppose G acts nontrivially on $\mathbb{Z}$-cohomology. Let G act on the unit interval I by $t \mapsto 1 - t$. If (X, A) is an F_2-n-disk, then $(I, \partial I) \times (X, A)$ is an F_2-$(n+1)$-disk and G acts trivially on cohomology. Conclude that $n - r$ is odd in this case.

5. Borel-Smith functions.

In the previous section, we have seen that group actions on spheres must satisfy certains conditions: The fixed point sets cannot be arbitrary. As an application of these results, we analyse actions on spheres where all fixed point sets are spheres again. We shall see with the help of representation theory that certain strong conditions have to be satisfied. In particular, actions of p-groups on spheres resemble representation spheres in some respect.

Let $p \in \mathbb{Z}$ be a prime and G a finite p-group. Suppose X is a finite-dimensional G-complex which is an F_p-cohomology sphere. Then, for each subgroup $H \subset G$, the fixed point set X^H is an F_p-cohomology sphere of some dimension $d(H)$. The integer $d(H)$ only depends on the conjugacy class of H. Thus we obtain a function

$$n\colon \psi(G) \to \mathbb{Z},\ (H) \mapsto d(H) + 1$$

from the set $\psi(G)$ of conjugacy classes of subgroups to the integers; this is called the **dimension function** of X.

For each finite group G, let $C(G)$ denote the ring of functions $\psi(G) \to \mathbb{Z}$. The next definition assembles the conditions on dimension functions which follow from (4.23), (4.31), (4.32), and (4.40) if the appropriate homological conditions are satisfied.

(5.1) Definition. A function $n \in C(G)$ is said to satisfy the **Borel-Smith relations** and is called a **Borel-Smith function** if the following holds:

(i) If $H \lhd K \subset G$, $K/H \cong \mathbb{Z}/p$, and p is odd, then $n(H) - n(K)$ is even.

(ii) If $H \lhd K \subset G$, $K/H \cong \mathbb{Z}/p \times \mathbb{Z}/p$, H_i/H the subgroups of order p in K/H, then $n(H) - n(K) = \sum_{i=0}^{p} (n(H_i) - n(K))$.

(iii) If $H \lhd L \lhd K \subset G$ and $L/H \cong \mathbb{Z}/2$, then $n(H) - n(L)$ is even if

$K/H \cong \mathbb{Z}/4$, and $n(H) - n(L)$ is divisible by 4 if K/H is a generalized quaternion group of order 2^k, $k \geq 3$.

The function n is called **monotone** if the following holds:

(iv) For all $K \subset G$, $n(K) \geq 0$. If $H \subset K$ and K/H is a p-group, then $n(H) \geq n(K)$.

We let $C_b(G) \subset C(G)$ be the additive subgroup of Borel-Smith functions. Let H' denote the commutator subgroup of H. We put

$$B = \{(H) \in \psi(G) \mid H/H' \text{ cyclic}\}.$$

Note that B contains the conjugacy classes of cyclic subgroups.

(5.2) Proposition.
(i) *The rank of $C_b(G)$ is equal to the cardinality of B.*
(ii) *B is the set of cyclic conjugacy classes if and only if G is nilpotent.*

Proof. (i) We only show that rank $C_b(G) \leq |B|$. For equality, see tom Dieck-Petrie [1982], (10.3). Conditions (i) and (iii) in (5.1) do not affect the rank of $C_b(G)$. The linear relation (ii) implies that, for a Borel-Smith function n, the value $n(K)$ can be computed from the values of smaller subgroups provided K has a quotient of the form $\mathbb{Z}/p \times \mathbb{Z}/p$, i.e. provided K/K' is not cyclic (5.3).

(ii) Suppose G is nilpotent. This means: G is the direct product of its p-Sylow subgroups $G(p)$. We use the following lemma, see Huppert [1967], III. 7.1.

(5.3) Lemma. *If a p-group G is not cyclic, then it has a quotient of the form $\mathbb{Z}/p \times \mathbb{Z}/p$.* □

Using this, we see that B contains only cyclic conjugacy classes.

Now assume that B consists of the cyclic conjugacy classes.

If G/G' is cyclic, then $(G) \in B$ hence G is cyclic. Otherwise, G has at least two distinct maximal normal subgroups M and N (of prime index). By induction, M and N are nilpotent. Therefore, $G = MN$ is nilpotent by Fitting's lemma, see Huppert [1967], III. 4.1. □

Let $RO(G)$ denote the real representation ring of the finite group G. The dimension function of a real representation V is the function $\mathrm{Dim}\, V\colon (H) \mapsto \dim_{\mathbb{R}} V^H$. The assignment $V \mapsto \mathrm{Dim}\, V$ extends to an additive homomorphism $\mathrm{Dim}\colon RO(G) \mapsto C(G)$. We denote its kernel by $RO_0(G)$ and its image by $C_r(G)$. Note that $C_r(G) \subset C_b(G)$. The converse inclusion holds for nilpotent groups.

(5.4) Theorem. *Let G be nilpotent and $n \in C(G)$ a Borel-Smith function. Then* *there exists an element x in the real representation ring $RO(G)$ of G such that* $n = \operatorname{Dim} x$.

In other words, this theorem says that for finite nilpotent groups the Borel-Smith functions are precisely those which are the dimension functions of linear virtual representations. In particular, homotopy representations of nilpotent groups have stably the dimension function of a linear representation; unstably, this is not true in general.

The proof of (5.4) needs some facts from representation theory which we are now going to collect in (5.5) to (5.12).

In the next lemma, μ denotes the Möbius function of elementary number theory.

(5.5) Lemma. *Let V be a representation of G over the field $K \subset \mathbb{C}$. Then*

$$|G| \dim_K V^G = \sum_C \Big(\sum_{D \subset C} \mu(|C/D|)\, |D| \dim_K V^D \Big).$$

The sum is taken over the cyclic subgroups C of G.

Proof. Since $\dim_{\mathbb{C}} (V \otimes_K \mathbb{C})^G = \dim_K V^G$, if suffices to look at complex representations. Let $V\colon G \to \mathbb{C}$ also denote the character of V. Starting point is the orthogonality relation

$$(5.6) \qquad |G| \dim V^G = \sum_{g \in G} V(g).$$

If C^* denotes the set of generators of the cyclic group C, we can rewrite the sum as

$$\sum_C \Big(\sum_{g \in C^*} V(g) \Big).$$

Since $\sum_C V(g) = \sum_{D \subset C} \Big(\sum_{g \in D^*} V(g) \Big)$,

we obtain

$$\sum_{g \in C^*} V(g) = \sum_{D \subset C} \mu(|C/D|) \sum_{g \in D} V(g)$$

by Möbius inversion. But the inner sum over $g \in D$ equals $|D| \dim V^D$. □

From (5.5), we see that $\operatorname{Dim} x = \operatorname{Dim} y$ if and only if $\dim x^C = \dim y^C$ for all cyclic subgroups C. In other words: The homomorphism

$$(5.7) \qquad C_r(G) \to \prod_C C_r(C)$$

induced by the restriction to $C \subset G$, C cyclic, is injective. We also see that $C_r(G)$ is free abelian of rank at most the number of conjugacy classes of cyclic subgroups of G.

Recall that the representation ring admits Adams operations $\psi^k\colon RO(G) \to RO(G)$; see Atiyah-Tall [1969], Serre [1971] p. 86. If V is an irreducible G-module, then $\psi^k V$, for k prime to $|G|$, is again an irreducible real G-module. This follows from the fact that $\psi^k V$ can be interpreted as a Galois-conjugate of V: Write V as a matrix representation with entries in a suitable cyclotomic field $Q(\zeta)$ and apply the automorphism $\zeta \mapsto \zeta^k$; see Atiyah-Tall [1969]. Galois automorphisms obviously map irreducible representations to irreducible ones. Let $V_1, \ldots, V_t$ be a complete set of non-conjugate irreducible real G-modules.

Let $RO_\psi(G) \subset RO(G)$ be the subgroup generated by elements of the form $V - \psi^k V$ for k prime to $|G|$. Write

$$RO(G)_r = RO(G)/RO_\psi(G).$$

Then $RO(G)_r$ is free abelian with basis $V_1, \ldots, V_t$.

(5.8) Lemma. $RO_\psi(G) \subset RO_0(G)$.

Proof. The Adams operation ψ^k has the following effect on characters

$$\psi^k V(g) = V(g^k).$$

Using (5.6), we see $\dim V^G = \dim(\psi^k V)^G$ because, for $(k, |G|) = 1$, the map $g \mapsto g^k$ is a permutation of G. □

(5.9) Proposition. $RO_\psi(G) = RO_0(G)$.

Proof. The map $RO(G) \to RO(G)$, $x \mapsto \sum_{(k,|G|)=1} \psi^k x$, where the sum is taken over all $k \in (\mathbb{Z}/|G|)^*$, the units in the ring $\mathbb{Z}/|G|$, induces a homomorphism $\mathrm{tr}\colon RO(G)_r \to RO(G)$. If we compose with $RO(G) \to RO(G)_r$, we obtain multiplication by $|\mathbb{Z}/|G|^*|$. Therefore, tr is injective. The restriction homomorphism $RO(G) \to \prod_C RO(C)$ is injective since representations are detected by characters; hence

$$RO(G)_r \to \prod_C RO(G)_r,$$

the analogue of (5.7), is injective, too.

By (5.8), we have a commutative diagram

$$\begin{array}{ccc}
RO(G)_r & \longrightarrow & \prod_C RO(C)_r \\
\downarrow & & \downarrow \\
C_r(G) & \longrightarrow & \prod_C C_r(C)
\end{array}$$

with surjective vertical arrows. Therefore, it suffices to show that $RO(C)_\Gamma \to C_r(C)$ is injective for cyclic C. This is done by inspection: The Galois-conjugacy classes for irreducible representations V are determined by the kernel of V and for each $D \subset C$ there exists exactly one class with kernel D. Therefore, the classes are detected by their dimension functions and the dimension functions are linearly independent. □

(5.10) Proposition. *The rank of $C_r(G)$ equals the number of conjugacy classes of cyclic subgroups of G.*

Proof. We have already seen that $C_r(G)$ has at most this rank. Let $C \subset G$ be cyclic and let $\mathbb{R}(G/C)$ be the permutation representation of the G-set G/C over $\mathbb{R}$. The character of $\mathbb{R}(G/C)$ is given by $g \mapsto |G/C^g|$. Therefore, the characters of the $\operatorname{tr} \mathbb{R}(G/C)$, where C runs through a set of non-conjugate cyclic subgroups, are linearly independent (compare I (2.18)). Therefore, $RO(G)_\Gamma$ has at least the rank in question. Now use (5.9). □

(5.11) Proposition. *Let G be a p-group. Suppose G does not have a normal subgroup isomorphic to $\mathbb{Z}/p \times \mathbb{Z}/p$. Then G is cyclic if p is odd and G is cyclic, dihedral, quaternion, or semi-dihedral if $p = 2$.*

Proof. Huppert [1967], III. 7.5, III. 7.6, I. 14.9. □

(5.12) Proposition. *Let G be nilpotent. Let G have a normal subgroup $N \cong \mathbb{Z}/p \times \mathbb{Z}/p$ and an irreducible faithful (real or complex) representation V. Let $H_0, \ldots, H_p$ be the subgroups of order p in N. Then one of the H_i, say H_0, is normal in G while the $N_i = NH_i$ for $i > 0$ are normal subgroups of index p. The restriction $\operatorname{res}_{N_i} V$ splits into irreducible representations $V_1 \oplus \ldots \oplus V_p$ where V_i has kernel H_i. If W is irreducible and not isomorphic to V, then $\operatorname{res}_{N_i} W$ does not contain a summand which is isomorphic to one of the V_j.*

Proof. The representation $\operatorname{res}_N V$ decomposes into a direct sum $W \oplus V_0 \oplus \ldots \oplus V_p$ where W is trivial and V_i has kernel H_i. If H_i is a normal subgroup, then V_i is a G-direct summand of V. Since V is irreducible and faithful, we have $W = \{0\}$ and $V_i \neq \{0\}$ implies that H_i is not normal in G. Therefore, not all H_i are normal and G acts non-trivially on the set $H_0, \ldots, H_p$ by conjugation. Since G is nilpotent, only the p-Sylow subgroup can act non-trivially, consequently, there exists an orbit of length p, say $\{H_1, \ldots, H_p\}$. Hence H_0 is normal and H_i for $i > 0$ is not normal. Hence $V_0 = \{0\}$ and the N_i, being of index p in G, are normal. This shows $N_1 = \ldots = N_p$ and the $H_1, \ldots, H_p$ are normal subgroups therein. Since V_j is the fixed point set of H_j, the subspace V_j is N_i-invariant and

$$\operatorname{res}_{N_j} V = V_1 \oplus \ldots \oplus V_p$$

as N_i-representation. The group G acts transitively on the set $\{V_1, \ldots, V_p\}$; hence $V \cong \operatorname{ind}_{N_i}^G V_j$. Suppose W is irreducible and not isomorphic to V. Then, by Frobenius reciprocity,

$$\begin{aligned} 0 = \langle V, W \rangle_G &= \langle \operatorname{ind}_{N_i}^G V_j, W \rangle_G \\ &= \langle V_j, \operatorname{res}_{N_i} W \rangle_{N_i} \end{aligned}$$

with $\langle V, W \rangle_G = \dim \operatorname{Hom}_G(V, W)$. This shows that $\operatorname{res}_{N_i} W$ cannot contain V_j as direct summand. (See Huppert [1967], V.17 for such considerations.) □

Proof of (5.4). We know that $C_r(G) \subset C_b(G)$ and (5.4) asserts that equality holds. From (5.2) and (5.10) we see that both groups have the same rank. Therefore, given $n \in C_b(G)$, there exist integers $a, a_1, \ldots, a_t$ such that

$$an = \sum a_i V_i$$

where $V_1, \ldots, V_t$ is the basis of $C_r(G)$ given by (the dimension functions of) a complete set of non-conjugate irreducible real G-modules. We assume, by induction, that the theorem holds for all groups of order less than $|G|$. We want to show that the a_i are divisible by a.

Look at a particular V_i and assume that there exists a proper subgroup H of G with the following property: The restriction $\operatorname{res}_H V_i$ splits into non-conjugate irreducible real H-modules which are not isomorphic to direct summands of $\operatorname{res}_H V_j$ for $i \neq j$. Then, considering $\operatorname{res}_H n$ and applying the induction hypothesis, we see that a divides a_i. Propositions (5.11) and (5.12) describe situations where such subgroups H can be found. We note that it suffices to look at faithful representations V_i. For, if V_i has kernel L, we consider the function $n' \colon K/L \mapsto n(K)$ in $C(G/L)$; it is contained in $C_b(G/L)$.

Thus assume that no normal subgroups of type $\mathbb{Z}/p \times \mathbb{Z}/p$ exist in G. In order to apply (5.12), we note that $V_1, \ldots, V_p$ are not Galois conjugate because they have different kernels and hence different dimension functions; moreover, one uses the fact that Galois conjugation commutes with induction and restriction.

We now look at the groups of (5.11) separately according to the four different types of their 2-Sylow subgroup.

G_2 cyclic, hence G cyclic. The faithful irreducible representations are Galois conjugate. Hence

$$an = a_0 V_0 + \sum a_i V_i.$$

Only V_0 is faithful and therefore $a | a_i$ by induction. Hence $an' = a_0 V_0$ with suitable $n' \in C_b(G)$. If $G = \mathbb{Z}/2$, then $C_r(G) = C(G)$ and there is nothing to prove. Otherwise, there exists a proper subgroup H of G with $n'(1) - n'(H)$ even on account of (5.1), (i) and (iii). Hence

$$an'(1) = 2a(2^{-1} n'(1)) = a_0 V_0(1).$$

Since V_0 is complex, $V_0(1) = 2$ and hence $a | a_0$.

G_2 generalized quaternion. The faithful irreducible representations are again Galois conjugate. As above, we reduce to $an = a_0 V_0$ with V_0 faithful. By (5.1, iii), $n(1) \equiv 0(4)$. Since V_0 is quaternionic, $V_0(1) = 4$ and $a | a_0$.

G_2 semi-dihedral. Let

$$G_2 = S_m = \langle a, b \,|\, a^{2^{m-1}} = 1 = b^2,\ bab^{-1} = a^{-1+2^{m-2}} \rangle.$$

One checks that S_m contains the quaternion subgroup $\langle A, B \,|\, A^{2^{n-3}} = B^2,\ BAB^{-1} = A^{-1} \rangle$ with $A = a^2$ and $B = ab$. The faithful irreducible representations of G are Galois conjugate and complex 2-dimensional. Their restriction to a maximal proper subgroup H with 2-Sylow subgroup $H_2 = Q_{m-1}$ are faithful, hence quaternionic and irreducible. This reduces the situation to the previous case: G_2 quaternion.

G_2 dihedral. The faithful irreducible representations are Galois conjugate. By induction, we reduce again to a situation $an = a_0 V_0$ with V_0 faithful. Let H be a non-central subgroup of order two in G. Then $V_0(H) = 1$ and therefore $an(H) = a_0$, i.e. $a | a_0$.

This completes the proof of Theorem (5.4). □

(5.13) Theorem. *Let $n \in C(G)$ be a monotone Borel-Smith function for the p-group G. Then there exists a real representation V of G such that* $n = \operatorname{Dim} V$.

Proof. Let $V_1, \ldots, V_t$ be as above. We know from (5.4) that there exist integers a_i such that $n = \sum a_i V_i$. We have to show that the a_i are non-negative. By induction on $|G|$, we have only to look at the coefficients of faithful V_i. Using (5.11) and (5.12) as in the proof of (5.4), we reduce to the case that G does not possess normal subgroups of type $\mathbb{Z}/p \times \mathbb{Z}/p$. Let V_0 be faithful. If G is cyclic, quaternion, or dihedral, let H be the unique central subgroup of order p. Then $0 \leq n(1) - n(H) = a_0 V_0(1)$ hence $a_0 \geq 0$. If G is semi-dihedral, we restrict to the quaternion subgroup of index 2 as in the proof of (5.4). □

There are results for complex representations which are analogous to (5.4) and (5.13).

(5.14) Theorem. (i) *Let G be nilpotent. Suppose $n \in C_b(G)$ takes only even values. Then there exist complex representations V and W such that* $n = \operatorname{Dim} V - \operatorname{Dim} W$.

(ii) *Let G be a p-group. Suppose $n \in C_b(G)$ takes only even values and is monotone. Then there exists a complex representation V such that* $n = \operatorname{Dim} V$.

The proof is analogous to (5.4) and (5.13) and is left to the reader. □

References: Dotzel-Hamrick [1981], Dotzel [1984], tom Dieck [1979a], [1982], [1982a].

For compact Lie groups G, one has to take into account the relation (4.40) for a torus. If one considers only continuous functions $\psi(G) \to \mathbb{Z}$, then the torus relations follow by continuity from the relations for finite subquotients. The Borel-relations for dimension functions can be used to give a homotopical definition of the representation ring of a torus.

Let T be a torus. Recall from II. 10 that we have canonical homomorphisms

(5.15) $RO(T) \to V(T) \to V^{\infty}(T) \to \tilde{V}^{\infty}(T)$.

(5.16) Theorem. *The homomorphisms* (5.15) *are isomorphisms (*tom Dieck [1982a]*).*

Proof. (i) Suppose X is a homotopy representation and Y a generalized homotopy representation. Then X and Y are T-homotopy equivalent if and only if they have the same dimension function.

Suppose they have the same dimension function. Then, by II. 4, there exists a T-map $f: X \to Y$ with degree $f^T = 1$ if $X^T \neq \emptyset$. From (4.45), Ex. 5, we see that, for all $H \subset T$, degree $f^H = \pm 1$ so that f is a T-homotopy equivalence.

(ii) Let Dim: $\tilde{V}^{\infty}(T) \to C'(T)$ be the dimension function map II (10.9). Then Dim $RO(T) \to C'(T)$ and Dim: $\tilde{V}^{\infty}(T) \to C'(T)$ have the same image. For $T = S^1$, this follows by inspection since, by the Smith relation, $\operatorname{Dim} X(\mathbb{Z}/n) \equiv \operatorname{Dim} X(S^1) \bmod 2$ and all functions having this property come from $RO(S^1)$.

For a general torus and a generalized homotopy representation, one derives from the Borel relation that

$$[X] - [X^T] \quad \text{and} \quad \sum_S ([X^S] - [X^T])$$

have the same dimension function (the sum taken over the tori S of codimension one in T). Now use the fact that the dimension function of $[X^S] - [X^T]$ comes from the T/S-space X^S, $T/S \cong S^1$.

(iii) By (i), the map Dim: $V^{\infty}(T) \to C'(T)$ is injective. By (ii), any generalized homotopy representation has stably the dimension function of a linear homotopy representation. Hence (i) also tells us that Dim: $\tilde{V}^{\infty}(T) \to C'(T)$ is injective, too. By (ii), $RO(T)$ maps onto each of the groups $V(T)$, $V^{\infty}(T)$, $\tilde{V}^{\infty}(T)$, and by II (5.16), the map Dim: $RO(T) \to C'(T)$ is injective. □

Bauer [1986] has shown among other things that for an extension $1 \to T \to G \to S \to 1$ of a torus T by a finite p-group S, a Borel-Smith function is the dimension function of some $x \in RO(G)$.

(5.17) Exercises.

1. Determine the faithful irreducible representations of the dihedral, semidihedral, and quaternion groups and show that they are Galois conjugate.

2. Verify the statement in the proof of (5.4) that the semidihedral group contains a quaternion group of index 2.
3. Give a proof of (5.14).
4. Let $R(G)$ be the complex representation ring of a nilpotent group G. Define a homomorphism $d_1: R(G) \to C(G)$ by assigning to each irreducible representation V the function $(H) \mapsto \frac{1}{2} \dim_{\mathbb{C}} V^H$ if V is of quaternionic type and $(H) \mapsto \dim_{\mathbb{C}} V^H$ if V is of real or complex type. Show that the image of d_1 is a direct summand (tom Dieck [1979a]). Similarly, show that one obtains an embedding $d_2: RO(G) \to C(G)$ as a direct summand by mapping each irreducible V to $d(V)^{-1} \operatorname{Dim} V$ where $d(V) = \dim_{\mathbb{R}} \operatorname{Hom}_G(V, V)$.
5. Show that for complex representations the analogue $R_\psi(G) = R_0(G)$ of (5.9) holds.
6. Let X be a homotopy representation for the torus. There exists a real representation V of T such that:
   (i) For each p-group $H \subset T$, $\operatorname{Dim} X(H) = \operatorname{Dim} SV(H)$.
   (ii) There exists a T-map $X \to SV$ of degree one.
7. Let $f: X \to Y$ be a T-map of degree one between simply-connected generalized homotopy representations for the torus T. Then the orbit map $f/T: X/T \to Y/T$ is a homotopy equivalence.
8. Let G be a finite cyclic group. Let $RO_h(G) \subset RO(G)$ be the subgroup generated by all $V - W$ such that SV and SW are G-homotopy equivalent. Show that the natural map $RO(G)/RO_h(G) \to V(G)$ is an isomorphism. Hint: Use the Smith relations and II (4.12).
   Remark: $V(G) \to V^\infty(G)$ is an isomorphism in this case, too.

6. Further results for cyclic groups. Applications.

We continue in the style of section 4 and derive cohomological results for actions of cyclic groups. Again, we make the general assumptions

(6.1) (i) $(p) \subset \mathbb{Z}$ is a fixed prime ideal.
(ii) $K = F_p$ is the prime field of characteristic p used as coefficient group for cohomology.
(iii) $G = \mathbb{Z}/p$ for $p \neq 0$ and $G = S^1$ for $p = 0$.
(iv) (X, A) is a pair of finite-dimensional G-complexes of finite orbit type.
(v) $\dim_K \bigoplus_i H^i(X, A)$ is finite.

It is useful to consider the following model for EG: The direct limit of inclusions of unit spheres

$$\ldots \subset S(\mathbb{C}^n) \subset S(\mathbb{C}^{n+1}) \subset \ldots$$

with action of $G \subset S^1 \subset \mathbb{C}$ given by scalar multiplication. Denote this limit by

S^∞. We shall consider the groups $H^*(S^q \times_G (X, A))$ for finite and infinite q. In the following, we suppress A from the notation.

Let us look at the commutative diagram induced by the inclusions $i\colon X^G \subset X$ and $s_q\colon S^q \subset S^\infty$ for odd q.

$$\begin{array}{ccccc}
 & & H^m(S^\infty \times_G X) & \xrightarrow[(S^\infty \times_G i)^*]{} & H^m(S^\infty \times_G X^G) \\
 & & \Big\downarrow {\scriptstyle (s_q \times_G X)^*} & & \Big\downarrow {\scriptstyle (s_q \times_G X^G)^*} \\
H^m(S^q \times_G (X, X^G)) & \longrightarrow & H^m(S^q \times_G X) & \xrightarrow[(S^q \times_G i)^*]{} & H^m(S^q \times_G X^G)
\end{array}$$

Let m be chosen large enough so that $(S^\infty \times_G i)^*$ is an isomorphism (4.9). The map $(s_q \times_G X^G)^*$ is always an epimorphism: Using the Künneth formula, it suffices to show that $s_q^*\colon H^*(S^\infty/G) \to H^*(S^q/G)$ is surjective. This we leave to the reader, see exercise (2.7.2).

We conclude from the diagram that $(S^q \times_G i)^*$ is surjective so that the cohomology sequence at the bottom of the diagram is short exact. In terms of dimensions over K, we thus obtain the relation

(6.2) $$\dim H^m(S^q \times_G X) = \dim H^m(S^q \times_G X^G) + \dim H^m(S^q \times_G (X, X^G)).$$

We can apply the Künneth formula to $H^m(S^q \times_G X^G)$ and obtain, using the abbreviation $k = m - q + \dim G$,

(6.3) $$\dim H^m(S^q \times_G X^G) = \begin{cases} \sum\limits_{k \le i} \dim H^i(X^G), & p \neq 0 \\ \sum\limits_{\substack{k \le i \\ k \equiv i(2)}} \dim H^i(X^G), & p = 0. \end{cases}$$

From the spectral sequence of the fibration $S^q \times_G X \to S^q/G$ we see

(6.4) $$\dim H^m(S^q \times_G X) \le \begin{cases} \sum\limits_{k \le i} \dim H^i(X)^G, & p \neq 0 \\ \sum\limits_{\substack{k \le i \\ k \equiv i(2)}} \dim H^i(X)\,, & p = 0. \end{cases}$$

Finally, we have, for m large enough,

(6.5) Lemma. $\dim H^m(S^q \times_G (X, X^G)) \ge \dim H^{k-d}(X/G, X^G)$ *with* $d = \dim G$.

Proof. For $p \neq 2$, this follows immediately from the spectral sequence of the relative fibration $S^q \times_G (X, X^G) \to (X/G, X^G)$. For $p = 0$, the same argument works for semi-free actions. The general case can be reduced to the semi-free case by the remarks at the end of II.9. □

Now we put (6.2)–(6.5) together and obtain

(6.6) Proposition. *Under the hypothesis of* (6.1), *we have for each integer* $k \geq 0$

$$\dim H^{k-d}(X/G, A/G \cup X^G) + \sum_{k \leq i} \dim H^i(X^G, A^G) \leq \sum_{k \leq i} \dim H^i(X, A)^G.$$

If $p = 0$, then a similar inequality holds if the two sums are replaced by the sum over $k \leq i$, $k \equiv i \mod 2$. In particular, $\dim \bigoplus H^i(X/G, A/G)$ and $\dim \bigoplus H^i(X^G, A^G)$ are finite ($d = \dim G$). □

(6.7) Theorem. *Under the hypothesis* (6.1), *the Euler characteristics with K-coefficients of* (X, A), (X^G, A^G), *and* $(X/G, A/G)$ *are defined. The following relations exist between these Euler characteristics:*

(i) $\chi(X, A) + (p-1)\chi(X^G, A^G) = p\chi(X/G, A/G)$, *for* $p \neq 0$.
(ii) $\chi(X, A) = \chi(X^G, A^G)$, *for* $p = 0$.

Proof. Let us consider the case $p \neq 0$ first. Suppose G acts freely on $X \setminus A$. For each $K[G]$-module M, we have the equivariant cohomology groups $\mathfrak{H}_G^*(X, A; M)$ introduced in II.3.

If $M = K[G]$ is the group ring, then

(6.8) $\quad \mathfrak{H}_G^*(X, A; M) \cong H^*(X, A)$

and if $M = K$ is the trivial module, then

(6.9) $\quad \mathfrak{H}_G^*(X, A; M) \cong H^*(X/G, A/G)$

(see II (9.7)).

If $0 \to M_1 \to M_2 \to M_3 \to 0$ is an exact sequence of $K[G]$-modules, then we have an exact sequence

$$\ldots \to \mathfrak{H}_G^*(X, A; M_1) \to \mathfrak{H}_G^*(X, A; M_2) \to \mathfrak{H}_G^*(X, A; M_3) \to \ldots.$$

If the $\mathfrak{H}_G^*(X, A; M)$ are finite-dimensional vector spaces over K, we define an Euler characteristic $\chi(X, A; M) = \sum_i (-1)^i \dim_K \mathfrak{H}_G^i(X, A; M)$.

From the exact sequence just mentioned, we obtain

(6.10) If $\chi(X, A; M_2)$ and $\chi(X, A; M_3)$ are defined, then $\chi(X, A; M_2)$ is defined and $\chi(X, A; M_2) = \chi(X, A; M_1) + \chi(X, A; M_3)$.

We shall use the following algebraic fact

(6.10) Lemma. *There exists a filtration by* $K(G)$*-submodules*

$$0 = M_0 \subset M_1 \subset M_2 \subset \ldots \subset M_p = K[G]$$

such that $M_{i+1}/M_i \cong K$ *is the trivial module.*

Proof. If M is any finite-dimensional $K[G]$-module, then $M^G \neq \{0\}$; because if $M^G = \{0\}$, then $G = \mathbb{Z}/p$ would act freely on $M \setminus \{0\}$ and this contradicts the fact that $|M| - 1$ is not divisible by p. □

We continue with the proof of (6.7). Since $\chi(X, A; K)$ is defined, we see by induction that the $\chi(X, A; M_i)$, M_i from (6.11), are all defined. From (6.8)–(6.10) we conclude $\chi(X, A) = p\chi(X/G, A/G)$.

If G does not act freely on $X \setminus A$, we can apply the preceding argument to $(X, A \cup X^G)$; note that $\chi(X^G, A^G)$ is defined and, because of the exactness of $H^*(X, A \cup X^G) \to H^*(X, A) \to H^*(A \cup X^G, A) = H^*(X^G, A^G)$, we see that $\chi(X, A \cup X^G)$ is defined and the relation

$$\chi(X, A \cup X^G) = \chi(X, A) - \chi(X^G, A^G)$$

holds. Similarly,

$$\chi(X/G, A/G \cup X^G) = \chi(X/G, A/G) - \chi(X^G, A^G).$$

Applying these relations to the relatively free case $(X, A \cup X^G)$ already proved, we obtain (6.7, i).

We turn to the case $p = 0$. Here, we use another method: We follow the Euler characteristic through the spectral sequence for H_G^* and use the localization theorem.

We consider $H_G^{(*)}(X, A)$ as $\mathbb{Z}/2$-graded $H^{(*)}(BG)$-module. Let L be the quotient field of $H^{(*)}(BG)$; it is concentrated in degree zero. Therefore, $H_G^{(*)}(X, A) \otimes_{H^*(BG)} L$ is a $\mathbb{Z}/2$-graded L-module. The localization theorem yields an isomorphism of $\mathbb{Z}/2$-graded L-modules

$$M(X, A) := H_G^*(X, A) \otimes_{H^*(BG)} L \cong H_G^*(X^G, A^G) \otimes_{H^*(BG)} L = M(X^G, A^G).$$

From the Künneth formula we see that

$$\chi(X^G, A^G) = \dim_L M(X^G, A^G)^0 - \dim_L M(X^G, A^G)^1.$$

The spectral sequence $(E_r^{*,q})$ of the fibration $X_G \to BG$ consists of $H^*(BG)$-modules. We tensor with L and consider this as a spectral sequence of L-modules. The module $M(X, A)$ has a filtration by submodules $F_q M = (F_q M^0, F_q M^1)$ and

$$F_q M^i / F_{q-1} M^i \cong \bigoplus_m E_\infty^{2m+i-q,q}, \quad i = 0,1.$$

Therefore,

$$\dim_L M^i = \sum_q \dim_L \bigoplus_m E_\infty^{2m+i-q,q}.$$

Since $E_r^{s,t} \neq 0$ implies that s is even, we see that

$$\chi(X^G, A^G) = \dim_L M^0 - \dim_L M^1 = \sum_q (-1)^q \dim_L (E_\infty^{(*),q}).$$

Since the Euler characteristic of a chain complex and its homology are equal, it follows that the right hand sum equals

$$\sum_q (-1)^q \dim_L (E_2^{(*),q})$$

which is, by the Künneth formula, equal to $\chi(X, A)$. □

(6.12) Theorem. *Let G be a finite group and X a finite-dimensional G-complex. If* $\tilde{H}^*(X; \mathbb{Z}) = 0$, then $\tilde{H}^*(X/G; \mathbb{Z}) = 0$.

Proof. Let G be $\mathbb{Z}/p$, p a prime. Since $\tilde{H}^*(X; \mathbb{Z}/p) = 0$, we conclude from (4.10) and (6.6) that $\tilde{H}^*(X/G; \mathbb{Z}/p) = 0$. Let G be a p-group. Since, for $H \lhd G$, we have $X/G = (X/H)/(G/H)$, we see by induction that $\tilde{H}^*(X/G; \mathbb{Z}/p) = 0$ in this case. Let G be a general group and P its p-Sylow group. By II.9, there exist homomorphisms

$$\tilde{H}^*(X/G; \mathbb{Z}/p) \to \tilde{H}^*(X/P; \mathbb{Z}/p) \to \tilde{H}^*(X/G; \mathbb{Z}/p)$$

whose composition is multiplication by $|G/P|$. Thus $\tilde{H}^*(X/G; \mathbb{Z}/p) = 0$ for all primes p. The exact sequence

$$\tilde{H}^*(X/G; \mathbb{Z}) \xrightarrow{\ p\ } \tilde{H}^*(X/G; \mathbb{Z}) \to \tilde{H}^*(X/G; \mathbb{Z}/p) = 0$$

shows that $\tilde{H}^*(X/G; \mathbb{Z})$ is divisible. By II.9, there is a composition

$$\tilde{H}^*(X/G; \mathbb{Z}) \to \tilde{H}^*(X; \mathbb{Z}) \to \tilde{H}^*(X/G; \mathbb{Z})$$

which is multiplication by $|G|$. This proves the assertion. □

We mention a few related results. For the proof of the next theorem, we refer to Bredon [1972], p. 91.

(6.13) Theorem. *Let G be a compact Lie group, X a G-space, $f\colon [0,1] \to X/G$ a path, and $q\colon X \to X/G$ the orbit map. Then there exists a path $F\colon [0,1] \to X$ such that $qF = f$.* □

(6.14) Corollary. *If X is arcwise connected and if there exists a connected orbit, then $q_*\colon \pi_1(X) \to \pi_1(X/G)$ is surjective.* □

(6.15) Theorem. *Let X be a finite-dimensional G-complex and let G be finite. If X is contractible, then X/G is contractible.*

Proof. By (6.12), we know that $\tilde{H}^*(X/G; \mathbb{Z}) = 0$. If we can show that $\pi_1(X/G) = 0$, then contractibility of X/G follows from the Whitehead theorem of algebraic topology. If G is a p-group, the $X^G \neq \emptyset$ by (3.14) and therefore

$\pi_1(X/G) = 0$ by (6.14). By induction on $|G|$, we conclude that X/G is contractible whenever G is solvable. Now let G be arbitrary and let $f: I \to X/G$ be a loop and $F: I \to X$ a lifting as in (6.13). Then F joins two points x and gx where $g \in G$. Let C be the cyclic group generated by g. We already know that X/C is contractible. But F yields a loop in X/C which projects to f. Therefore, f is nullhomotopic. □

The most general results of this type are due to Oliver [1976a]. In particular, using also earlier results of Conner [1960], he proves

(6.16) Theorem. *Any action of a compact Lie group on a Euclidean space has contractible orbit space.* □

We mention some further results on bundle cohomology and localization.

In certain cases the localization theorem can be reformulated in terms of deformations of algebras. This important viewpoint has been introduced by V. Puppe [1978], [1979]. It leads to the use of commutative algebra and algebraic geometry. V. Hauschild [1983] has successfully applied this method.

Localization methods can also be used in rational homotopy theory. Related is the use of bundle cohomology on the cochain level. For this method, see Allday [1978], [1979], Allday-V. Puppe [1985], Carlsson [1983], Atiyah-Bott [1984].

For actions on Poincaré duality spaces, see Allday-Skjelbred [1974], Chang-Skjelbred [1972], Bredon [1973], Stieglitz [1978].

The bundle cohomology can be used to define a topological substitute for the weight system of a compact Lie group. This viewpoint is due to W.C. Hsiang and W.Y. Hsiang. See W.Y. Hsiang [1975] for a general introduction and further references. See also Allday-Skjelbred [1974], Chang-Skjelbred [1974]. For estimations of the size of the fixed point cohomology $H^*(X^G)$ in terms of the size of $H^*(X)$, see Bredon [1964], V. Puppe [1974], Chang [1976]. Further references: Duflot [1981], [1983], Chang-Skjelbred [1976], Skjelbred [1974], [1978], [1978a].

For localization methods in cobordism theory, see tom Dieck [1970], [1970a], [1970b], [1972a], [1972b], [1974a].

For localization in K-theory, see Atiyah-Segal [1968].

(6.17) Exercises.

1. Let X be a G-ENR. Consider Alexander-Spanier cohomology with compact support (isomorphic to sheaf – or presheaf cohomology with compact support; see Spanier [1966], Chapter 6; Bredon [1967a], Chapter II).
   Let G be a finite p-group acting freely on X. Suppose $H^*(X; F_p)$ is finite-dimensional. Show: $\chi(X/G; F_p)$ is defined and

$$\chi(X; F_p) = |G|\chi(X/G; F_p).$$

Hints: If $H \lhd G$, then G/H acts freely on the G/H-ENR X/H. Thus (induction) it suffices to consider $G = \mathbb{Z}/p$. There is an isomorphism

$$H^i(X; F_p) \cong H^i(X/G; A)$$

where A is the local coefficient system with stalks $H^0(\pi^{-1}(x); F_p)$, $\pi: X \to X/G$ the quotient map. Use (6.11) in order to filter $A = A_p \supset \ldots \supset A_0 = 0$ by F_pG-modules such that A_i/A_{i-1} is the trivial module. Use the Cartan spectral sequence of a covering (Bredon [1967a], p. 154) to show that $H^i(X/G; F_p)$ is finite-dimensional. Finally use the additivity of the Euler characteristic $\chi(X; A_i) = \chi(X; A_{i-1}) + \chi(X; A_i/A_{i-1})$.

2. Let X be as in the previous exercise. Suppose $H^*(X; \mathbb{Z})$ is a finitely generated abelian group. Let the finite group G act freely on X. Show that

$$\chi_G(X; \mathbb{Q}) = \chi(X/G; \mathbb{Q}) \cdot \mathbb{Q}G \in R(G; \mathbb{Q}).$$

Here, $\mathbb{Q}G$ denotes the regular representation of G over $\mathbb{Q}$. Hints: Show that the two elements in question have the same character. Thus one has to show that

$$\chi(X) = |G|\chi(X/G)$$

and

$$\chi(X)(g) = 0 \quad \text{for} \quad g \neq 1.$$

Prove the first equation for cyclic groups as follows: Since $H^*(X; \mathbb{Z})$ is finitely generated, the universal coefficient formula for cohomology with compact support (Spanier [1966], p. 338) shows that $\chi(X; \mathbb{Q}) = \chi(X; F_p)$. The Cartan spectral sequence of a covering shows that $H^*(X/G; \mathbb{Z})$ is finitely-generated. Now use exercise 1 and induction on $|G|$.
From II.9 one has $H^i(X; \mathbb{Q})^G \cong H^i(X/G; \mathbb{Q})$ if one assumes that the transfer exists for the cohomology theory in question. From the character formula $|G| \dim \psi^G = \sum_{g \in G} \psi(g)$ for each character ψ and from $\chi(X) = |G|\chi(X/G)$ one obtains, for cyclic G, that

$$\sum_{g \neq 1} \chi_G(X)(g) = 0.$$

Prove $\chi_G(X)(g) = 0$, $g \neq 1$, by induction on the order of g as follows. Start with

$$H^i(X; \mathbb{C}) = H^i(X/G; A),$$

G generated by g and A a coefficient system with typical stalk $\mathbb{C}G$. Decompose A according to the irreducible $\mathbb{C}G$-modules

$$A = \oplus A_j, \quad 0 \leq j < m = |G|$$

where g acts on A_j through multiplication with $\zeta^j = \exp(2\pi i j/m)$. The equalities of traces

$$\begin{aligned}\operatorname{Tr}(g^k, H^i(X; \mathbb{C})) &= \sum_j \operatorname{Tr}(g^k, H^i(X/G; A_j)) \\ &= \sum_j \zeta^{jk} \dim H^i(X/G; A_j)\end{aligned}$$

yield

$$L(g^k, X) = \sum_j \zeta^{jk} \chi(X/G; A_j)$$

for the Lefschetz number of the g^k-action on X.
But, for $(k, m) = 1$, $L(g^k, X) \in \mathbb{Z}$ is obtained from $L(g, X)$ by applying a Galois automorphism of $\mathbb{Q}(\zeta)$ over $\mathbb{Q}$. Therefore, $L(g^k, X) = L(g, X)$ for $(k, m) = 1$. Thus obtain

$$0 = \sum_{(k,m)=1} L(g^k, X) + \sum_{(k,m)\neq 1} L(g^k, X).$$

By induction, the second sum in zero and since the summands of the first sum are all equal to each other, one obtains $L(g, X) = 0$. Show that the result for cyclic G implies the statement for general G.

3. Let X be a compact G-ENR. Suppose G is cyclic with generator g. Show that the Lefschetz number

$$L(g, X) = \sum_i (-1)^i \operatorname{Tr}(g, H^i(X; \mathbb{Q}))$$

is equal to the Euler characteristic $\chi(X^g)$. Hints: Let $X_1 = X^g, X_2, \ldots, X_r$ be the orbit bundles of X. Then $H^*(X_i, \mathbb{Z})$ is a finitely generated abelian group (cohomology with compact support) and $L(g, X_j) = 0$ for $j > 1$ by the previous exercise. Hence

$$L(g, X) = \sum_j L(g, X_j) = L(g, X_1) = \chi(X^g).$$

4. Let G be a finite group and let X be a compact G-ENR. Show that

$$|G|\,\chi(X/G) = \sum_{g\in G} \chi(X^g).$$

Hint: Use $H^i(X/G) \cong H^i(X)^G$ and the previous exercise.

Chapter IV: The Burnside Ring

This chapter comprises the following sections.

1. The Burnside ring and generalizations as universal additive invariants.
2. The Burnside ring and its elementary properties.
3. The space of subgroups of a compact Lie group in the Hausdorff metric.
4. The prime ideal spectrum of the Burnside ring.
5. Description of the Burnside ring by congruences between Euler characteristics of fixed point sets.
6. Finiteness theorems for compact Lie groups G; in particular, the boundedness of the orders of $\pi_0(NH/H)$, H subgroup of G.
7. Description of the idempotent elements of the Burnside ring in terms of the subgroup structure.
8. Mackey and Green functors for compact Lie groups.
9. Dress induction for compact Lie groups and the hyperelementary induction theorem.
10. Localization of stable homology theories at prime ideals of the Burnside ring. Splitting of homology theories.

1. Additive invariants.

The Euler characteristic for finite cell complexes is the universal additive invariant. This section gives a treatment of additive invariants for finite G-complexes and reduces its computation to the study of Euler characteristics of fixed point sets.

An **additive invariant** for finite G-complexes consists of an abelian group A and a map b which assigns to each finite G-complex X an element $b(X) \in A$ such that the following axioms hold:

(1.1) (i) If X and Y are G-homotopy equivalent, then $b(X) = b(Y)$.

(ii) If X and Y are subcomplexes of Z, then

$$b(X) + b(Y) = b(X \cap Y) + b(X \cup Y).$$

(iii) $b(\emptyset) = 0$.

An additive invariant can essentially be calculated by counting cells.

(1.2) Proposition. *Let $n(X, H, i)$ be the number of i-cells of type (H) of X and put*

$n(X, H) = \sum_{i \geq 0} (-1)^i n(X, H, i)$. *Then*

$$b(X) = \sum_{(H)} n(X, H)\, b(G/H).$$

The sum is taken over conjugacy classes of subgroups.

Proof. Induction on the number of cells and on the dimension. Let $Z = X \cup (G/H \times D^n)$ be obtained from X by attaching an n-cell of type (H). Let $Y = G/H \times D^n(\frac{1}{2})$ be the closed cell in $G/H \times D^n$ of radius $\frac{1}{2}$ about the center. If we remove the interior $\mathring{Y}$ of Y from Z, then the resulting space is G-homotopy equivalent to X. Therefore, by (1.1), (i) and (ii),

(1.3) $$\begin{aligned} b(Z) &= b(Z \setminus \mathring{Y}) + b(Y) - b((Z \setminus \mathring{Y}) \cap Y) \\ &= b(X) + b(G/H \times D^n) - b(G/H \times S^{n-1}). \end{aligned}$$

(With a little good will, $Z \setminus \mathring{Y}$ and Y are subcomplexes of the space Z in some cellular structure.) One shows by induction

(1.4) $$b(G/H \times S^n) = (1 + (-1)^n)\, b(G/H), \quad n \geq -1.$$

Namely, let D_+ (resp. D_-) be the upper (lower) hemisphere of S^n. Then

$$\begin{aligned} b(G/H \times S^n) &= b(G/H \times D_+) + b(G/H \times D_-) - b(G/H \times S^{n-1}) \\ &= 2b(G/H) - (1 + (-1)^{n-1})\, b(G/H) \\ &= (1 + (-1)^n)\, b(G/H). \end{aligned}$$

The induction starts with $n = -1$ by (1.1, iii). Using (1.3) and (1.4), we obtain $b(Z) = b(X) + (-1)^n b(G/H)$, the induction step. Again (1.1, iii) is used to start the induction. □

It is apparent from (1.2) that a universal additive invariant is obtained if the $b(G/H)$ are linearly independent. In order to deal with the universal situation, it is slightly more convenient to work with pointed G-complexes. Therefore, let $\mathfrak{C}(G)$ denote the category of pointed finite G-complexes. An **additive invariant** for $\mathfrak{C}(G)$ is a pair (B, b), B an abelian group and b an assignment, which associates to X an element $b(X) \in B$ such that the following axioms hold:

(1.5) (i) If X and Y are pointed G-homotopy equivalent, then $b(X) = b(Y)$.
(ii) If A is a pointed subcomplex of X, then $b(A) - b(X) + b(X/A) = 0$.

An additive invariant (U, u) for $\mathfrak{C}(G)$ is called **universal** if each additive invariant (B, b) is obtained from (U, u) as $b(X) = \varphi u(X)$ for a unique homomorphism $\varphi\colon U \to A$. As usual, a universal additive invariant is uniquely determined. The existence of a universal invariant is shown by a construction of a Grothendieck group: Let F be the free abelian group generated by the pointed G-homotopy classes of finite complexes. Write $[X]$ for the basis element of F

defined by the complex X. Let N be the subgroup of F generated by all elements $[A]-[X]+[X/A]$ for pointed subcomplexes A of X. Put $U(G)=F/N$ and let $u(X)$ be the class of $[X]$ in $U(G)$. Then $(U(G), u)$ is a universal additive invariant. Let X^+ be X with a separate base point added.

For each subgroup H of G, we obtain an additive invariant $(\mathbb{Z}, \chi_H)$ as follows: Let $\chi(A)$ denote the Euler characteristic of A and put

$$(1.6)\qquad \chi_H(X)=\chi(X^H/NH)-1\in\mathbb{Z}.$$

Elementary properties of the Euler characteristic (e.g. Dold [1972], V. 5) show that this defines an additive invariant. By universality, $X\mapsto\chi_H(X)$ extends to a homomorphism from $U(G)$, also denoty by

$$(1.7)\qquad \chi_H\colon U(G)\to\mathbb{Z}.$$

(1.8) Proposition. *$U(G)$ is the free abelian group on elements $u(G/H^+)$ where G/H runs through the isomorphisms types of homogeneous G-spaces (equivalently: H runs through a complete set of conjugacy classes of closed subgroups H of G).*

Proof. For each additive invariant, (1.5, ii) yields the relation $b(P)=0$ for a point P. If CX denotes the cone on X and $S^1\wedge X=CX/X$ the suspension, then $b(CX)=b(P)=0$ by (1.5, i); then $b(X)=-b(S^1\wedge X)$ by (1.5, ii). Moreover, $b(X\vee Y)=b(X)+b(Y)$ by (ii). We show that $U(G)$ is generated by the $u(G/H^+)$. A zero-dimensional $X\in\mathfrak{C}(G)$ has the form $X=\vee_j G/H_j^+$. Hence $b(X)=\sum_j b(G/H_j^+)$ by the previous remarks. If X_n is the n-skeleton of X, then $X_n/X_{n-1}\cong\vee_j S^n\wedge G/H_j^+$. Hence

$$\begin{aligned}u(X_n)&=u(X_{n-1})+u(X_n/X_{n-1})=u(X_{n-1})+\textstyle\sum_j u(S^n\wedge G/H_j^+)\\&=u(X_{n-1})+(-1)^n\textstyle\sum u(G/H_j^+).\end{aligned}$$

Thus, by induction on the dimension of X, $u(X)$ is contained in the subgroup generated by the $u(G/H^+)$. We show that the $u(G/H^+)$ are linearly independent. Suppose $0=\sum_{(H)}a(H)u(G/H^+)$ in $U(G)$. Let K be maximal with respect to inclusion such that $a(K)\neq 0$. Then $(G/H^+)^K=P$, a point, if H is smaller than K, whence

$$0=\chi_K(0)=\sum a(H)\chi_K u(G/H^+)=\sum a(H)(\chi((G/H)^+)^K-1)=a(K),$$

a contradiction. □

(1.9) Corollary. *The family (χ_H) of homomorphisms χ_H yields an injective homomorphism $\chi\colon U(G)\to\prod_{(H)}\mathbb{Z}$.* □

(1.10) Proposition. *The assignment $(X,Y)\mapsto X\wedge Y$ induces a product $U(G)\times U(G)\to U(G)$ which makes $U(G)$ into a commutative ring with unit.*

Proof. If $L: \mathfrak{C}(G) \to \mathfrak{C}(G)$ is a functor which preserves homotopy types and cofibration sequences, then $X \mapsto u(L(X))$ is an additive invariant which, by universality, yields a homomorphism $u_L: U(G) \to U(G)$ such that $u_L(u(X)) = u(L(X))$. Each X defines such a functor via $Y \mapsto X \wedge Y$. Thus $u(Y) \mapsto u(X \wedge Y)$ is a well defined homomorphism $U(G) \to U(G)$. By interchanging the role of X and Y, it is seen that this homomorphism only depends on $u(X)$. Moreover, $(u(X), u(Y)) \mapsto u(X \wedge Y)$ is bilinear. □

(1.11) Remark. In general, the maps χ_H of (1.7) are not ring homomorphisms.

We give another description of the ring $U(G)$ which is based on more general spaces. Define the following equivalence relation on the set of compact Euclidean G-neighbourhood retracts (compact G-ENR): $X \sim Y$ if and only if $\chi(X^H/NH) = \chi(Y^H/NH)$ for all $H \subset G$. Let $U'(G)$ be the set of equivalence classes. Disjoint union (addition) of representatives induces the structure of an abelian group on $U'(G)$. Let $u'(X)$ be the class of the space X in $U'(G)$. If K is a compact ENR with trivial G-action and Euler characteristic $\chi(K) = -1$, then $X \times K$ represents the additive inverse of $u'(X)$.

(1.12) Proposition. (i) *$U'(G)$ is the free abelian group with basis $u'(G/H)$ where H runs through a complete set of conjugacy classes of subgroups of G.*
(ii) *For each compact G-ENR X, the relation $u'(X) = \sum_{(H)} \chi_c(X_{(H)}/G)\, u'(G/H)$ holds. Here, $\chi_c(X_{(H)}/G) = \chi(GX^H/G, GX^{>H}/G)$.*

Proof. The linear independence of the $u'(G/H)$ is shown as in the proof of (1.8). Relation (ii) implies that the $u'(G/H)$ are generators. Thus it remains to verify this relation. Elements of $U'(G)$ are detected by the well-defined homomorphisms $\chi'_K: u'(X) \mapsto \chi(X^K/NK)$. We show that both sides of the desired equality yield the same result when χ'_K is applied.

By additivity of the Euler characteristic, we have

$$\chi(X^K/NK) = \sum_{(H)} \chi_c(X^K_{(H)}/NK).$$

The projection $X_{(H)} \to X_{(H)}/G$ is a fibre bundle with fibre G/H. This implies that $X^K_{(H)}/NK \to X_{(H)}/G$ is a fibre bundle with fibre $(G/H^K)/NK$. From II (8.13), Ex. 4, we obtain $\chi_c(X^K_{(H)}/NK) = \chi((G/H^K)/NK)\,\chi_c(X_{(H)}/G)$ from which the desired equality follows. □

A similar definition, applied to finite G-complexes, yields a group $U''(G)$. Since finite G-complexes are compact G-ENR, II (8.12), the inclusion of categories induces a homomorphism $U''(G) \to U'(G)$ which, using (1.12), is seen to be an isomorphism. The map

$$\iota: U''(G) \to U(G),\ u''(X) \mapsto u(X^+)$$

is a well-defined homomorphism, which, by (1.8), is seen to be an isomorphism. Moreover, $\chi_{H}\iota = \chi''_{H}$.

The multiplication in $U(G)$ corresponds to cartesian product in $U'(G)$.

(1.13) Proposition. *Cartesian product of representatives induces a multiplication on $U'(G)$ and $U''(G)$. The homomorphism ι is an isomorphism of rings.*

Proof. We only have to show that cartesian product is compatible with the equivalence relation, i.e. we must show that $\chi((X\times Y)^K/NK)$ can be computed from the $\chi(X^H/NH)$ and $\chi(Y^H/NH)$ or, equivalently, from the $\chi_c(X_H/NH)$ and $\chi_c(Y_H/NH)$. We begin with the relation

$$\chi((X\times Y)^K/NK) = \sum_{(H)} \chi_c((X\times Y)^K_{(H)}/NK).$$

The map $(X\times Y)^K_{(H)}/\mathrm{NK} \to Y_{(H)}/G$ is a fibre bundle with fibre $(X^K \times_G H^K)/NK$. Now we use the fact that G/H^K consists of a finite number of NK-orbits, I (5.10), say $G/H^K = \sum_U NK/U$ as NK-space. This implies

$$\chi_c((X\times Y)^K/NK) = \sum_U \chi_c(Y_{(H)}/G)\,\chi(X^K/U).$$

Finally, we use

$$\chi(X^K/U) = \sum_{(H)} \chi_c(X_{(H)}/G)\,\chi((G/H^K)/U)$$

and conclude that $\chi((X\times Y)^K/NK)$ can indeed be computed from the $\chi_c(X_H/NH)$ and $\chi_c(Y_H/NH)$. □

We use the isomorphisms $U'(G) \cong U''(G) \cong U(G)$ obtained above as identifications.

(1.14) Proposition. *Suppose NH/H is not finite. Then $[G/H] \in U(G)$ is nilpotent.*

Proof. By the descending chain property for subgroups of G, the spaces $(G/H)^k = G/H \times \ldots \times G/H$, k factors, $k \geq 1$, altogether have only finitely many conjugacy classes of isotropy groups. We expand $[G/H]^k \in U(G)$ in terms of the basis, $[G/H]^k = \sum_{(K)} a_K[G/K]$. Choose a maximal (L) with $a_L \neq 0$. We claim that the expansion of $[G/H]^{k+1}$ does not contain $[G/L]$ with a non-zero coefficient. Multiplying the expansion of $[G/H]^k$ by $[G/H]$, we see that $[G/L]$ can only occur in the expansion of $[G/H][G/L]$. We use (1.12): $(G/H\times G/L)_L = G/H^L \times NL/L$ and hence $\chi_c((G/H\times G/L)_L/NL) = \chi(G/H^L)$. Since NH/H acts freely on G/H^L and since $\dim NH/H > 0$, we have a free circle action on G/H^L which implies $\chi(G/H^L) = 0$. □

We now generalize the preceding constructions by looking at G-maps $f: Z \to X$ of compact G-ENR's Z into the G-space X. Recall that we have the component category $\pi_0(G, X)$, see I (10.3), where the objects are the G-homotopy classes $[x]: G/H \to X$ and where morphisms from $[x]: G/H \to X$ to $[y]: G/K \to X$ are the G-maps $\sigma: G/H \to G/K$ such that $y\sigma \simeq x$. An object $\alpha: G/H \to X$ may be identified with the path component X^H_α into which G/H is mapped by the morphism α. The automorphism group $\mathrm{Aut}(\alpha)$ of $\alpha = [x]$ consists of those $\sigma: G/H \to G/H$ such that $x\sigma \simeq x$. We have $\sigma \in NH/H$ by identifying σ with $\sigma(eH) = nH$, $n \in NH$. We have an action of NH/H on X^H and $\pi_0(X^H)$; and $N_\alpha H/H$, the isotropy group of $\alpha \in \pi_0(X^H)$, is isomorphic to $\mathrm{Aut}(\alpha)$.

Given $f: Z \to X$ as above and $\alpha: G/H \to X$, we let $Z(f, \alpha) = Z^H \cap f^{-1}(X^H_\alpha)$ be the subspace of Z^H which is mapped under f into X^H_α. The action of NH/H on Z^H induces an action of $N_\alpha H/H$ on $Z(f, \alpha)$ or, in other terms, $Z(f, \alpha)$ is an $\mathrm{Aut}(\alpha)$-space.

Two maps $f_i: Z_i \to X$ for $i = 0, 1$ are called equivalent if and only if for each $\alpha: G/H \to X$ the Euler characteristics $\chi(Z(f_i, \alpha)/\mathrm{Aut}(\alpha))$ are equal. Let $U(G; X)$ be the set of equivalence classes. Let $[f: Z \to X]$ or $[f]$ denote the class of f. We define the structure of an abelian group on $U(G; X)$ by using as composition law the disjoint union

$$[f_0: Z_0 \to X] + [f_1: Z_1 \to X] = [f_0 + f_1: Z_0 + Z_1 \to X].$$

The neutral element is represented by $\emptyset \to X$.

Let A be a compact ENR with trivial G-action and Euler characteristic $\chi(A) = -1$. Then one verifies that $f \circ \mathrm{pr}: Z \times A \to Z \to X$ represents an additive inverse for $[f: Z \to X]$. Each α defines a homomorphism

$$\chi_\alpha: U(G; X) \to \mathbb{Z}, \ [f: Z \to X] \mapsto \chi(Z(f, \alpha)/\mathrm{Aut}(\alpha))$$

and if α ranges over the isomorphism classes $\mathrm{Is}\,\pi_0(G, X)$ of objects of $\pi_0(G, X)$, we obtain an injective homomorphism

$$U(G; X) \to \prod_\alpha \mathbb{Z}, \quad \alpha \in \mathrm{Is}\,\pi_0(G, X).$$

If X is a point, then $U(G; X) = U(G)$, the previously defined object.

We record for later use the following properties of this construction. The proof is left as an exercise.

(1.15) Proposition. (i) *Suppose $f_i: Z_i \to X$ are given and $\sigma: Z_0 \to Z_1$ is a G-homotopy-equivalence such that $f_1\sigma$ is G-homotopic to f_0. Then $[f_0] = [f_1]$.*

(ii) *Let*

$$\begin{array}{ccc} Z_0 & \longrightarrow & Z_1 \\ {\scriptstyle i}\downarrow & & \downarrow \\ Z_2 & \longrightarrow & Z \end{array}$$

be a push-out of compact G-ENR's such that i is a G-cofibration. Let $f: Z \to X$ be given and let $f_i: Z \to X$ be the resulting elements. Then

$$[f_0] + [f_1] = [f] + [f_2] \quad \textit{in } U(G; X).$$

By (1.15, i), each object $\alpha: G/H \to X$ yields a well-defined element $[\alpha] \in U(G; X)$ and this depends only on the isomorphism class of the object. □

If $f: Z \to X$ and $\beta \in \mathrm{Is}\,\pi_0(G, X)$ are given, we let $Z(\beta) \subset Z_{(H)}$ be the G-subspace consisting of the orbits $C \subset Z_{(H)}$ such that $f|C: C \to X$ defines β.

(1.16) Proposition. *$U(G; X)$ is the free abelian group on the elements $[\alpha]$ for $\alpha \in \mathrm{Is}\,\pi_0(G, X)$. For each $f: Z \to X$, the equality*

$$[f] = \sum_\beta \chi_c(Z(\beta)/G)\,[\beta], \quad \beta \in \mathrm{Is}\,\pi_0(G, X)$$

holds in $U(G; X)$.

Proof. We verify that the $[\alpha]$ are linearly independent. We denote objects of $\pi_0(G, X)$ in the following manner: $\alpha: G/H_\alpha \to X$. Suppose $x = \sum n_\alpha [\alpha: G/H_\alpha \to X] = 0$ in $U(G; X)$. Let $H_\gamma = K$ be a maximal subgroup such that $n_\gamma \neq 0$. Consider

$$\chi_\gamma(\beta: G/H_\beta \to X) = \chi(G/H_\beta^K \cap \beta^{-1}(X_\gamma^K)/\mathrm{Aut}(\gamma)).$$

If this is non-zero, then H_γ is subconjugate to H_β. By maximality of H_γ, only the summands $n_\beta \chi_\gamma [\beta: G/H_\beta \to X]$ with $H_\gamma \sim H_\beta$ can contribute to $\chi_\gamma x$. In this case,

$$gH_\beta \in G/H_\beta^K \cap \beta^{-1}(X_\gamma^K)$$

is equivalent to

$$g^{-1} H_\gamma g = H_\beta \quad \text{and} \quad \beta(gH_\beta) \simeq \gamma(eH_\gamma) \quad \text{in } X^K$$

and therefore $\sigma \in \mathrm{Hom}(G/H_\gamma, G/H_\beta) = G/H_\beta^K$ corresponding to gH_β is an isomorphism from γ to β in $\pi_0(G; X)$. Thus only the summand $n_\gamma[\gamma]$ remains and $\chi_\gamma[\gamma] \neq 0$ then contradicts $\chi_\gamma(x) = 0$. Therefore, the $[\alpha]$ must be linearly independent.

The second assertion of (1.16) shows that the $[\alpha]$ generate $U(G; X)$. Thus it remains to prove this assertion. We show that both sides of the asserted equality have the same value under χ_α. The additivity of the Euler characteristic yields

$$\chi_\alpha(f) = \sum_\beta \chi_c(Z(\beta)^K \cap f^{-1}(X_\alpha^K)/\mathrm{Aut}(\alpha))$$

if $\alpha: G/K \to X$. Now one verifies

(1.17) (i) $Z(\beta) \to Z(\beta)/G$ is a fibre bundle with fibre G/H.

(ii) $Z(\beta)^K \cap \beta^{-1}(X_\alpha^K) \to Z(\beta)/G$ is a fibre bundle with fibre $G/H_\beta^K \cap \beta^{-1}(X_\alpha^K)$.

(iii) $\mathrm{Aut}(\alpha)$ acts as automorphism group of the bundle in (ii) and, taking $\mathrm{Aut}(\alpha)$-orbit spaces, again induces a fibre bundle.

Using (1.17) and the multiplicativity of the Euler characteristic in fibre bundles, one obtains

$$\chi_c(Z(\beta)^K \cap f^{-1}(X_\alpha^K)/\mathrm{Aut}(\alpha)) = \chi_c(Z(\beta)/G)\,\chi(G/H_\beta^K \cap \beta^{-1}(X_\alpha^K)/\mathrm{Aut}(\alpha))$$

and this shows that χ_α yields the same number when applied to both sides of the equation in (1.16). □

If one uses G-maps $f: Z \to X$ of finite G-complexes Z into X and proceeds as above, one obtains a group which is isomorphic to $U(G; X)$. If $G/H \times E^n \subset Z$ is an (open) n-cell of Z, then the restriction of f to $G/H \times E^n$ defines a certain basis element $[\alpha]$ of $U(G; X)$. We therefore call this cell an n-cell of type α. Let $n(\alpha, i)$ be the number of i-cells of type α and let $n(\alpha) = \sum(-1)^i n(\alpha, i)$. Using (1.15), a similar proof as for (1.2) yields

(1.18) Proposition. $[f: Z \to X] = \sum_\alpha n(\alpha)\,[\alpha]$. □

We can also characterize $U(G; X)$ through a universal property. An additive invariant for finite G-complexes $f: Z \to X$ over X is an abelian group A and an assignment $a[f] \in A$ to each $f: Z \to X$ such that the following holds:

(1.19) (i) If $f_0: Z_0 \to X$ and $f_1: Z_1 \to X$ and a G-homotopy equivalence σ with $f_1\sigma \simeq_G f_0$ are given, then $a[f_0] = a[f_1]$.

(ii) If

$$\begin{array}{ccc} Z_0 & \longrightarrow & Z_1 \\ {\scriptstyle j}\big\downarrow & & \big\downarrow \\ Z_2 & \longrightarrow & Z \end{array}$$

is a pushout of finite G-complexes and j a G-cofibration and $f: Z \to X$ a G-map with restrictions $f_i: Z_i \to X$, then

$$a[f] + a[f_0] = a[f_1] + a[f_2].$$

(iii) $a[\emptyset \to X] = 0$.

An additive invariant (A, a) is called universal if every other additive invariant (B, b) is obtained from (A, a) by composing with a unique homomorphism $A \to B$. From (1.15) we see that $U(G; X)$ and $u[f: Z \to X] = [f]$ is an additive invariant. By a Grothendieck construction, there exists a universal additive invariant (A, a) and hence a uniquely defined homomorphism $\varphi: A \to U(G; X)$. The proof of (1.18) shows that A is generated by the $a[G/H \to X]$ and that φ is surjective. Using (1.16), one sees that φ is an isomorphism. Thus we have

(1.20) Proposition. *$U(G; X)$ and $(Z \to X) \mapsto [Z \to X]$ is a universal additive invariant for finite G-complexes over X.* □

We list further properties of the groups $U(G; X)$.

(1.21) Proposition. *Let H be a subgroup of G and X an H-space. There is a natural isomorphism $U(H; X) \to U(G; G \times_H X)$ which is induced on representatives by mapping $[f: Z \to X]$ to $[G \times_H f: G \times_H Z \to G \times_H X]$.*

Proof. We use the definition of the groups via finite complexes (but see exercise 4). One has to show that the map in question is well-defined. This follows with the universal property (1.20) because it is easily seen that the functor $Y \mapsto G \times_H Y$ preserves the relation given by (1.19). An inverse homomorphism is obtained by restricting $Z \to G \times_H X$ to $H \times_H X \cong X$; this is again well-defined by the universal property. □

If $\varphi: H \to G$ is a homomorphism between compact Lie groups, then regarding a G-space via φ as H-space induces a homomorphism

(1.22) $\quad U(\varphi): U(G; X) \to U(H; X).$

The universal property again shows this to be well-defined.

(1.23) Proposition. *Cartesian product induces a bilinear map*

$$U(G; X) \times U(H; Y) \to U(G \times H; X \times Y)$$
$$(a: A \to X, b: B \to Y) \mapsto (a \times b: A \times B \to X \times Y).$$

If $G = H$, one can compose with the diagonal $G \to G \times G$, using (1.22), *to obtain an internal bilinear pairing*

(1.24) $\quad U(G; X) \times U(G; Y) \to U(G; X \times Y).$

If X and Y are points, this induces the ring structure in $U(G)$. □

The groups $U(G; X)$ can be constructed via pointed complexes. The following presentation is related to Oliver-Petrie [1982]. Suppose X is a fixed G-space and $\pi_0(G, X) = \pi(X)$ the component category. A $\pi(X)$-space consists of a pointed G-space Z together with pointed subspaces Z_α for each $\alpha \in \mathrm{Ob}\, \pi(X)$ such that the following holds:

(1.25) $\quad$ (i) $Z^H = \bigvee_{\varrho(\alpha) = G/H} Z_\alpha$ for all $H \subset G$.

(We let $\varrho(\alpha) = G/H$ if $\alpha: G/H \to X$.)

(ii) For each morphism

$$\sigma\colon(\alpha\colon G/H \to X) \to (\beta\colon G/K \to X),$$

the induced map $\sigma^*\colon Z^K \to Z^H$ satisfies $\sigma^*(Z_\beta) \subset Z_\alpha$.

In particular $\alpha \mapsto Z_\alpha$ is a functor $\pi(X) \to \mathrm{Top}_0$ from $\pi(X)$ into the category of pointed spaces Top_0. We denote a $\pi(X)$-space by Z and usually suppress the system (Z_α) from the notation.

Suppose $Z(1)$ and $Z(2)$ are $\pi(X)$-spaces. A pointed map $f\colon Z(1) \to Z(2)$ is a $\pi(X)$-map if for all α the inclusion $f(Z(1)_\alpha) \subset Z(2)_\alpha$ holds. Then f is a natural transformation between the corresponding functors.

Let $\alpha\colon G/H \to X$ be given. Let S be a space with trivial G-action. We equip $(G/H \times S)^+$ with the structure of a $\pi(X)$-space by setting

$$(G/H \times S)^+_\beta = (\mathrm{Hom}_G(\beta, \alpha) \times S)^+$$

where $\mathrm{Hom}_G(\beta, \alpha) \subset \mathrm{Hom}_G(G/K, G/H)$ for $\beta\colon G/K \to X$ carries the compact-open-topology. This certainly defines a contravariant functor $\pi(X) \to \mathrm{Top}_0$ and (1.25, i) is easily verified.

We use this construction to define $\pi(X)$-complexes. They are obtained by successive attachment of $\pi(X)$-cells. For a $\pi(X)$-space Z and a $\pi(X)$-map $(G/H, \alpha) \times S^{n-1} \to Z$, we form the pushout diagram

(1.26)
$$\begin{array}{ccc} ((G/H, \alpha) \times S^{n-1})^+ & \longrightarrow & Z \\ \big\downarrow \cap & & \big\downarrow \\ ((G/H, \alpha) \times D^n)^+ & \longrightarrow & Z' \end{array}$$

and say that Z' is obtained from Z by attaching an n-cell of type α. The space Z' inherits the structure of a $\pi(X)$-space in a canonical way such that (1.26) becomes a diagram of $\pi(X)$-spaces. If we start with a base point, this notion of cell-attachment leads to the notion of a pointed $\pi(X)$-complex.

Now we briefly consider additive invariants for finite $\pi(X)$-complexes and G-maps between $\pi(X)$-complexes. There is a notion of homotopy for $\pi(X)$-maps: Pointed G-homotopy which maps the Z_α-subspaces into the corresponding spaces. The universal group for finite $\pi(X)$-complexes $U(G; \pi(X))$ is defined using the relations

(1.27) (i) If Z and Z' are homotopy-equivalent $\pi(X)$-complexes, then $[Z] = [Z']$ in $U(G; \pi(X))$.

(ii) If $A \to X \to C$ is a cofibration sequence of $\pi(X)$-complexes, then $[A] - [X] + [C] = 0$ in $U(G; \pi(X))$.

The group $U(G; \pi(X))$ is isomorphic to $U(G; X)$. The basis element

$\alpha\colon G/H \to X$ corresponds to $(G/H^+, \alpha)$ where we define $(G/H^+, \alpha)_\beta = \mathrm{Hom}_G(\beta, \alpha)^+$ as above.

Finally, we consider a new type of additive invariants for spaces which was introduced by Lück [1986]. Each space of a suitable category will have its own value group for the invariants.

Let C be a full subcategory of the category of G-spaces such that C is equivalent to a small category. We assume that

(1.28) (i) If $X \in \mathrm{Ob}(C)$ and $Y \simeq_G X$, then C contains a space which is G-homeomorphic to Y.

(ii) If X is the pushout of $X_2 \leftarrow X_0 \xrightarrow{k} X_1$ and k is a G-cofibration and X_2, X_0, X_1 are objects of C, then X is G-homeomorphic to an object of C.

(iii) $\emptyset \in \mathrm{Ob}(C)$.

A **functorial additive invariant** (B, b) for C consists of a covariant functor B from C to the category of abelian groups and an assignment b which associates an element $b(X) \in B(X)$ to each $X \in \mathrm{Ob}(C)$. The following axioms are assumed to hold:

(1.29) (i) If $f\colon X \to Y$ is a G-homotopy equivalence in C, then $B(f)(b(X)) = b(Y)$. If $f \simeq_G g$, then $B(f) = B(g)$.

(ii) Let the following square be a pushout of G-spaces between objects in C.

$$\begin{array}{ccc} X_0 & \xrightarrow{\;k\;} & X_1 \\ \downarrow & \searrow^{j_0} & \downarrow{\scriptstyle j_1} \\ X_2 & \xrightarrow[\;j_2\;]{} & X\,. \end{array}$$

Let k be a G-cofibration. Then

$$b(X) = B(j_1)(b(X_1)) + B(j_2)(b(X_2)) - B(j_0)(b(X_0)).$$

(iii) $b(\emptyset) = 0$.

A functorial additive invariant (U, u) is called **universal** if for each functorial additive invariant (B, b) there exists a unique natural transformation $F\colon U \to B$ such that $F(X)(u(X)) = b(X)$ for $X \in \mathrm{Ob}(C)$.

An additive invariant for C is defined as the special case where B is a constant functor (see (1.1) for the category of finite G-complexes).

Let (B, b) be a functorial additive invariant for C. Assume the point space P is in C and let $p_X\colon X \to P$. We define an additive invariant (A, a) for C by setting $A = B(P)$ and $a(X) = B(p_X)(b(X))$. The following is left as an exercise:

(1.30) Proposition. *If (B, b) is universal, then (A, a) is universal.* □

As always, universal (functorial) additive invariants are determined up to unique isomorphism. Existence is shown by a Grothendieck construction.

(1.31) Proposition. *There exists a universal functorial additive invariant.*

Proof. Suppose C is a small category. Given $Y \in \mathrm{Ob}(C)$, let $F(Y)$ be the free abelian group on the set of G-homotopy classes $[f]$ of G-maps $f: X \to Y$ in C. Let $R(Y) \subset F(Y)$ be the subgroup generated by elements of the form:

(i) $[f_1] - [f_2]$ if there exists a G-homotopy equivalence h such that $[f_1 h] = [f_2]$.
(ii) $[f] - [f_1] - [f_2] + [f_0]$ if there exists a pushout diagram as in (1.19) and $f: X \to Y$ such that $f_i = fj_i$ for $i = 0, 1, 2$.
(iii) $[\emptyset \to Y]$.

Let $U(Y) = F(Y)/R(Y)$ and let $u(Y) \in U(Y)$ be the element represented by the identity. A morphism $h: Y \to Z$ in C induces $U(h): U(Y) \to U(Z)$ by composition with h. The reader should check that this construction yields a universal invariant. □

Formal properties of universal (functorial) additive invariants lead to product and restriction formulas. We explain this below.

Suppose G_1 and G_2 are compact Lie groups and C_1, C_2, resp. C categories of G_1-, G_2-, resp. $G_1 \times G_2$-spaces such that the category $C_1 \times C_2$ is contained in C via $(X_1, X_2) \mapsto X_1 \times X_2$. Let (U_1, u_1), (U_2, u_2), resp. (U, u) be universal functorial additive invariants for the categories in question.

For each $Y_2 \in \mathrm{Ob}(C_2)$, let $T(Y_2)$ be the abelian group of natural transformations $U_1(?) \to U(? \times Y_2)$. A morphism $f: Y_1 \to Y_2$ induces a homomorphism $T(f): T(Y_1) \to T(Y_2)$ by composition with $U(\mathrm{id} \times f)$. Since $(U(? \times Y_2), u(? \times Y_2))$ is a functorial additive invariant for C_1, there exists a natural transformation $t(Y_2): U_1(?) \to U(? \times Y_2)$ uniquely determined by the condition that $t(Y_2)(Y_1)$ sends $u_1(Y_1)$ to $u(Y_1 \times Y_2)$ for all $Y_1 \in \mathrm{Ob}(C_1)$. Then (T, t) is a functorial additive invariant for C_2 so that there exists exactly one natural transformation $F: U_2 \to T$ with $F(Y_2)(u_2(Y_2)) = t(Y_2)$ for all $Y_2 \in \mathrm{Ob}(C_2)$. The transformation F can be interpreted as a natural pairing

(1.32) $P(Y_1, Y_2): U_1(Y_1) \otimes U_2(Y_2) \to U(Y_1 \times Y_2)$

which is uniquely determined by the condition that it sends $u_1(Y_1) \otimes u_2(Y_2)$ to $u(Y_1 \times Y_2)$.

The pairing (1.32) is an **external product formula**: Its significance is that the invariant $u(Y_1 \times Y_2)$ only depends on the invariants $u_1(Y_1)$ and $u_2(Y_2)$ and not on any other properties of the spaces involved.

We now consider the restriction. Let H be a closed subgroup of G. Let C be a category of G-spaces and D a category of H-spaces with universal functorial additive invariants (U, u) and (V, v), respectively. Suppose that the restriction $\operatorname{res}_H X$ of $X \in \operatorname{Ob}(C)$ lies in D so that we have a restriction functor $\operatorname{res}: C \to D$. By universality, there exists a unique natural transformation $R: U \to V \circ \operatorname{res}$ which sends $u(Y)$ to $v(\operatorname{res} Y)$ for all $Y \in \operatorname{Ob}(C)$. We call R a **restriction formula**; its significance is that the v-invariant of $\operatorname{res} Y$ only depends on the u-invariant of Y.

We can combine the external product formula and the restriction formula to obtain an **internal product formula**. Suppose the category C of G-spaces with universal functorial additive invariant (U, u) is closed with respect to products $Y_1 \times Y_2$ of objects $Y_i \in \operatorname{Ob}(C)$ (with diagonal action). Then there is a unique natural pairing

(1.33) $P(Y_1, Y_2): U(Y_1) \otimes U(Y_2) \to U(Y_1 \times Y_2)$

which sends $u(Y_1) \otimes u(Y_2)$ to $u(Y_1 \times Y_2)$.

The ring $U(G)$ is used in Becker-Schultz [1978] and Oliver [1976b].

(1.34) Exercises.

1. Let X be a compact G-ENR and $H \subset G$ a subgroup. Show that the Euler characteristic $\chi(X^H/NH)$ is defined. Show that the relation
$$\chi(X^K/NK) = \sum_{(H)} \chi_c(X^K_{(H)}/NK)$$
holds. Show that $X^K_{(H)}/NK \to X_{(H)}/G$ is a fibre bundle with fibre $(G/H^K)/NK$.
2. Show that the spaces G/H^k, $k \geq 1$, have only finitely many conjugacy classes of isotropy groups.
3. Give a proof of (1.15), (1.17), and (1.18).
4. Show via computations with Euler characteristics for G-ENR's that the assignments used in (1.21), (1.22), and (1.23) yield well-defined maps when applied to H-ENR's and G-ENR's, respectively. In particular, show that the equivariant ENR-property is preserved under these constructions.
5. Establish the isomorphism $U(G; X) \cong U(G; \pi(X))$.
6. Show that composition with a G-map $f: X \to Y$ induces a homomorphism $f_*: U(G; X) \to U(G; Y)$.
7. Let $\mathfrak{A}$ be a category of pointed G-complexes with the following properties: (i) $\mathfrak{A}$ contains a point. (ii) If $f: X \to Y$ is in $\mathfrak{A}$, then the canonical map $Y \to C(f)$ into the mapping cone of f is also in $\mathfrak{A}$. An additive invariant for $\mathfrak{A}$ is a pair (K, k) consisting of an abelian group K and an assignement k: Ob $(\mathfrak{A}) \to K$ such that: (i) If X and Y are G-homotopy equivalent, then $k(X) = k(Y)$. (ii) If $X \to Y \to Z$ is a cofibration sequence, then $k(X) - k(Y) + k(Z) = 0$. If the G-homotopy classes of objects in $\mathfrak{A}$ form a set, then $\mathfrak{A}$ has a universal additive invariant $(K(\mathfrak{A}), k)$. If $\mathfrak{A}$ is closed under $\wedge$, then $(kX, kY) \mapsto$

$k(X \wedge Y)$ is a well-defined product on $K(\mathfrak{A})$ which turns $K(\mathfrak{A})$ into a commutative ring. Let $\mathfrak{A}_1$ be the full subcategory of $\mathfrak{A}$ consisting of $X \in \mathrm{Ob}(\mathfrak{A})$ such that all X^H are simply-connected. If $\mathfrak{A}_1$ possesses a universal additive invariant, then so does $\mathfrak{A}$ and the natural map $K(\mathfrak{A}_1) \to K(\mathfrak{A})$ induced by inclusion of categories is an isomorphism.
A suitable category of this type is $\mathfrak{A}_G$ consisting of G-complexes X with the following properties:
(i) X is finite-dimensional of finite orbit type.
(ii) For each $H \subset G$, the homology $H_*(X^H; \mathbb{Z})$ is finitely generated.
8. For $H \subseteq G$, let $\mathfrak{A}_G(H)$ be the full subcategory of $\mathfrak{A}_G$ (defined in the previous exercise) consisting of objects which, apart from the base point, have only orbits of type G/H. There is a functor $\varrho_H: \mathfrak{A}_G \to \mathfrak{A}_G(H)$ which maps X to $GX^H/GX^{>H}$. Let $(K(\mathfrak{A}_G(H)), k_H)$ be a universal additive invariant for $\mathfrak{A}_G(H)$. Then $X \mapsto k_H(\varrho_H(X))$ is an additive invariant for $\mathfrak{A}_G$. Put $K = \bigoplus_{(H)} K(\mathfrak{A}_G(H))$ and $k(X) = (k_H \varrho_H X | (H))$. Show that (K, k) is a universal additive invariant for $\mathfrak{A}_G$. Show that $X \mapsto X^H$ induces an isomorphism $K(\mathfrak{A}_G(H)) \to K(\mathfrak{A}_{NH/H}(1))$.
Remark: One can show that $K(\mathfrak{A}_G(1)) \cong K_0(\mathbb{Z}(G/G_0))$ where $K_0(\mathbb{Z}\pi)$ denotes the Grothendieck group of projective finitely generated $\mathbb{Z}\pi$-modules.

2. The Burnside ring.

For a finite group G, we have defined the Burnside ring $A(G)$ in I(2.18). We now give a different definition of $A(G)$ for compact Lie groups (see also II.8).

Consider the following equivalence relation on the set of finite G-complexes: $X \sim Y$ if and only if for all $H \subset G$, the spaces X^H and Y^H have the same Euler characteristic. Let $A(G)$ be the set of equivalence classes and let $[X] \in A(G)$ be the class of X. Disjoint union and cartesian product of G-complexes are compatible with this equivalence relation and induce composition laws, denoted as addition and multiplication, respectively, on $A(G)$. It is easy to verify that $A(G)$ together with these composition laws is a commutative ring with identity. The zero element is represented by a complex X such that the Euler characteristic $\chi(X^H) = 0$ for each $H \subset G$. If K is a space with trivial G-action and $\chi(K) = -1$, then $X \times K$ represents the additive inverse of X in $A(G)$.

Let $\phi(G)$ denote the set of conjugacy classes (H) such that NH/H is finite. For each $H \subset G$, the assignment $X \mapsto \chi(X^H)$ induces, by definition of $A(G)$, a ring homomorphism

(2.1) $\quad \varphi_H: A(G) \to \mathbb{Z}.$

2.2) Proposition. *Additively, $A(G)$ is the free abelian group on $[G/H]$, $(H) \in \phi(G)$. For each G-complex X, the following relation holds* $[X] = \sum_{(H) \in \phi(G)} \chi_c(X_{(H)}/G)[G/H]$.

Proof. The elements $[G/H]$, $(H) \in \phi(G)$, are linearly independent. For suppose there were a relation $\sum a_H[G/H] = 0$. Let (K) be maximal such that $a_K \neq 0$. Then we obtain the contradiction

$$0 = \varphi_K(\sum a_H[G/H]) = a_K \chi(G/K^K) = a_K |WK| \neq 0.$$

In order to prove the second assertion, we note that the assignment $X \to [X] \in A(G)$ is an additive invariant in the sense of (1.1). From (1.12), we obtain the relation $[X] = \sum_{(H)} \chi_c(X_{(H)}/G)[G/H]$. But if (H) is not in $\phi(G)$, i.e. if WH is not finite, then $[G/H] = 0$ in $A(G)$ because G/H^K carries a free WH-action and hence a free S^1-action. The latter fact implies $\chi(G/H^K) = 0$; see II(8.13), Ex. 4. □

From (2.2) one concludes immediately that, for finite G, the ring $A(G)$ is isomorphic to the ring defined in I(2.18). In general, we have a canonical map

(2.3) $\quad v: U(G) \to A(G),$

the identity on representatives, and v is a surjective ring homomorphism.

(2.4) Proposition. *For finite G, the homomorphism v is an isomorphism. In general, the kernel of v is the nilradical of U(G) (= set of nilpotent elements of U(G)).*

Proof. By means of the (φ_H) of (2.1), $A(G)$ can be considered as a subring of $\prod_{(H)} \mathbb{Z}$. Therefore, $A(G)$ has nilradical 0. Now use (1.14). □

The following observation is due to Schwänzl [1977].

(2.5) Proposition. *The multiplication table of the $[G/H]$ has non-negative coefficients, i.e. if $[G/H][G/K] = \sum_{(L)} n_L[G/L]$, then $0 \leq n_L$.*

Proof. We have $n_L = \chi_c((G/H \times G/K)_{(L)}/G)$. Moreover,

$$\begin{aligned}(G/H \times G/K)_{(L)}/G &\cong (G/H \times G/K)_L/NL \\ &\subset (G/H \times G/K)^L/NL.\end{aligned}$$

By I(5.10), the space G/H^L consists of finitely many NL/L-orbits. Since NL/L is finite, the set $(G/H \times G/K)_{(L)}/G$ is finite and therefore its Euler characteristic is non-negative. □

We give some applications of the geometric definition of $A(G)$. Let $R(G)$ be the complex representation ring of G. The action of G on X induces an action of G on $H_i(X; \mathbb{C})$. Let $[H_i(X; \mathbb{C})] \in R(G)$ be the element corresponding to this representation. Define the equivariant Euler characteristic of X to be

$$\chi_G(X) = \sum_{i \geq 0} (-1)^i [H_i(X; \mathbb{C})] \in R(G).$$

The character value of $\chi_G(X)$ at $g \in G$ is the alternating sum of the traces of $H_i(l_g)\colon H_i(X; \mathbb{C}) \to H_i(X; \mathbb{C})$ where l_g is the left translation by g. This alternating sum of traces is called the Lefschetz index $L(l_g)$ of the map l_g. It turns out that, in the present case, $L(l_g)$ is the Euler characteristic of the l_g-fixed points:

(2.6) Lemma. $L(l_g) = \chi(X^g) = \chi(X^S)$. *Here, S is the closed subgroup generated by g.*

Proof. It can be shown (by thickening) that a finite G-complex X has the G-homotopy type of a compact differentiable G-manifold U with boundary. Thus it suffices to prove the assertion for U. The fixed point set U^S has an equivariant tubular neighbourhood W in U (Bredon [1972], VI.2); consequently, the inclusion $U^S \to W$ is a G-homotopy equivalence. The Lefschetz index only depends on a neighbourhood of the fixed point set and can therefore be computed on W. Now use Dold [1972], VI.6.6. See also III(6.17). □

From (2.6) we see that the assignment $X \mapsto \chi_G(X)$ is compatible with the equivalence relation defining $A(G)$ and therefore induces a homomorphism

(2.7) $\quad \chi_G\colon A(G) \to R(G)$

called **equivariant Euler characteristic**. Note that the component G_0 of $e \in G$ acts trivially on $H_i(X; \mathbb{C})$ so that $\chi_G(X)$ may be considered as an element of $R(G/G_0)$. Also one may use rational coefficients in which case $\chi_G(X)$ is an element of the rational representation ring $R(G/G_0; \mathbb{Q})$.

If G is finite and S a finite G-set, then $\chi_G(S)$ is represented by the permutation representation $\mathbb{Q}(S)$ with S as basis. Thus

(2.8) $\quad \chi_G\colon A(G) \to R(G; \mathbb{Q})$

may also be called the permutation representation invariant.

We recall the following fact from representation theory.

(2.9) Proposition. *Let G be a finite p-group. Then* (2.8) *is surjective.*

Proof. Ritter [1972], Segal [1972], tom Dieck [1979]. □

For another application, let $A(G)^*$ denote the group of units of $A(G)$. Let $S(V)$ be the unit sphere of a real representation V of G. The element $e(V) = 1 - [S(V)]$ satisfies

$$\varphi_H e(V) = 1 - \chi(S(V)^H) = (-1)^{\dim V^H}.$$

Therefore, $e(V \oplus W) = e(V)e(W)$ and $e(V)^2 = 1$. We obtain a homomorphism

(2.10) $\quad \nu\colon R(G;\mathbb{R}) \to A(G)^*, V \mapsto e(V)$

from the additive group of the real representation ring into the multiplicative group $A(G)^*$. Without proof, we mention the following result of J. Tornehave [1984]

(2.11) Theorem. *Let G be a finite 2-group. Then ν is surjective.* □

The definition of $A(G)$ shows that elements of $A(G)$ may be identified with integer valued functions on conjugacy classes; more precisely, $[X]$ is identified with the function $(H) \mapsto \chi(X^H)$. From (2.2) we see that it suffices to consider those H with finite NH/H. In the next section we show that it suffices to look at functions which are continuous in a suitable sense. Then it is interesting to identify $A(G)$ as a subring in the ring of all functions. This will be achieved by describing congruences among Euler characteristics of fixed point sets (section 5).

The Burnside ring is a contravariant functor on the category of compact Lie groups. If $\alpha\colon G_1 \to G_2$ is a homomorphism between Lie groups, then, via α, a G_2-space X can be viewed as a G_1-space $\operatorname{res}_\alpha X$ and $X \mapsto \operatorname{res}_\alpha X$ induces a well-defined ring homomorphism

(2.12) $\quad \alpha^* = A(\alpha)\colon A(G_2) \to A(G_1).$

The equivariant Euler characteristic is then a natural transformation $\chi_G\colon A(?) \to R(?)$.

If we consider the H-fixed point set X^H as WH-space, then the assignment $X \mapsto X^H$ induces a well-defined ring homomorphism

(2.13) $\quad \tilde{\varphi}_H\colon A(G) \to A(WH).$

There is an analogous homomorphism $\varrho_H\colon R(G) \to R(WH)$ which sends the G-representation V to the WH-representation V^H. But $\tilde{\varphi}_H$ and ϱ_H are **not** compatible via the equivariant Euler characteristic. For instance, if we look at the permutation representation $\mathbb{C}(S)$ of a finite G-set S, then $\dim \mathbb{C}(S^H) = |S^H|$ and $\dim \mathbb{C}(S)^H = |S/H|$.

(2.14) Proposition. *Let H be a subgroup of G. The assignment $X \mapsto G \times_H X$ induces an additive homomorphism*

$$e_H^G\colon A(H) \to A(G).$$

(X a compact H-ENR.) If WH is infinite, then $e_H^G = 0$.

Proof. $G \times_H X$ is a compact G-ENR. Given $K \subset G$, $(G \times_H X)^K \neq \emptyset$ implies

$(K) \leq (H)$. Assume $K \subset H$. We have to show that $\chi((G \times_H X)^K)$ can be computed from Euler characteristics of fixed point sets X^L. We look at the fibration

$$(G \times_H X)^K \to G/H^K.$$

The fibre over gH is homeomorphic to $\mathrm{Fix}(gKg^{-1} \cap H, X)$. If WH is infinite, then the components of G/H^K carry a free S^1-action and therefore have Euler characteristic zero; hence $\chi((G \times_H X)^K) = 0$ in this case; and we see that e_H^G is the zero map. If $(K) \in \phi(G)$, then G/H^K is finite and therefore

$$\chi((G \times_H X)^K) = \sum_{gH \in G/H^K} \chi(X^{gKg^{-1} \cap H}). \quad \square$$

Recall that we have obtained an isomorphism $I_G: A(G) \to \omega_G^0$ in section II.8. If $d_H: \omega_G^0 \to \mathbb{Z}$ sends $x \in \omega_G^0$, represented by $f: S^V \to S^V$, to the degree of f^H, then $d_H I_G = \varphi_H$. The isomorphisms I_G are compatible with the maps (2.12) and (2.13) and their obvious analogues for stable homotopy groups ω_G^0. There is also an extension homomorphism $\tilde{e}_H^G: \omega_H^0 \to \omega_G^0$ defined as follows: We have an isomorphism $\omega_H^0 \cong \omega_G^0(G/H^+)$, see II(6.9). In II.8, we have constructed transfer maps $W^c \to G/H^+ \wedge W^c$ and these induce homomorphisms $\omega_G^0(G/H^+) \to \omega_G^0$. The composition of these two homomorphisms is $\tilde{e}_H^G$. The reader may show as an exercise that

(2.15) $I_G e_H^G = \tilde{e}_H^G I_H$.

Let K be a subgroup of finite index in G. For each K-space X, we have the multiplicative induction I(4.10)

$$m_K^G X = \mathrm{Hom}_K(G, X).$$

(2.16) The assignment $X \mapsto m_K^G(X)$ induces a map $m_K^G: A(K) \to A(G)$. In general, this map is not additive but preserves products.

Proof. If X is a K-ENR, then $m_K^G(X)$ is a G-ENR. Given $H \subset G$, we compute $\chi(\mathrm{Hom}_K(G, X)^H)$. We use I(4.11). Since K has finite index in G, the K-space G/H is K-homeomorphic to a finite disjoint union $\coprod_i K/K(i)$ of homogeneous spaces. The canonical bijections

$$\begin{aligned} \mathrm{Hom}_K(G, X)^H &= \mathrm{Hom}_G(G/H, \mathrm{Hom}_K(G, X)) \\ &= \mathrm{Hom}_K(G/H, X) \\ &= \mathrm{Hom}_K(\textstyle\coprod_i K/K(i), X) \\ &= \textstyle\prod_i X^{K(i)} \end{aligned}$$

show that $\chi(\mathrm{Hom}_K(G, X)^H)$ can be computed from Euler characteristics of fixed point sets X^L $(L \subset K)$. $\square$

On the level of homotopy groups, the multiplicative induction $\tilde{m}_H^G: \omega_0^H \to \omega_0^G$ is given by the construction of I(4.14), Ex. 8. The reader may show as an exercise that

(2.17) $I_G m_H^G = \tilde{m}_H^G I_H$.

(2.18) Proposition. *Let L be a finite normal subgroup of G. The assignment $X \mapsto X/L$ induces a map $A(G) \to A(G/L)$.*

Proof. Let X be a G-ENR. Let $p: X \to X/L$ be the quotient map. Fix $H \subset G/L$ and set $B = p^{-1}(X/L^H)$. Let P be the inverse image of H in G. We consider X and B as P-spaces. A P-orbit of X, which is isomorphic to P/U, is contained in B if and only if $P = LU$. Hence B is a union of P-orbit bundles of X. From (2.2) we see that the class $[B] \in A(P)$ only depends on $[X] \in A(P)$. From III(6.17), Ex. 4 we obtain

$$\chi(X/L^H) = \chi(B/L) = |L|^{-1} \sum_{g \in L} \chi(B^g).$$

Hence $\chi(X/L^H)$ can be computed from $[B] \in A(P)$, which only depends on $[X]$.

We still have to show that X/L is a G/L-ENR. By II(8.10), it suffices to verify that $X/L^H = B/L$ is an ENR. Since B is a union of P-orbit bundles of X, it is a P-ENR by II(8.10). Hence B/L is an ENR by II(8.9). □

We now discuss symmetric powers. Let S_r denote the symmetric group on r symbols. If X is a G-space, then the diagonal action of G on X^r and the permutation action of S_r commute. So we can view X^r as $(S_r \times G)$-space. Let M be an $(S_r \times G)$-space with trivial G-action. Then we obtain an $(S_r \times G)$-space $M \times X^r$ and a G-space $(M \times X^r)/S_r$. With these notations we can state

(2.19) Proposition. *The assignment $(M, X) \mapsto (M \times X^r)/S^r$ induces a map*

$$A(S_r) \times A(G) \to A(G).$$

Proof. We assume, of course, that X is a compact G-ENR and M a compact S_r-ENR. Then X^r is a compact $(S_r \times G)$-ENR. We begin by showing that $X \mapsto X^r$ induces a map $w: A(G) \to A(S_r \times G)$. The standard embedding $S_{r-1} \subset S_r$ yields an embedding of $S_{r-1} \times G$ as a subgroup of finite index in $S_r \times G$. If we view X as an $(S_{r-1} \times G)$-space via the projection $S_{r-1} \times G \to G$, then the $(S_r \times G)$-space X^r is obtained from X using the multiplicative induction corresponding to $S_{r-1} \times G \subset S_r \times G$. Thus w is well-defined by (2.16). Now consider the following composition of maps

$$A(S_r) \times A(G) \xrightarrow[p \times w]{} A(S_r \times G) \times A(S_r \times G)$$

$$\xrightarrow[m]{} A(S_r \times G) \xrightarrow[q]{} A(G)$$

in which w is given as above, p is induced by the projection $S_r \times G \to S_r$, the map m is the ring multiplication, and q is the quotient map from (2.18). One checks that the composition above is given on representatives by $(M, X) \mapsto (M \times X^r)/S_r$. □

Let $\Pi \subset S_r$ be a subgroup. Then the Π-symmetric power of a G-space X is defined to be the G-space X^r/Π. Since $(S_r/\Pi \times X^r)/S_r \cong X^r/\Pi$, we obtain from (2.19)

(2.20) Proposition. *The assignment* $X \mapsto X^r/\Pi$ *induces a map* $A(G) \to A(G)$. □

The elements of $A(S_r)$ can be interpreted as operations on the Burnside ring. We are going to explain this. Write the map (2.19) as follows:

$$A(S_r) \times A(G) \to A(G), (x, y) \mapsto x \cdot y.$$

Let X resp. Y be an S_r-resp. S_t-space. We write

$$X \circ Y = S_{r+t} \times_{S_r \times S_t} (X \times Y)$$

using the standard embedding $S_r \times S_t \subset S_{r+t}$. Let $G \int S_r$ be the wreath-product of G with S_r. This is the set $S_r \times G^r$ with group-law

$$(s^{-1}; g_1, \ldots, g_r)(t; h_1, \ldots, h_r) = (s^{-1}t; g_1 h_{s(1)}, \ldots, g_r h_{s(r)}).$$

If M is a G-space, then M^r becomes an $G \int S_r$-space with action

$$(s^{-1}; g_1, \ldots, g_r)(m_1, \ldots, m_r) = (g_1 m_{s(1)}, \ldots, g_r m_{s(r)}).$$

We consider $S_t \int S_r$ as a subgroup of S_{rt} in the following way: If $M = S_t$ as S_t-space, then $S_t \int S_r$ acts as a group of permutations on M^r; now identify M^r suitably with $\{1, \ldots, rt\}$. (The conjugacy class of $S_t \int S_r$ in S_{rt} is then uniquely determined.) For X and Y as above, we write

$$X \circledast Y = S_{rt} \times_{S_t \int S_r} (X \times Y^r)$$

(2.21) Proposition. *The assignments* $(X, Y) \mapsto X \circ Y$ *and* $(X, Y) \mapsto X \circledast Y$ *induce maps*

$$A(S_r) \times A(S_t) \to A(S_{r+t}), (x, y) \mapsto x \circ y$$
$$A(S_r) \times A(S_t) \to A(S_{rt}), (x, y) \mapsto x * y,$$

respectively. The graded additive group

$$A = \bigoplus_{r \geq 0} A(S_r)$$

becomes a graded ring with multiplication $\circ$. *Moreover, one has*

$$
\begin{aligned}
(a+b)*c &= a*c+b*c \\
(a\circ b)*c &= (a*c)\circ(b*c) \\
(a*b)*c &= a*(b*c) \\
b*1 &= b \qquad \textit{for} \quad 1\in B(S_0).
\end{aligned}
$$

For $a_1, a_2 \in A$ *und* $b \in A(G)$, *one has*

$$
\begin{aligned}
(a_1\circ a_2)\cdot b &= (a_1\circ b)\cdot(a_2\circ b) \\
(a_1 * a_2)\cdot b &= a_1\cdot(a_2\cdot b).
\end{aligned}
$$

Proof. See exercise 9. □

The Burnside ring (of a finite group) carries λ-operations and induced Adams operations ψ. A λ-structure based on exterior or symmetric powers was considered by Knutson [1973] and further studied by Siebeneicher [1976] and Gay-Morris-Morris [1983]. These λ-structures turn $A(G)$ into a special λ-ring only for cyclic G. Another λ-structure, suggested by Knutson [1973], was studied by Blass [1979] and Ochoa [1984]. It was shown in tom Dieck [1984b] that $A(G)$ can be turned into a special λ-ring essentially for regular p-groups.

The Burnside ring for general compact groups was studied in Gordon [1975]. Gordon [1977] treats the Burnside ring of a finite torus extension.

Permutation representations of finite G-sets can be defined over the integers. One obtains an equivariant Euler characteristic $A(G) \to R(G; \mathbb{Z})$ which has better properties than χ_G; see Oliver [1978].

For computations with Euler characteristics and orbit bundles which are similar to the methods of this section, see K. S. Brown [1974], [1975], [1982], [1982a]. The Burnside ring of a compact Lie group was introduced in tom Dieck [1975a].

(2.22) Exercises.

1. The Burnside ring $A(G)$ can be defined using compact Euclidean G-neighbourhood retracts instead of finite G-complexes. (Compare the analogous definition of $U'(G)$ in section 1.)
2. Define an equivalence relation on the set of G-maps $f: X \to X$ of finite G-complexes (resp. compact G-ENR) by

$$(f_1: X_1 \to X_1) \sim (f_2: X_2 \to X_2)$$

if and only if the Lefschetz numbers $L(f_1^H)$ and $L(f_2^H)$ are equal for all $H \subset G$. Show that the set of equivalence classes is canonically isomorphic to $A(G)$; see Ulrich [1983]. For related material, compare Okonek [1983] and Dold [1984]. Extend (2.12)–(2.21) to this more general situation.
3. Show that the analogue of (2.5) for $U(G)$ holds true. Prove a similar result for the element in $U(G; G/H_1 \times \ldots \times G/H_r)$ which is represented by the identity.

4. Let $G \subset \mathrm{SO}(3)$ be a finite cyclic group. Show that $u(\mathrm{SO}(3)/G)$ has square zero in $U(\mathrm{SO}(3))$.
5. Let $e_g\colon R(G) \to \mathbb{C}$ be the evaluation of characters at $g \in G$. Let $S \subset G$ be the closed subgroup generated by g. Show that the diagram

$$\begin{array}{ccc} A(G) & \xrightarrow{\chi_G} & R(G) \\ \varphi_S \downarrow & & \downarrow e_g \\ \mathbb{Z} & \xrightarrow{\subset} & \mathbb{C} \end{array}$$

   is commutative.
6. Show that $\chi_G\colon A(G) \to R(G/G_0; \mathbb{Q})$ is an isomorphism if and only if G is isomorphic to the direct product of a torus and a finite cyclic group.
7. Verify (2.15) and (2.17).
8. Show that $m_K^G(X)$ is a G-ENR if X is a K-ENR.
9. Give a proof of (2.21). Hint: The identities are verified by considering representatives. It remains to be shown that $\circ$ and $*$ are well-defined; they can be written as suitable compositions of maps which we have shown to be well-defined earlier.

3. The space of subgroups.

We begin by recalling some notions from point set topology. Let E be a metric space with bounded metric d. Let $F(E)$ be the set of non-empty subsets of E. Put $r(A, B) = \sup(d(x, B) \mid x \in A)$ for $A, B \in F(E)$ and define

$$h(A, B) = \max(r(A, B), r(B, A)).$$

Then r is a metric on $F(E)$, called **Hausdorff metric**. If E is complete, then $F(E)$ is complete. If E is compact, then $F(E)$ is compact.

The convergence in $F(E)$ of a sequence X_i to the limit X can be expressed as follows: For each $\varepsilon > 0$, there exists n_0 such that for $n > n_0$:

(3.1) (i) For $x \in X_n$, there exists $y \in X$ with $d(x, y) < \varepsilon$.
(ii) For $x \in X$, there exists $y \in X_n$ with $d(x, y) < \varepsilon$.

If Y_n is the closure of $\bigcup_{p \geq 0} X_{n+p}$, then X is the intersection of the Y_n.

We want to use this metric on the set $S(G)$ of closed subgroups of the compact Lie group G.

(3.2) Proposition.

(i) *$S(G)$ is a closed (hence compact) subset of $F(G)$.*

(ii) *The action $G \times S(G) \to S(G)$, $(g, H) \mapsto gHg^{-1}$ is continuous. The orbit space $\psi(G)$ is countable, hence a totally disconnected compact Hausdorff space.*

(iii) *$\phi(G) \subset \psi(G)$ is a closed subspace.*

Proof.
(i) We start with a bi-invariant metric d on G, i.e. with a metric satisfying $d(x, y) = d(gx, gy) = d(xg, yg)$ for all $x, y, g \in G$ (see exercise 1). Let $X = \lim H_i$, $H_i \in S(G)$. Given $x, y \in X$ and $\varepsilon > 0$, choose n_0 such that for $n > n_0$ there exist $x_n, y_n \in H_n$ with $d(x, x_n) < \varepsilon/2$, $d(y, y_n) < \varepsilon/2$. Then, using the bi-invariance, $d(xy^{-1}, x_n y_n^{-1}) \leq d(xy^{-1}, x_n y^{-1}) + d(x_n y^{-1}, x_n y_n^{-1}) < \varepsilon$. If $xy^{-1} \notin X$, then $X \cup \{xy^{-1}\}$ would satisfy (3.1), (i) and (ii), in contradiction to $X = \lim H_i$.
(ii) Let $\lim g_i = g$ in G and $\lim H_i = H$ in $S(G)$. Bi-invariance of d and the triangle inequality yield $d(g_n x_n g_n^{-1}, gxg^{-1}) \leq 2d(g, g_n) + d(x, x_n)$. Using this, one verifies that gHg^{-1} is precisely the set of points satisfying (3.1), (i) and (ii) for the sequence $g_n H_n g_n^{-1}$. The space $\psi(G)$ is countable, see I(5.18), exercise 8.
(iii) We show that $T(G) = \{H \mid WH \text{ finite}\}$ is closed in $S(G)$. Let $H = \lim H_i$, $H_i \in S(G)$. By I(5.9), there exists $\varepsilon > 0$ such that each subgroup in the ε-neighbourhood of H is conjugate to a subgroup of H. Hence the H_i are eventually conjugate to subgroups of H. But if $K \in T(G)$ and $K \subset H$, then $H \in T(G)$; this follows from I(5.10) because G/H^K consists of finitely many WK-orbits and hence is a finite set with free WH-action. The proof of (iii) will be completed once we have shown that convergence in $S(G)$ and $\psi(G)$ is equivalent in the sense of the following proposition. □

(3.3) Proposition. *Let $(H) = \lim (H_i)$ in $\psi(G)$. Then there exist n_0 and $K_n \in S(G)$ for $n \geq n_0$ such that $(K_n) = (H_n)$, $K_n \subset H$, $\lim K_n = H$.*

Proof. By I(5.9), for each $\varepsilon > 0$ there is an integer $n_0(\varepsilon)$ such that for $n > n_0(\varepsilon)$ there exists u_n with $d(u_n, e) < \varepsilon$ and $u_n H_n u_n^{-1} \subset H$. Therefore, we can find a sequence $g_n \in G$ converging to e such that for almost all n the inclusion $g_n H_n g_n^{-1} \subset H$ holds. □

(3.4) Proposition. *Suppose* $\lim(H(i)) = G$. *Let X have finite orbit type. Then $X^{H(i)} = X^G$ for almost all i.*

Proof. Put $\varepsilon = \min h(K, G)$ where (K) runs through the finite set of isotropy types of X different from (G). Then $\varepsilon > 0$. Since $(L) \subset (K)$ implies $h(L, G) \geq h(K, G)$, we see that $h(L, G) < \varepsilon$ implies that L is not subconjugate to any isotropy group different from G. Thus if $h(H(i), G) < \varepsilon$, then

$$X^{H(i)} = \bigcup_{(K)} X_{(K)}^{H(i)} = X_{(G)}^{H(i)} = X^G. \quad \square$$

(3.5) Proposition. *Let X be a compact G-ENR. Then the mapping $\psi(G) \to \mathbb{Z}$, $(H) \mapsto \chi(X^H)$ is continuous ($\mathbb{Z}$ carries the discrete topology).*

Proof. Let $(H) = \lim(H_i)$. By (3.3), we can assume that $H(i) \subset H$ and $H = \lim H(i)$. The assertion follows from (3.4) applied to the H-space X. $\square$

(3.6) Proposition. *Suppose $1 \to T \to P \to F \to 1$ is an exact sequence of compact Lie groups. Suppose T is a torus and F is finite. Then P is the limit of finite subgroups.*

Proof. Let m be an integral multiple of the order of F. Then there exists a finite subgroup $P(m)$ of P such that $P(m) \cap T$ is the subgroup of those elements of T whose orders divide m and such that $P(m)$ projects onto F. One verifies that P is the limit of these $P(m)$. $\square$

(3.7) Proposition. *A compact Lie group G is a limit of proper subgroups if and only if G_0, the component of $e \in G$, is not semi-simple.*

Proof. Suppose $G = \lim H(n)$, $H_n \neq G$. Set $K(n) = G_0 \cap H(n)$. Then $G_0 = \lim K(n)$. Hence it is no essential restriction to assume that G is connected. By passing to a subsequence, we can assume that the semi-simple groups $[H(n)_0, H(n)_0]$ converge to a limit H and therefore eventually have the same dimension as H. Using (3.3), we thus may assume that we have the following situation: $G = \lim L(n)$, $L = [L(n)_0, L(n)_0]$ for all n, $L \neq G$. Since $L \lhd L(n)$, we have $L \lhd G$ and G/L is the limit of finite torus extensions $L(n)/L$. Suppose the $L(n)/L$ are finite. We now invoke the theorem of Jordan, which states that there exists an integer j such that each finite subgroup of G/L has a normal abelian subgroup of index less than j (see (6.4)). Choose an abelian normal subgroup $A(n)$ of $L(n)/L$ according to this theorem. Then the limit A of the $A(n)$ is an abelian normal subgroup of index less than j in G/L. Since G/L is connected, $G/L = A$ is a torus and therefore G is not semi-simple. We leave the case of infinite $L(n)/L$ to the reader.

Conversely, if G is not semi-simple, we can find a normal subgroup L of G_0 such that G_0/L is a non-trivial torus (Hochschild [1965], XIII Theorem 1.3). One shows that L is a characteristic subgroup of G_0. Therefore, G/L is a finite extension of a torus. Now use (3.6). $\square$

(3.8) Exercises.

1. Show that a compact Lie group G carries a bi-invariant metric. Hint: Use integration over G.
2. Complete the proof of (3.6).
3. Show that an extension of a torus by a p-group is a limit of finite p-groups.
4. Let $S \subset S(G)$ be a closed subset. Suppose S is closed with respect to intersections and contains G. For each $H \in S(G)$, the set $\{K | K \supset H, K \in S\}$ contains a unique minimal element $m(H) = m_S(H)$. Show that $m: S(G) \to S$ is continuous. If S is closed with respect to conjugation, then m induces a continuous map on the corresponding spaces of conjugacy classes. If $T \subset S$ is a subset with similar properties, then $m_T m_S = m_T$.
5. Let $H = \lim H_i$ and $K = \lim K_i$ in $S(G)$. Suppose $H_i \rhd K_i$ for all i. Show that $H \rhd K$.
6. For $H \in S(G)$, let $\tau(H) \subset NH$ be the pre-image of the maximal torus of NH/H. Show that $\tau: S(G) \to S(G)$ is continuous. Thus there exist continuous retractions $S(G) \to T(G)$ and $\psi(G) \to \phi(G)$.
7. Show that $\phi(G)$ is an inverse limit of continuous retractions $\phi(G) \to F_i$ onto finite sets $F_1 \subset F_2 \subset F_3 \subset \ldots$. Similarly for $\psi(G)$.

4. Prime ideals.

Recall from (2.1) the definition of the ring homomorphisms $\varphi_H: A(G) \to \mathbb{Z}$, $X \mapsto \chi(X^H)$. If $(p) \subset \mathbb{Z}$ is a prime ideal, then

(4.1) $\quad q(H,p) = \varphi_H^{-1}(p) \subset A(G)$

is a prime ideal of $A(G)$. We shall show that all prime ideals of $A(G)$ arise in this manner.

(4.2) Theorem. *Let $q \subset A(G)$ be a prime ideal and let p be the characteristic of $A(G)/q$. There exists a unique $(K) \in \phi(G)$ with $q = q(K,p)$ and $\varphi_K(G/K) \not\equiv 0 \bmod (p)$.*

Proof. (See Dress [1969a] for finite G). Let $T(q) = \{(H) \in \phi(G) | [G/H] \notin q\}$. Then $T(q)$ is non-empty because $(G) \in \phi(G)$ and $[G/G] = 1 \notin q$. Let (H) be minimal with respect to inclusion and conjugation in $T(q)$. Such (H) exist by the descending chain condition for subgroups of compact Lie groups. For each $x \in A(G)$, there exists a relation of type

(4.3) $\quad [G/H]x = \varphi_H(x)[G/H] + \sum a_K [G/K]$

where the sum is taken over $(K) < (H)$, $(K) \neq (H)$. In order to see this, take a finite G-complex X that represents x and look at the orbits of $G/H \times X$. From

(2.2), we see that a relation like (4.3) must exist with some constant c in place of $\varphi_H(x)$. In order to determine c, apply φ_H to both sides of the equation and use that $(H) \in \phi(G)$ implies $\varphi_H(G/H) \neq 0$. Now (4.4) and the minimality of (H) imply $[G/H]x \equiv \varphi_H(x)[G/H] \bmod q$. Dividing by $[G/H] \notin q$, we get $x \equiv \varphi_H(x) \bmod q$ or $q = q(H, p)$ with p being the characteristic of $A(G)/q$. If K is any subgroup of G with $q = q(K, p)$ and $\varphi_K(G/K) \not\equiv 0 \bmod p$ (p the characteristic of $A(G)/q$), then $\varphi_K(G/K) \equiv \varphi_H(G/K) \not\equiv 0 \bmod p$ for an (H) as in the beginning of the proof. In particular, $G/H^K \neq \emptyset$; and similarly, $G/K^H \neq \emptyset$. This can only happen if $(H) = (K)$. □

(4.4) Proposition. *Every homomorphism $f\colon A(G) \to R$ into an integral domain R has the form $f(x) = \varphi_K(x) \cdot 1$ for a suitable $K \subset G$.*

Proof. The kernel of f is a prime ideal $q(K, p)$. Therefore, $f\colon A(G) \to A(G)/q(K, p) \to R$ must be the map $x \mapsto \varphi_K(x) \cdot 1$ because of the fact that there is a unique isomorphism $A(G)/q(K, p) \cong \mathbb{Z}/(p)$. □

(4.5) Proposition.

(i) *Suppose $H \lhd K \subset G$. Assume that K/H is an extension of a torus by a finite p-group. Let $p = 0$ if K/H is a torus. Then $q(H, p) = q(K, p)$.*

(ii) *If $q(K, 0) = q(L, 0)$ and $(K) \in \phi(G)$, then, up to conjugation, $L \lhd K$ and K/L is a torus.*

(iii) *Given $L \subset G$, there exists $K \in \phi(G)$ such that $L \lhd K$ and K/L is a torus. Moreover, in this case $\varphi_L = \varphi_K$.*

Proof.
(i) There exists L such that $H \lhd L \lhd K$, L/H is a torus, K/L a finite p-group. Let X be a finite G-complex. The group K/L acts on X^L with fixed point set X^K. Hence $\chi(X^K) \equiv \chi(X^L) \bmod p$ and $\chi(X^L) = \chi(X^H)$ by III(6.7).
(ii) Since $q(K, 0) = q(L, 0)$, we have $\varphi_K = \varphi_L$. From $\chi(G/K^L) = \varphi_L(G/K) = \varphi_K(G/K) = |WK| \neq 0$ we see that G/K^L is non-empty and hence $(L) \leq (K)$. We can assume that $L \subset K$. Let T be a maximal torus in WL and let P be its pre-image in NL. By (i), we have $q(P, 0) = q(L, 0)$. We show that $(P) \in \phi(G)$; then, by (4.2), we conclude that $(P) = (K)$. Assume that $(P) \notin \phi(G)$. Then WP contains a non-trivial maximal torus S. We denote by Q its pre-image in NP. We claim that L is still normal in Q. Let $q \in Q$ induce the conjugation automorphism c_q on P. Since Q/P is a torus, c_q is homotopic to an inner automorphism, hence an inner automorphism itself by I(5.18), exercise 1; thus it preserves the normal subgroup L. From the exact sequence $1 \to P/L \to Q/L \to S \to 1$ and $P/L = T$ we conclude that Q/L is a torus and hence T is not maximal; a contradiction.
(iii) Use the proof of (i) and (ii). □

Let $C(G) = C(\phi(G), \mathbb{Z})$ be the ring of continuous functions from $\phi(G)$ into the discrete space $\mathbb{Z}$. Then we have

(4.6) Proposition. *The map* $\varphi: A(G) \to C(G)$, $x \mapsto ((H) \mapsto \varphi_H(x))$ *is well-defined and an injective ring homomorphism.*

Proof. By (3.5), $(H) \mapsto \varphi_H(x)$ is continuous and therefore an element of $C(G)$. By (4.5), only those φ_H with $(H) \in \phi(G)$ are needed to detect elements of $A(G)$; therefore, φ is injective. □

(4.7) Proposition. *Suppose* $q(H,p) = q(K,p)$, $H \in \phi(G)$, $K \in \phi(G)$, $|WH| \not\equiv 0 \bmod p$, $|K/K_0| \not\equiv 0 \bmod p$ *where* K_0 *is the component of e in K, and* $p \neq 0$. *Then, up to conjugation,* $K \lhd H$ *and* H/K *is a finite p-group.*

Proof. Choose P such that $K \subset P \subset NK$ and P/K is a p-Sylow group of NK/K. We claim that $NP \subset NK$. Take $x \in NP$ and let K^x be the x-conjugate of K. Then $K/(K \cap K^x) \subset P/K^x$; hence $K/(K \cap K^x)$ is a finite p-group. On the other hand, K, K^x, and P have the same component K_0 of the identity; hence $K/(K \cap K^x)$ is a quotient of K/K_0 which has order prime to p by assumption. Therefore, $K = K \cap K^x = K^x$ and $x \in NK$. But then $|NP/P| \not\equiv 0 \bmod p$ since P/K is a p-Sylow group of NK/K. Now (4.2) and (4.5) imply $(P) = (H)$ and hence the assertion. □

Let G be finite and $H \subset G$. Then there exists a unique smallest normal subgroup H_p of H such that H/H_p is a p-group.

(4.8) Proposition. *Let G be finite,* $H \subset G$, $|WH| \not\equiv 0 \bmod p$. *Then* $q(H,p) = q(K,p)$ *if and only if* $(H_p) \leq (K) \leq (H)$.

Proof. Use (4.5) and (4.7). □

Let $A(G)_0 = A(G) \otimes_{\mathbb{Z}} \mathbb{Q}$ and let Spec $A(G)_0$ be the set of prime ideals of $A(G)_0$ equipped with the Zariski topology. (See Boubaki [1961b], Ch. II for definition and elementary properties of the Zariski topology on the prime ideal spectrum of a commutative ring.) We have a map

(4.9) $q: \phi(G) \to \operatorname{Spec} A(G)_0$

which assigns to (H) the prime ideal $q(H,0) = \text{kernel}\,(\varphi_H \otimes \mathbb{Q})$. By (4.2) and (4.4), this map is bijective. The embedding $\varphi: A(G) \to C(G)$ yields an embedding

(4.10) $\varphi: A(G)_0 \to C(\phi(G), \mathbb{Q})$.

Of course, $\mathbb{Q}$ carries the discrete topology. There is a map

(4.11) $F\colon \phi(G) \to \operatorname{Spec} C(\phi(G), \mathbb{Q})$

defined by $F(x) = \{f \mid f(x) = 0\}$. The diagram

(4.12)
$$\begin{array}{ccc} & \phi(G) & \\ {}^{q}\swarrow & & \searrow{}^{F} \\ \operatorname{Spec} A(G)_0 & \xleftarrow[\operatorname{Spec}\varphi]{} & \operatorname{Spec} C(\phi(G), \mathbb{Q}) \end{array}$$

is commutative.

(4.13) Proposition.
(i) *The map F is a homeomorphism.*
(ii) *The map q is a homeomorphism.*
(iii) *The map φ in* (4.10) *is an isomorphism.*

Proof.
(i) For this assertion, $\phi(G)$ can be replaced by any compact, totally disconnected space X. We construct an inverse to F. Let $b \in \operatorname{Spec} C(X, \mathbb{Q})$ be given. We claim that $P_b = \bigcap_{f \in b} f^{-1}(0)$ consists of a single element of X. Suppose $x, y \in P_b$ are different. Choose $f \in b$ with $f(x) = 0$. Since X is totally disconnected, we can find a closed and open U with $x \in U$, $y \notin U$. Let $K(U)$ be the characteristic function of U. Consider the functions $f_1 = fK(U) + (1 - K(U))$ and $f_2 = f(1 - K(U)) + K(U)$. Then $f_1 f_2 = f \in b$. Since $f_2(x) = 1$, we have $f_2 \notin b$ and hence $f_1 \in b$. But $f_1(y) = 1$, hence $y \in P_b$, a contradiction.

Suppose $P_b = \emptyset$. Then for each $x \in X$ there is a function $g_x \in b$ with $g_x(x) \neq 0$. By normalizing, we can assume that $g_x(x) = 1$. The sets $g_x^{-1}(1) = U(x)$ form a cover of X by closed and open sets. Choose a finite subcover $U(x_i)$, $1 \le i \le n$. Let $V_i = U(X_1) \cup \ldots \cup U(x_i)$. Then one shows by induction on i, using the identity

$$K(V_{i+1}) = K(V_i) + K(U(x_{i+1})) - K(V_i)K(U(x_{i+1})),$$

that $K(V_i)$ is contained in b. But $K(V_n)$ is the constant 1 which is not in b, a contradiction.

A map $d\colon \operatorname{Spec} C(X, \mathbb{Q}) \to X$ is now defined by mapping b to the unique element of P_b. One verifies that F and d are inverse to each other. The map d is continuous: Let V be closed and open in X. Then $d^{-1}(V) = \{b \mid K(V) \notin b\}$, which is open by definition of the Zariski topology. The closed and open subsets form a base for the topology of X. The map F is continuous: Let $U = \{b \mid f \notin b\}$, $f \in C(X, \mathbb{Q})$, be a basic open set of $\operatorname{Spec} C(X, \mathbb{Q})$. Let $u = F(x) \in U$ be given. Then $f(x) \neq 0$. The set $V = f^{-1}f(x)$ is open and closed in X and contains x. Moreover, $u \in F(V) \subset U$.

(ii) An element $x \in C(\phi(G), \mathbb{Z})$ is a locally constant function, hence an integral linear combination of idempotent functions. Therefore, this ring is integral over any subring. Consequently, $C(\phi(G), \mathbb{Z}) \otimes \mathbb{Q} \cong C(\phi(G), \mathbb{Q})$ is integral over $A(G)_0$. It follows that the induced mapping Spec φ is closed (Atiyah-Mac Donald [1969], p. 67, Exercise 1). Since it is continuous and bijective, it is a homeomorphism.
(iii) Let $U \in \phi(G)$ be an open and closed subset and $q(U) \in \operatorname{Spec} A(G)_0$ the corresponding set. By Bourbaki [1961b], II.4.3, Prop. 15, there is a unique idempotent element $e(U) \in A(G)_0$ such that $q(U) = \{\mathfrak{p} \in \operatorname{Spec} A(G)_0 | A(G)_0(1 - e(U)) \subset \mathfrak{p}\}$ or, equivalently, $U = \{(H) \in \phi(G) | \varphi_H(A(G)_0(1 - e(U)) = 0\}$. From this we see that $\varphi(U) \in C(\phi(G), \mathbb{Q})$ is the characteristic function of U. Since characteristic functions of this type generate $C(\phi(G), \mathbb{Q})$ as an algebra, we conclude that φ is surjective. □

(4.14) Proposition. *$C(G)$ is the free abelian group with basis $(x_H = |WH|^{-1} \varphi(G/H) | (H) \in \phi(G))$.*

Proof. A priori, the x_H are contained in $C(G) \otimes \mathbb{Q}$. But since WH acts freely on G/H^K, we see that the numbers $\chi(G/H^K)$ are divisible by $|WH|$ so that $x_H \in C(G)$. By (4.13, iii), each element x of $C(G)$ is a linear combination $x = \sum r_H x_H$ with rational coefficients r_H. Let (K) be maximal such that $r_K \neq 0$. Then $x(K) = \varphi_K(\sum r_H x_H) = r_K |WK|^{-1} \varphi_K(G/K) = r_K \in \mathbb{Z}$. Now apply the same reasoning to $x - r_K x_K$ to obtain the result by induction.

(4.14) says in particular that $C(G)/A(G)$ is a torsion group. Actually, it has bounded exponent by Theorem (6.9).

For prime ideals in $\omega_G^0(X)$ and their relation to the prime ideals of $A(G)$, see tom Dieck [1979], 8.6.

(4.15) Exercises.

1. Let O(2) be the group of orthogonal (2,2)-matrices. Show that the groups in $\phi(\mathrm{O}(2))$ are O(2), SO(2) and the dihedral groups D_m of order $2m$. Show that $ND_m = D_{2m}$. Show that $q(D_m, 2) = q(D_n, 2)$ if $n = 2^j m$.
2. Let Spec R denote the prime ideal spectrum of a commutative ring R with Zariski topology (see Bourbaki [1961b], Ch. II).
   (i) Let X be a compact, totally disconnected space. Then $(x, (p)) \mapsto \{f | f(x) \in (p)\}$ defines a homeomorphism $F: X \times \operatorname{Spec} \mathbb{Z} \to \operatorname{Spec} C(X, \mathbb{Z})$ where $C(X, \mathbb{Z})$ is the ring of continuous functions $X \to \mathbb{Z}$.
   (ii) Show that $A(G) \subset C(G)$ is an integral extension and conclude that $\operatorname{Spec} C(G) \to \operatorname{Spec} A(G)$ is closed and surjective.
   (iii) Show that the map $q: \phi(G) \times \operatorname{Spec} \mathbb{Z} \to \operatorname{Spec} A(G)$, $((H), (p)) \mapsto q(H, p)$ is continuous, closed, and surjective. (See tom Dieck [1979], 5.7.10).

3. Show that the inclusion $A(G) \subset C(G)$ can be recovered from the algebraic structure of $A(G)$ as follows: $A(G) \otimes \mathbb{Q}$ is the total quotient ring of $A(G)$ (all non-zero-divisors made invertible); and $C(G)$ is the integral closure of $A(G)$ in its total quotient ring.
4. Let T be a maximal torus of the compact Lie group G. Show that $\chi(G/T) = \chi(G/T^T) = |NT/T|$. Use the conjugation theorem for the maximal tori to show that $NT \to G \to G/G_0$, G_0 component of e, is surjective. Show that $\chi(G/NT) = 1$ and $|NT/T| = |G/G_0||W|$, W Weyl group of G_0. Verify $[G/T]^2 = |NT/T|[G/T]$ in $A(G)$. Let N_pT denote the pre-image of the p-Sylow group of NT/T. Show that $\chi(G/N_pT) \not\equiv 0 \bmod p$. Let $H \subset G$ be an extension $1 \to S \to H \to F \to 1$ with S a torus and F a p-group. Show that $\chi(G/N_pT^H) \not\equiv 0 \bmod p$ and conclude that H is conjugate to a subgroup of N_pT.
5. Let G be a compact Lie group and H a subgroup. Show that $\chi(G/H) \neq 0$ if and only if H has maximal rank (i.e. contains a maximal torus).

5. Congruences.

We have seen that the Burnside ring $A(G)$ may be identified with a subring of $C(G)$, the ring of continuous (= locally constant) functions from the space $\phi(G)$ of conjugacy classes (H), with WH finite, into $\mathbb{Z}$. In this section, our aim is to identify this subring.

We begin by considering finite groups G. Let S be a finite G-set and let $V(S)$ be the complex vector space spanned by the elements of S. The G-action on the basis S of $V(S)$ induces a linear action on $V(S)$. The resulting G-module $V(S)$ is called the **permutation representation** associated to S. The character of $V(S)$ is a function on G; it will be denoted by the same symbol. The orthogonality relations for characters (Serre [1971], p. 28) say in particular that for each complex representation V the relation

(5.1) $$\sum_{g \in G} V(g) = |G| \dim_{\mathbb{C}} V^G$$

holds. By definition, $V(g)$ is the trace of the left translation $l_g\colon V \to V$. We apply this to $V(S)$ and obtain

$$V(S)(g) = \operatorname{Trace} l_g = |S^g|.$$

Therefore, applying (5.1) to this case yields

(5.2) $$\sum_{g \in G} \varphi_{\langle g \rangle}(x) \equiv 0 \bmod |G|$$

for each $x \in A(G)$. We have denoted by $\langle g \rangle$ the cyclic group generated by g. Let H^* denote the set of generators of the cyclic group H. The number of elements g for which $\langle g \rangle$ is conjugate to H equals $|H^*||G/NH|$. Therefore, (5.2) can be rewritten as

$$(5.3)\qquad \sum_{(H)\,\text{cyclic}} |H^*||G/NH|\varphi_H(x) \equiv 0 \bmod |G|.$$

We now apply the same argument to $V(S^H)$ considered as NH/H-module and obtain

$$(5.4)\qquad \sum_{(K)} |NH/NH \cap NK||K/H^*|\varphi_K(x) \equiv 0 \bmod |NH/H|;$$

this time, the sum is taken over all NH-conjugacy classes K such that H is normal in K and K/H is cyclic. Congruence (5.4) has the form

$$(5.5)\qquad \sum_{(K)} n(H,K)\varphi_K(x) \equiv 0 \bmod |NH/H|;$$

the $n(H,K)$ are integers, $n(H,H) = 1$, the sum is taken over all G-conjugacy classes (K) such that H is normal in K and K/H is cyclic. It was first observed by A. Dress that the congruences of type (5.5) characterize elements of $A(G)$ in $C(G)$.

We now treat the case of a compact Lie group G. Let X be a finite G-complex. Let $H \in \phi(G)$. We look at X^H with the finite group NH/H acting on it. The element $\chi_{NH/H}(X^H) = \sum(-1)^i H_i(X^H; \mathbb{C}) \in R(NH/H)$ in the complex representation ring has character $n \mapsto \chi_{NH/H}(X^H)(n) = \chi(X^K)$; here, K is the group generated by n and H. Thus we obtain a congruence of the type

$$(5.6)\qquad \sum_{(K)} n(H,K)\chi(X^K) \equiv 0 \bmod |NH/H|;$$

the $n(H,K)$ are again integers, $n(H,H) = 1$, and the sum is taken over all G-conjugacy classes (K) such that H is normal in K and K/H is cyclic.

(5.7) Theorem. *The congruences* (5.6) *are a complete set of congruences for the image of* $\varphi\colon A(G) \to C(G)$, *i.e. a function* $z \in C(G)$ *is contained in* $\varphi A(G)$ *if and only if for all* $(H) \in \phi(G)$ *the congruence* (5.6)

$$\sum_{(K)} n(H,K)z(K) \equiv 0 \bmod |NH/H|$$

is satisfied.

Proof. According to (4.14), we can write z as an integral linear combination $z = \sum n_K x_K$. If n_K is divisible by $|WK|$, then $n_K x_K \in \varphi A(G)$. Therefore, modulo the image of φ, we have to consider a function $z \neq 0$ where no n_K is divisible by $|WK|$. Choose (H) maximal among the (K) with $n_K \neq 0$. Consider the congruence belonging to H. The only non-zero term in the sum is $n(H,H)z(H) = n_H$ which has to be zero modulo $|WH|$, a contradiction. □

One can also consider the embedding

$$\varphi' : A(G) \to C(\psi(G), \mathbb{Z}) =: C'(G)$$

into the ring of continuous functions on all conjugacy classes. The image of φ' is given by

(5.8) Proposition. *An element $z \in C'(G)$ is contained in the image of φ' if and only if the congruences* (5.7) *hold and, moreover, $z(H) = z(K)$ whenever $H \triangleleft K$ and K/H is a torus.*

Proof. We know already by (4.5, iii) and (5.7) that $z \in \operatorname{Im} \varphi'$ satisfies the stated conditions. On the other hand, if $z' \in C'(G)$ is given, consider the restriction $z = z' \mid \phi(G)$. We apply (5.7) to z and conclude that there exists $x \in A(G)$ such that $\varphi(x) = z$. We claim that $\varphi'(x) = z'$. Let H be a subgroup of G. If $(H) \notin \phi(G)$, then, by (4.5, ii), there exists a subgroup K such that $H \triangleleft K$, K/H a torus and $(K) \in \phi(G)$. Thus

$$\varphi_K(x) = \varphi_H(x) = z(H) = z(K). \quad \square$$

(5.9) Remark. We usually identify $A(G)$ with its image in $C(G)$ or $C'(G)$.

Of course, the congruences appearing in (5.7) are not uniquely determined. Occasionally, different sets of congruences are useful. We only consider the case of finite groups.

Let $WH(p)$ be a p-Sylow subgroup of WH. We can consider $V(S^H)$ above as $WH(p)$-module and obtain a congruence modulo $|WH(p)|$ for each (H) and (p). Of course, the set of these congruences also describes $A(G) \subset C(G)$ because elements of $A(G)$ satisfy these congruences and they describe a subgroup of index $\prod_{(H)} |WH|$, which is the index of $A(G) \subset C(G)$. These congruences are primary. Thus they are useful when the localization $A(G)_{(p)}$ of $A(G)$ at the prime $p \in \mathbb{Z}$ is considered. In order to describe the inclusion $A(G)_{(p)} \subset C(G)_{(p)}$, only p-primary congruences are necessary.

Still another set of congruences is obtained by combinatorial considerations.

Let μ be the Möbius function of the subgroup lattice (see Aigner [1975], IV.2). Then one has

(5.10) Proposition. *Let G be a finite group. Then, for $x \in A(G)$, the congruence*

$$\sum \mu(H) \varphi_H(x) \equiv 0 \bmod |G|$$

holds; the sum is taken over all subgroups H of G.

Proof. For a G-set S, let $S_K = \{s \in S \mid G_s = K\}$. Then $S^H = \coprod_{H \subset K} S_K$. Thus we can apply Möbius inversion to obtain

$$|S_H| = \sum_{H \subset K} \mu(H, K) |S^K|.$$

Put $\mu(H) = \mu(1, H)$ and note that $|S_1| \equiv 0 \bmod |G|$ because S_1 is the set of free orbits of S. □

For a p-group G, one has $\mu(H) \neq 0$ if and only if H is elementary abelian and $\mu(H) = (-1)^d p^{\binom{d}{2}}$ for $|H| = p^d$ if H is elementary abelian (see Kratzer-Thévenaz [1984], p. 431).

At this point we mention the stable version of the results obtained in II.5. Suppose V and W are two complex representations such that $\dim V^H = \dim W^H$ for all $H \subset G$. Denote by $\omega(V, W)$ the stable module $\lim [S^U S^V, S^U S^W]$; compare II.6. By the equivariant Hopf theorem II(4.11), we obtain an embedding

$$\omega(V, W) \subset C(G)$$

by looking at degree functions. This subgroup of $C(G)$ only depends on $\alpha = V - W \in R(G)$ and will therefore be denoted by C_α. It is an $A(G)$-module via the embedding. It turns out that C_α consists of those functions $z \in C(G)$ which satisfy the congruences II(5.14). The $A(G)$-module C_α is a projective module of rank one; see tom Dieck-Petrie [1978]. The projective modules of rank one form the Picard group of the Burnside ring; its geometric significance is explained in tom Dieck-Petrie [1982]. The Picard group is computed in tom Dieck [1984a], [1985], [1978a].

(5.11) Exercises.

1. The orthogonal group SO(3) has the following conjugacy classes of subgroups

SO(3)	
$S^1 \cong SO(2)$	maximal torus
$NS^1 \cong O(2)$	normalizer of S^1
$I \cong A(5)$	icosahedral group
$O \cong S(4)$	octahedral group
$T \cong A(4)$	tetrahedral group
D_n, $n \geq 2$	dihedral group, order $2n$
$\mathbb{Z}/n$, $n \geq 1$	cyclic group

(compare Wolf [1967], 2.6). Show that $ND_n = D_{2n}$, $n \neq 2$; $ND_2 = S(4)$, $NA(4) = S(4)$, $NS(4) = S(4)$, $NA(5) = NA(5)$, $NO(2) = O(2)$. The cyclic groups have infinite index in their normalizers.

2. The ring ASO(3) is the set of functions $z \in C(G)$ such that

(i) $z(H)$ arbitrary for $H = SO(3)$, $A(5)$, $S(4)$, $O(2)$.
(ii) $z(D_n) \equiv z(D_{2n}) \bmod 2$, $n \neq 2$.
(iii) $z(A(4)) \equiv z(S(4)) \bmod 2$.

(iv) $z(\mathrm{O}(2)) \equiv z(S^1) \bmod 2$.
(v) $z(D_2) + 2z(A(4)) + 3z(D_4) \equiv 0 \bmod 6$.
The continuity of z gives $\lim_j z(D_{2^j n}) = z(\mathrm{O}(2))$.

3. The group SO(3) has a unique complex irreducible representation of dimension $d = 2n+1$, $n = 0, 1, 2, \ldots$ (see Bröcker-tom Dieck [1985]). Verify the following table of fixed point dimensions.

H	$\dim V^H$	n
$\mathbb{Z}/m$	$1+2k$	$km \leq n \leq (k+1)m-1$
D_m	$k+\frac{1}{2}+(-\frac{1}{2})^n$	$km \leq n \leq (k+1)m-1$
T	k	$n \equiv 1, 2, 5 \bmod 6$
	$k+1$	$n \equiv 0, 3, 4 \bmod 6$
O	k	$n \equiv 1, 2, 3, 5, 7, 11$
	$k+1$	$n \equiv 0, 4, 6, 8, 9, 10$ modulo 12
I	k	$n \equiv 1, 2, 3, 4, 5, 7, 8, 9, 11, 13, 14, 17, 19, 23, 29$ modulo 30
	$k+1$	otherwise

4. $A\mathrm{SO}(3)$ contains the following idempotent elements $\neq 0,1$: Let $G = \mathrm{SO}(3)$. Then $x = G/I - G/T - G/D_5 - G/D_3$, $y = G/\mathrm{O}(2) + G/O - G/D_4 - G/D_3$, $x+y$, $1-x$, $1-y$, $1-x-y$ are the idempotents $\neq 0,1$. See also section 7: G contains 3 conjugacy classes of perfect subgroups: 1, I, and G. Therefore, there must be 2^3 idempotents. The element x corresponds to the function $z \in C(G)$ with $z(I) = 1$ and $z(H) = 0$ otherwise; and $1-x-y$ corresponds to the function z with $z(G) = 1$ and $z(H) = 0$ otherwise.
5. Show that a function $z \in C(\mathrm{SO}(3))$ with values in $\{\pm 1\}$ represents a unit of $A\mathrm{SO}(3)$ if and only if $z(D_2) = z(A(4))$. Show that there exist units which are not in the image of the homomorphism (2.10).

6. Finiteness theorems.

We have already shown in I(5.11)

(6.1) A compact differentiable G-manifold has only a finite number of orbit types.

Using this we show

(6.2) Proposition. *Let G be a compact Lie group. There is only a finite number of conjugacy classes of subgroups which are normalizers of connected subgroups.*

Proof. (Bredon, VII. Lemma 3.2, in Borel [1960].) Let L be the Lie algebra of G and E its exterior algebra. Recall that L is the tangent space of G at e. The differential of $c_g: G \to G$, $h \mapsto ghg^{-1}$ yields a linear map $l_g: L \to L$ and $\mathrm{Ad}: G \times L \to L$, $(g, v) \mapsto l_g(v)$ is called the adjoint representation of G. This action on L induces an action Ad on E and on the corresponding projective space $P(E)$ (compare I(2.8)). If h is a linear subspace of L with basis $h_1, \ldots, h_k$, then the exterior product $h_1 \wedge \ldots \wedge h_k$ determines a point ph of $P(E)$ which is independent of the choice of the basis (Plücker coordinates). A subgroup N of G leaves h invariant if and only if $ph \in P(E)$ is fixed under N. Thus if H is a subgroup with Lie algebra h, then:

$$gHg^{-1} = H \Leftrightarrow \mathrm{Ad}(g, h) = h \Leftrightarrow \mathrm{Ad}(g, ph) = ph.$$

Thus NH is the isotropy group G_{ph} of the G-action on $P(E)$. Now we can apply (6.1). □

(6.3) Proposition. *A compact Lie group G contains only a finite number of conjugacy classes (K) of subgroup K which are centralizers of closed subgroups.*

Proof. Consider the action $G \times G \to G$, $(g, h) \mapsto ghg^{-1}$. If $H \subset G$, then G^H is the centralizer ZH of H in G. Now apply (6.1). □

Next we state a classical finiteness theorem of Jordan. We denote the set of finite subgroups of G by $\mathrm{Fin}(G)$.

(6.4) Theorem. *There exists an integer j, depending only on the dimension and the number of components of G, with the following properties: Each $H \in \mathrm{Fin}(G)$ has a normal abelian subgroup A_H such that $|H/A_H| < j$. Moreover, the A_H can be chosen in such a way that $H \subset K$ implies $A_H \subset A_K$.*

Proof. Given integers k and d, there are only finitely many non-isomorphic groups G such that $\dim G = d$ and the number of components $|G/G_0| = k$; see (6.5). Therefore, these groups can be embedded into a fixed orthogonal group $\mathrm{O}(n)$ (Bröcker-tom Dieck [1985], III(4.1)). Hence it suffices to prove the theorem for $G = \mathrm{O}(n)$. For a proof in this case, we refer to Wolf [1967], p. 100–103. It is shown there that one can find a neighbourhood U of e in $\mathrm{O}(n)$ which has the following properties:

(i) $gUg^{-1} = U$ for all $g \in \mathrm{O}(n)$.

(ii) If $g, h \in U$ and g commutes with $[g, h] = ghg^{-1}h^{-1}$, then g commutes with h.
(iii) For $g, h \in U$, the sequence $[g, h], [g, [g, h]], [g, [g, [g, h]]], \ldots$ converges in U to e.

If $O(n)$ carries a normalized Haar measure μ such that $\mu(O(n)) = 1$, let U be a neighbourhood satisfying (i)–(iii) above. Let $\mu(U) > 1/j$. Given $H \subset \mathrm{Fin}(G)$, let A_H be the subgroup generated by $H \cap U$. Then A_H is a normal abelian subgroup and $|H/A_H| < j$. This construction also shows that $A_H \subset A_K$ if $H \subset K$. □

(6.5) Theorem. *There exist only finitely many non-isomorphic compact Lie groups with given dimension and number of components.*

Proof. The proof depends on various classical results. We describe the ingredients and leave the details to the reader.

We first look at connected groups G. Such groups have the form $G = (T \times H)/D$ where T is a torus, H a compact semi-simple Lie group, and D a finite central subgroup of $T \times H$ such that $D \cap T$ and $D \cap H$ are trivial (Hochschild [1965], XIII Theorem 1.3). Therefore, the projection of D into H is injective with image contained in the center ZH of H. This center is finite by a classical theorem of Weyl (see e.g. Bröcker-tom Dieck [1985], V(7.13)). Hence, given T and H, there is only a finite number of groups $(T \times H)/D$. By the classification theorem for compact semisimple Lie groups, there is only a finite number of groups H. This establishes the theorem for connected groups.

For the general case, one has to study extensions $1 \to G_0 \to G \to E \to 1$ of a compact group G_0 by a finite group E. One uses the general theory of group extensions (Mac Lane [1963], IV): Given an extension as above, conjugation in G induces a homomorphism $G \to \mathrm{Aut}(G_0)$. If we divide by the normal subgroup of inner automorphisms, we obtain an induced homomorphism $\gamma: E \to \mathrm{Aut}(G_0)/\mathrm{In}(G_0)$. This is an invariant of the extension. There is only a finite number of extensions with a given γ.

In order to see this, recall that the set of congruence classes of such extensions is in bijective correspondence to a cohomology group $H^2(E; C)$ with C = center of G_0 and E-module structure γ (see Mac Lane [1963], IV.8.8). The group C is an abelian compact Lie group, hence isomorphic to a product $C = T \times A$ of a torus T and a finite abelian group A. The exact sequence of E-modules $1 \to T \to C \to A \to 1$ gives the exact sequence

$$H^2(E; T) \to H^2(E; C) \to H^2(E; A).$$

Let $1 \to \mathbb{Z}^n \to \mathbb{R}^n \to T \to 1$ be a universal covering. Then $H^2(E; T) \cong H^3(E; \mathbb{Z}^n)$. The finiteness of the cohomology of finite groups with finitely generated coefficients implies $H^2(E; C)$ to be finite.

Thus it remains to study the set of homomorphisms $\gamma: E \to \mathrm{Aut}(G_0)/\mathrm{In}(G_0)$. In order to enumerate non-isomorphic groups G, it suffices to look at conjugation classes of homomorphisms γ. Suppose first that $G_0 = S$ is an n-dimensional torus, hence $\mathrm{Aut}(S)/\mathrm{In}(S) \cong \mathrm{GL}(n, \mathbb{Z})$. In this case, we have to enumerate isomorphism classes of $\mathrm{GL}(n, \mathbb{Z})$-modules of rank n. By representation theory and the Jordan-Zassenhaus-Theorem (Curtis-Reiner [1962], § 79), they are finite in number. For a general group G_0, one uses the additional fact that a semi-simple compact Lie group has finite outer automorphism group. □

Recall that the dimension of a maximal torus of G is called the rank of G. A subgroup H of G is called of maximal rank if $\operatorname{rank} H = \operatorname{rank} G$.

(6.6) Theorem. *A compact connected Lie group has only a finite number of conjugacy classes of connected subgroups of maximal rank.*

Proof. Borel-de Siebenthal [1949]. See also Wolf [1967], 8.10. □

We now consider solvable groups. The **derived group** $G^{(1)}$ of G is defined as the closure of the subgroup generated by all commutators. Inductively, we define $G^{(n)} = (G^{(n-1)})^{(1)}$. A group H is called **perfect** if $H = H^{(1)}$. A compact Lie group is called **solvable** if $G^{(n)} = \{e\}$ for some integer n (compare Tits [1983], IV § 1). If $1 \to A \to B \to C \to 1$ is an exact sequence of compact Lie groups, then B is solvable if and only if A and C are solvable. A compact Lie group is solvable if and only if it is an extension of a torus by a finite solvable group.

We have to use the following elementary facts.

(6.7) Proposition.

(i) *Each subgroup H of G has a unique minimal normal subgroup H_s such that H/H_s is solvable.*

(ii) *For each H, there exists an integer n such that $H^{(n)} = H_s$.*

(iii) *H_s is a perfect characteristic subgroup of H.*

(iv) *$H = H_s$ if and only if H is perfect.*

(v) $(H) = (K) \Rightarrow (H_s) = (K_s)$.

(vi) *$K \lhd H$, H/K solvable* $\Rightarrow K_s = H_s$.

Proof. (i) If $K \lhd H$, $L \lhd H$ and H/K, H/L are solvable, then $K \cap L \lhd H$ and $H/(K \cap L)$, being isomorphic to a subgroup of $H/K \times H/L$, is solvable. By the descending chain property for subgroups, there is a minimal group as stated.
(ii), (iii), and (iv). Since $H/H^{(1)}$ is abelian, $H/H^{(k)}$ is solvable by induction, hence $H_s \subset H^{(k)}$ for all k and $H^{(k)}/H_s$ is solvable. If $H^{(k)} \neq H_s$, then $H^{(k)}$ has a non-trivial abelian quotient, hence $H^{(k)} \neq H^{(k+1)}$. By the descending chain property, there is an integer n such that $H^{(n)} = H^{(n+1)}$; for this n, $H^{(n)} = H_s$ and $H^{(n)}$ is perfect. The $H^{(n)}$ are characteristic subgroups.
(v) and (vi) are obvious. □

(6.8) Theorem. *Let G be a compact Lie group. There exists an integer n such that the equality $H^{(n)} = H_s$ holds for all subgroups H.*

Proof. Note that $H^{(n)} = H_s$ if and only if $(H/H_s)^{(n)}$ is the trivial group. Therefore, we consider pairs H, K such that $H \lhd K \subset G$ and K/H is solvable. We show that there exists an integer n such that for all such pairs (K/H) the group $(K/H)^{(n)}$ is the trivial group.

For a solvable group L, let us call the smallest integer k such that $L^{(k)} = \{e\}$ the length $l(L)$ of L. Take a pair (K, H) as above. Since K/H is solvable, we have an exact sequence $1 \to T \to K/H \to F \to 1$ with a torus T and a finite solvable group F. Since $l(K/H) \leq l(T) + l(F) = 1 + l(F)$, we have only to show that the lengths of finite solvable subquotients are uniformly bounded.

Let K_0 denote the e-component of K. Then $K/H \to F$ induces a surjection $p: K/K_0 \to F$. There exists an integer $b(G)$ such that for every $K \subset G$ there exists a normal abelian subgroup A_K of K/K_0 such that $|K/K_0: A_K| < b(G)$. The existence of $b(G)$ is proved by induction on $\dim G$ and $|G/G_0|$. Given G, the bound exists for finite K by (6.4). Let K be a subgroup of positive dimension. Consider $K_0 \subset K \subset NK \subset N(K_0)$. Then K/K_0 is a finite subgroup of $N(K_0)/K_0 = U$ and $\dim U < \dim G$. By (6.2), only finitely many non-isomorphic groups U occur. By induction, this implies the required finiteness.

Returning to $p: K/K_0 \to F$ let $F_0 = pA_K$. Then $|F/F_0| < b(G)$. But $l(F) \leq l(F_0) + l(F/F_0) = 1 + l(F/F_0)$ since F_0 is abelian. But $l(F/F_0)$ is bounded because only a finite number of groups occurs. □

The next theorem is taken from tom Dieck [1977].

(6.9) Theorem. *There exists an integer b such that for each closed subgroup H of G the index $|WH: (WH)_0|$ is less than b.*

Proof. The proof proceeds in three steps: We first reduce to the case that WH is finite. Then we reduce to the case that H is finite. Finally, we show that for finite H with finite WH the order of WH is uniformly bounded.

The group $\mathrm{Aut}(H)/\mathrm{In}(H) = \mathrm{Out}(H)$ of automorphisms modulo inner automorphisms is discrete (see I(5.18), Ex. 1). Conjugation induces an injective homomorphism $NH/ZH \cdot H \to \mathrm{Out}(H)$ where ZH is the centralizer of H in G. Hence $NH/ZH \cdot H$ is finite. Therefore:

(6.10) WH is finite if and only if $ZH/ZH \cap H$ is finite.
Using (6.10) and the relation $Z(ZH \cdot H) \subset ZH \subset ZH \cdot H$, we obtain:
(6.11) For each $H \subset G$, the group $ZH \cdot H$ has finite index in its normalizer.

If $n \in G$ normalizes H, then also ZH and $ZH \cdot H$. We therefore have $NH/ZH \cdot H \subset N(ZH \cdot H)/ZH \cdot H$. Using (6.11) and the existence of an upper bound for the set

$$F(G) = \{|WH| \mid H \subset G, WH \text{ finite}\},$$

we obtain

(6.12) There exists an integer c such that $|NH/ZH \cdot H| < c$ for all $H \subset G$.

Now we can obtain the first reduction of our problem. From the exact sequence

$$1 \to ZH/ZH \cap H \to WH \to NH/ZH \cdot H \to 1$$

we see that the kernel of $WH/(WH)_0 \to NH/ZH \cdot H$ is a quotient of $ZH/(ZH)_0$. Now (6.3) and (6.12) show that $\{|WH/(WH)_0| \mid H \subset G\}$ is bounded.

We prove by induction on $|G/G_0|$ and $\dim G$ that $F(G)$ has an upper bound $a = a(G/G_0, \dim G)$. For finite G, we can take $a = |G|$. Suppose that an upper bound $a(K/K_0, \dim K)$ is given for all K with $\dim K < \dim G$. Let $T(G) = \{H \subset G \mid WH \text{ finite}\}$. Suppose $H \in T(G)$ is not finite. We consider the projection $p: N(H_0) \to N(H_0)/H_0 = U$. Let V be the normalizer of H/H_0 in U. Then $WH = V/(H/H_0)$ and therefore $H/H_0 \in T(U)$. Since $\dim U < \dim G$, we obtain, by induction hypothesis, $|WH| \leq a(U/U_0, \dim U)$. It follows from (6.2) that the possible values for $|U/U_0|$ are finite. Hence, for a given group G, the possible values of $|U/U_0|$ are bounded, say $|U/U_0| \leq m(G)$. Now use (6.5) to establish the induction step as far as non-finite H in $T(G)$ are concerned.

Now let $H \in T(G)$ be finite. Then $K = NH$ is finite and, by I(5.10), $K \in T(G)$. We choose $j = j(|G/G_0|, \dim G)$ and A_H, A_K according to (6.4). We have

$$|K/H| \leq |K/A_K| \cdot |A_K/H \cap A_K| \leq j \cdot |A_K/H \cap A_K|.$$

Hence it suffices to find a bound for the $|A_K/H \cap A_K|$. Consider the extension $1 \to A_H \to H \to S \to 1$. Conjugation $c(a)$ by $a \in A_K$ is trivial on A_H. Therefore, $c(a)$ induces an automorphism of S. Since $|S| \leq j$, this automorphism has order at most $J = j!$. Hence $c(a^r)$ is the identity on S and A_H for a suitable $r \leq J$. The group of such automorphisms modulo the subgroup of inner automorphisms by elements of A_H is isomorphic to $H^1(S; A_H)$, with S acting on A_H by conjugation. Since this group is annihilated by $|S|$, we see that $c(a^s)$ is an inner automorphism by an element of A_H for a suitable $s \leq J|S| \leq Jj$. In other words: $a^s h^{-1} \in ZH$. Hence it suffices to find a bound for the orders of the groups $(A_K \cap ZH)/(H \cap A_K \cap ZH)$.

The group $U_1 = A_K \cap ZH$ is contained in the normalizer NT of a suitable maximal torus of G. This follows from a general fact:

(6.13) Proposition. *Let A be a subgroup of a compact Lie group G which has a chain $\{e\} = A_n \subset A_{n-1} \subset \ldots \subset A_0 = A$ of normal subgroups such that each A_i/A_{i+1} is either a torus or finite cyclic. Then A is contained in the normalizer of a suitable maximal torus of G.*

Proof. Borel-Serre [1953], Théorème 1. □

Thus, if we set $U = U_1 \cap T$, we have $|U_1/U| \leq |G/G_0||wG_0|$ where wG_0 denotes the Weyl group of G_0, i.e. $wG_0 = (NT \cap G_0)/T$. We estimate the order of U. Since U is abelian, we have $U \subset ZU$; therefore, U is contained in the center $C = C(ZU)$ of ZU. We have $H \subset ZU$ as $U \subset ZH$ and therefore $C \subset NH$. Hence C is finite. We have

$$|ZU/(ZU)_0| \leq |G/G_0||N_0(ZU)_0/(ZU)_0|$$

where $N_0(ZU)_0 = N(ZU)_0 \cap G_0$ is the normalizer of $(ZU)_0$ in G_0. By (6.3), there is only a finite number of possible normalizers $N_0(ZU)_0$ and thus the possible component groups $ZU/(ZU)_0$ have bounded orders.

Since $U \subset C$ and C is a finite center of the set of groups ZU with bounded number of components, the proof of (6.9) will be complete if we can show

(6.14) Proposition. *There exists an integer c, depending only on $|G/G_0|$ and* $\dim G$, *such that a finite center $C(G)$ has order less than c.*

Proof. We let G/G_0 act by conjugation on $C(G_0)$. Then $C(G) \cap G_0$ is the fixed point set of this action. We have $C(G_0) = A \times T_1$ where A is a finite abelian group and T_1 is a torus. The group A is the center of a semi-simple group and therefore $|A|$ is bounded by a constant d_0 depending only on $\dim G$. The exact cohomology sequence associated to the universal covering $0 \to \pi_1 T_1 \to V \to T_1 \to 0$ shows that the fixed point set of the action of G/G_0 on $T_1 = C(G_0)_0$ is isomorphic to $H^1(G/G_0; \pi_1 T_1)$, hence its order is bounded by a constant d_1 depending only on G/G_0 and the rank of T_1. Hence $|C(G)| \leq |G/G_0| d_0 d_1$. □

As a corollary to (6.9), we can state (compare (4.14))

(6.15) Proposition. *Let n be the least common multiple of the numbers $|NH/H|$ for $(H) \in \phi(G)$. Then the cokernel of $A(G) \to C(G)$ has exponent n.* □

7. Idempotent elements.

We want to enumerate the idempotent elements of $A(G)$ in terms of the subgroup structure of G.

As in section 3, let $S(G)$ denote the space of closed subgroups of G and $\psi(G)$ the quotient space under the conjugation action. Let $H^{(1)}$ be the derived subgroup of H and H_s the smallest normal subgroup of H such that H/H_s is solvable. Let P be the space of perfect subgroups in $S(G)$. Let $C'(G)$ be the ring of continuous functions $\psi(G) \to \mathbb{Z}$. An element $e \in A(G)$ is idempotent, i.e. satisfies $e^2 = e$, if and only if $\varphi_H(e) \in \{0, 1\}$ for all H. A function $z \in C'(G)$ is idempotent if and only if it assumes values in $\{0, 1\}$.

(7.1) Proposition. *An idempotent $e \in C'(G)$ is contained in $A(G)$ if and only if the equality $e(H) = e(H_s)$ holds for all subgroups H of G.*

Proof. Suppose $e \in A(G)$. It suffices to show that $e(H) = e(H^{(1)})$. Since $H/H^{(1)}$ is abelian, there exists a chain $H^{(1)} = K_1 \lhd K_2 \lhd \ldots \lhd K_r = H$ such that K_{i+1}/K_i is either a torus or a cyclic group of prime order. In each case, by (5.8), $e(K_i) \equiv e(K_{i+1})$ modulo a suitable prime ideal of $\mathbb{Z}$. Since e assumes only the values 0 and 1, we have $e(K_i) = e(K_{i+1})$.

Suppose $e \in C'(G)$ satisfies $e(H) = e(H_s)$. We want to use (5.8) in order to show that $e \in A(G)$. If $H \lhd K$ and K/H is a torus, then $K_s = H_s$ and therefore $e(K) = e(H)$. If $(H) \in \phi(G)$ and $H \lhd K$, K/H cyclic, then $K_s = H_s$ and the congruences (5.7) are satisfied for a constant function. □

We are going to describe the set of idempotents with the help of commutative algebra.

(7.2) Proposition. *The maps $H \mapsto H^{(1)}$ and $H \mapsto H_s$ are continuous maps $S(G) \to S(G)$. The space P is closed in $S(G)$.*

Proof. In view of the compactness of $S(G)$ and (6.8), we have only to show that $H \mapsto H^{(1)}$ is continuous. Let $H_1, H_2, \ldots$ be a sequence of subgroups converging to H. Without loss of generality we can assume that the H_i are conjugate to subgroups of H. Moreover, by (3.3), we can find a sequence $g_1, g_2, \ldots$ of elements in G converging to e such that $K_i = g_i H_i g_i^{-1}$ is contained in H. We show that $\lim K_i^{(1)}$ exists and is equal to $H^{(1)}$. Fix $\varepsilon > 0$ and choose n such that, for $i \geq n$, $d(K_i, H) < \varepsilon$ in the Hausdorff metric. Let $c^k(K)$ be the closed subspace of a group K consisting of those elements which are products of at most k commutators. Then $d(K_i, H) < \varepsilon$ implies $d(c^k(K_i), c^k(H)) < 4k\varepsilon$. Choose k such that $d(c^k(H), H^{(1)}) < \varepsilon$. Then, for $n \leq i$, we have $d(c^k(H_i), H^{(1)}) < (4k+1)\varepsilon$ and, a fortiori, $d(K_i^{(1)}, H^{(1)}) < (4k+1)\varepsilon$. Finally, since $K_i^{(1)} = g_i H_i^{(1)} g_i^{-1}$, the sequence $H_1^{(1)}, H_2^{(2)}, \ldots$ converges and has the same limit $H^{(1)}$. □

As a corollary we obtain

(7.3) Proposition. *Let $H \subset G$ be perfect. Then $\{K \mid K_s = H\}$ and $\{K \mid K_s \sim H\}$ are closed subsets of $S(G)$.* □

In section 4, we obtained the closed quotient map

$$q: S(G) \times \operatorname{Spec} \mathbb{Z} \to \operatorname{Spec} A(G), (H, (p)) \mapsto q(H, p).$$

Let Π be the image of P in $\psi(G)$. Consider the composition

$$r\colon S(G) \times \operatorname{Spec} \mathbb{Z} \xrightarrow[\text{pr}]{} S(G) \xrightarrow[s]{} P \xrightarrow[c]{} \Pi$$

with projection pr and $s(H) = H_s$, $c(H) = (H)$. Then r is continuous by (7.2).

(7.4) Proposition. *The map r factors over q and thus induces a continuous surjective map $t\colon \operatorname{Spec} A(G) \to \Pi$, $q(H,p) \mapsto H_s$.*

Proof. Suppose $q(H,p_1) = q(K,p_2)$. Since p is the characteristic of $A(G)/q(H,p)$, we have $p_1 = p_2$. Let $p = p_1$. Let (H^*) be the unique conjugacy class such that $q(H,p) = q(H^*,p)$ and NH^*/H^* is finite of order prime to p; see (4.2). By (4.5), we can find a countable transfinite sequence $H \lhd H_1 \lhd H_2 \lhd \ldots \lhd H_\lambda \sim H^*$ such that H_{i+1}/H_i is solvable and H_j is the limit of the preceding subgroups if j is a limit ordinal. Using transfinite induction, it follows from (7.2) that $H_s = (H_\lambda)_s$.

The space Π is a countable compact metric space and therefore totally disconnected. Hence we get a unique continuous map $u\colon \Pi \to \pi_0 \operatorname{Spec} A(G)$ such that $u\pi_0 = t$ where $\pi_0\colon \operatorname{Spec} A(G) \to \pi_0 \operatorname{Spec} A(G)$ is the projection onto the space of components.

(7.5) Theorem. *The map u is a homeomorphism.*

Proof. $\pi_0 \operatorname{Spec} A(G)$ is compact as a quotient of the compact space $\operatorname{Spec} A(G)$. The space Π is a Hausdorff space. Therefore, it is enough to show that u is bijective. We already know from (7.3) that u is surjective. In order to show injectivity, let B and C be two components of $\operatorname{Spec} A(G)$ such that $e(B) = e(C)$. Choose elements $q(H,p) \in B$ and $q(K,l) \in C$. Then $(H_s) = tq(H,p) = tq(K,l) = (K_s)$. Since H/H_s is solvable, we can find a finite chain of subgroups $H_s = H_k \lhd H_{k-1} \lhd \ldots \lhd H_2 \lhd H_1 = H$ such that H_i/H_{i+1} is either a torus or finite cyclic of prime order. By (4.5), $q(H_i, p_i) = q(H_{i+1}, p_i)$ for a suitable prime. If $\bar{q}(H,p)$ denotes the closure of the point $q(H,p)$, then

$$q(H,p) \in \bar{q}_1(H_1, 0),\ \bar{q}(H_i, 0) \cap \bar{q}(H_{i+1}, 0) \neq \emptyset$$

and therefore $q(H_s, 0) \in B$. Similarly, $q(K_s, 0) \in C$ and therefore $B = C$. □

Using some commutative algebra, (7.5) leads to a description of the idempotent elements in $A(G)$. We recall the relevant facts.

Let U be an open and closed subset of $\operatorname{Spec} A(G)$. Then U is a union of components and projects to an open and closed subset of Π, denoted by $t(U)$. Let $e(U)$ be the idempotent element of $A(G)$ which corresponds to U (see Bourbaki [1961b], II.4.3, Proposition 15): The complement Z of U in $\operatorname{Spec} A(G)$ is given by

$$Z = V(A(G)e(U)) := \{q \in \operatorname{Spec} A(G) \mid A(G)e(U) \subset q\}.$$

Let $S(U) = \{H \subset G \mid \varphi_H e(U) = 1\}$. Then, by (7.1), $H \in S(U)$ if and only if $H_s \in S(U)$. The idempotent $e(U)$ is indecomposable if and only if U is a component. If the perfect subgroup H of G is not a limit of perfect subgroups, then, by (7.5),

$$U(H) := \{q(K, p) \mid (K_s) = (H)\}$$

is a component and thus H gives rise to an indecomposable idempotent $e_H := e(U(H))$.

(7.6) Proposition. *Let G be a finite group. The indecomposable idempotents of $A(G)$ correspond bijectively to the conjugacy classes of perfect subgroups of G.*

Proof. The bijection is given by assigning e_H to H. □

(7.7) Proposition. *The compact Lie group G is solvable if and only if 0 and 1 are the only idempotents in $A(G)$.*

Proof. If G is solvable, then $H_s = \{1\}$ for all subgroups H of G; and, by (7.1), an idempotent is a constant function.

If G is not solvable, then Π consists of more than one point and therefore there exist continuous non-constant functions $\Pi \to \{0, 1\}$ and, by (7.2), continuous non-constant functions $S(G) \to \{0, 1\}$ which yield, by (7.1), an idempotent different from 0 and 1. □

References for this section are tom Dieck [1977a], Schwänzl [1979]. Proposition (7.7) is due to Dress [1969a]. Explicit formulas for idempotents in terms of the standard basis $[G/H]$ can be found in D. Gluck [1981], Yoshida [1983], Kratzer-Thévenaz [1984].

Suppose X is a finite complex such that X^H is either empty or contractible, for all $H \subset G$. Then $[X] \in A(G)$ is an idempotent. It follows from Oliver [1976b] that each idempotent of $A(G)$ is the difference of two such geometrically realized idempotents.

(7.8) Proposition. *Let H and K be subgroups of G such that H_s and K_s are conjugate. Suppose $\chi(G/H^K) \neq 0$. Let*

$$G/H^K = M_1 + M_2 + \ldots + M_r$$

be a decomposition into connected components. Let d be the minimal dimension of the M_i. Then $\chi(M_i) \neq 0$ if and only if $\dim M_i = d$.

Proof. Since $\chi(G/H^K) \neq 0$, we see that K is subconjugate to H. Thus we can assume that $K \subset H$. It follows from $K/K \cap H_s \subset H/H_s$ that $K/K \cap H_s$ is solvable, hence $K_s \subset K \cap H_s \subset H_s$. Since K_s and H_s are conjugate, we have $H_s = K_s =: L$. Let

$$G/H^L = N_1 + \ldots + N_t$$

be the decomposition into the finite number of WL-orbits (see I(5.10)) and write $N_i \cong WL/A_i$. We claim that the A_i are solvable. In order to prove the claim, let N_i be the NL-orbit through $g_i H$. The isotropy group at $g_i H$ is $NL \cap g_i H g_i^{-1}$. Hence

$$A_i = (NL \cap g_i H g_i^{-1})/L.$$

But $g_i^{-1} L g_i = L$; this can be seen as follows. Let S be the image of $g_i^{-1} L g_i \to H \to H/L$. Then S is solvable and

$$S \cong g_i^{-1} L g_i / g_i^{-1} L g_i \cap L.$$

But since $g_i^{-1} L g_i$ is isomorphic to L, this group is perfect and cannot have a non-trivial solvable quotient. Hence $g_i^{-1} L g_i \cap L = g_i^{-1} L g_i$ and this implies the desired equality of groups. Therefore, A_i is a subgroup of the solvable group $g_i^{-1} H g_i / g_i^{-1} L g_i$ and this implies the claim.

Now let us consider

$$G/H^K \cong (G/H^L)^{K/L} \cong N_1^{K/L} + \ldots + N_t^{K/L}.$$

Let $U := K/L \subset WL$. The N_i^U decompose into NU-orbits, which are isomorphic to $NU/NU \cap w_j^{-1} A_i w_j$; thus they have the form NU/B with solvable B. By now we have shown that G/H^K has the form

$$NU/B_1 + \ldots + NU/B_k$$

with solvable $B_i \subset NU$. If NU/B_i consists of different components, then all these components have equal Euler characteristic. Therefore, $\chi(NU/B_i) \neq 0$ if and only if B_i has maximal rank in NU. Since B_i is solvable, it has maximal rank if and only if it has the dimension of the maximal torus of NU. □

(7.9) Exercises.

1. Let $1 \to T \to G \to F \to 1$ be an extension of a torus T by a finite group F. Let $\varphi: F \to \mathrm{Aut}(T)$ be the map induced by the conjugation action. Let F' be the kernel of φ. Show that the number of conjugacy classes of perfect subgroups of G is finite if and only if F/F' is solvable (tom Dieck [1979], 5.11.6).

8. Induction categories.

In order to prepare for the axiomatic induction theory of compact Lie groups, we construct the basic category which is used in this context.

Let G be a compact Lie group. We shall define a category $\Omega(G)$. The objects are the homogeneous spaces. The morphisms from G/H to G/K are the elements of the abelian group $U(G; G/H \times G/K)$, which was defined in section 1. We want the category to be additive: Composition of morphisms shall be a bilinear map

$$\text{(8.1)} \qquad U(G; G/H_1 \times G/H_2) \times U(G; G/H_2 \times G/H_3) \to U(G; G/H_1 \times G/H_3).$$

In order to define this pairing, we give a description via representatives. Suppose

$$(\alpha, \beta_1): A \to G/H_1 \times G/H_2$$
$$(\beta_2, \gamma): B \to G/H_2 \times G/H_3$$

are given. Consider the diagram

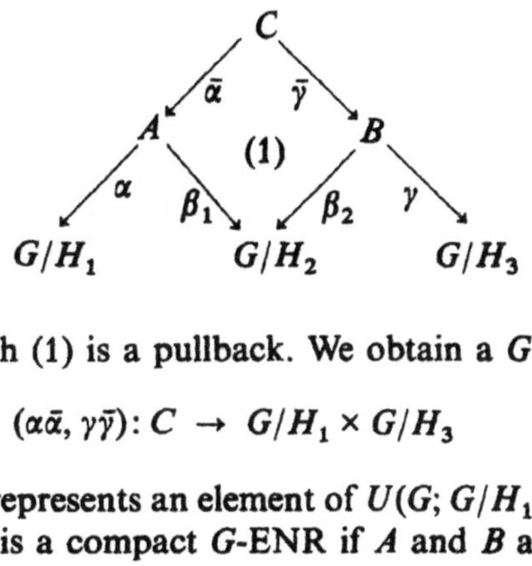

in which (1) is a pullback. We obtain a G-map

$$(\alpha\bar{\alpha}, \gamma\bar{\gamma}): C \to G/H_1 \times G/H_3$$

which represents an element of $U(G; G/H_1 \times G/H_3)$: This follows from the fact that C is a compact G-ENR if A and B are spaces of this type (exercise 2).

In order to see that this construction defines a pairing (8.1), we have to show that it induces a well-defined map. For this purpose, we give another construction. We use the canonical G-homeomorphism $G \times_H X \to G/H \times X$, $(g, x) \mapsto (g, gx)$ for G-spaces X without further mentioning. Consider the following composition of maps

$$
\begin{array}{ll}
\textbf{(8.3)} & U(G; G/H_1 \times G/H_2) \times U(G; G/H_2 \times G/H_3) \\
& \qquad\qquad (1) \downarrow \cong \\
& U(H_2; \mathrm{res}_{H_2} G/H_1) \times U(H_2; \mathrm{res}_{H_2} G/H_3) \\
& \qquad\qquad (2) \downarrow \\
& U(H_2; \mathrm{res}_{H_2} G/H_1 \times \mathrm{res}_{H_2} G/H_3) \\
& \qquad\qquad (3) \downarrow \cong \\
& U(G; G/H_2 \times (G/H_1 \times G/H_3)) \\
& \qquad\qquad (4) \downarrow \mathrm{pr}_* \\
& U(G; G/H_1 \times G/H_3).
\end{array}
$$

The map (1) uses the isomorphism (1.21), the map (2) is an application of (1.24), (3) is again (1.21), and (4) is induced by the projection (1.28, Ex. 6). Obviously, (8.3) is a well-defined bilinear pairing. We show that on suitable representatives (8.2) and (8.3) give the same result.

Suppose $(\alpha, \beta_1): A \to G/H_1 \times G/H_2$ and $(\beta_2, \gamma): B \to G/H_2 \times G/H_3$ are given. The map β_1 leads to a homeomorphism $A \cong G \times_{H_2} A(0)$ with $A(0) := \beta_1^{-1}(eH_2)$ and β_2 yields $B \cong G \times_{H_2} B(0)$ with $B(0) := \beta_2^{-1}(eH_2)$. Let $\alpha(0): A(0) \to \mathrm{res}_{H_2} G/H_1$ be the restriction of α to $A(0)$ and let $\gamma(0): B(0) \to \mathrm{res}_{H_2} G/H_3$ be the restriction of γ to $B(0)$. After the first step in (8.3) we have the element represented by $(\alpha(0), \gamma(0))$, after the second step the element represented by $\alpha(0) \times \gamma(0)$. If we apply the $G \times_{H_2}$-extension process to the pullback

$$
\begin{array}{ccc}
A(0) \times B(0) & \xrightarrow{\ \mathrm{pr}\ } & B(0) \\
\Big\downarrow{\scriptstyle \mathrm{pr}} & & \Big\downarrow \\
A(0) & \longrightarrow & P, \quad P = \text{Point},
\end{array}
$$

we obtain the pullback

$$
\begin{array}{ccc}
C = G \times_{H_2} (A(0) \times B(0)) & \xrightarrow{\ \bar{\gamma}\ } & G \times_{H_2} B(0) \cong B \\
\Big\downarrow{\scriptstyle \bar{\alpha}} & & \Big\downarrow{\scriptstyle \beta_2} \\
A \cong G \times_{H_2} A(0) & \xrightarrow{\ \beta_1\ } & G/H_2 .
\end{array}
$$

Thus if we apply the $G \times_{H_2}$-extension to $\alpha(0) \times \gamma(0)$ and project away the G/H_2-factor, we obtain $(\alpha\bar{\alpha}, \gamma\bar{\gamma})$. This shows the equivalence of the two definitions for the composition of morphisms.

Using the transitivity of pullbacks, we obtain from the first definition that composition of morphisms is associative. The diagonal $G/H \to G/H \times G/H$ represents the identity morphism of the object G/H. Thus we have established the category $\Omega(G)$.

The morphism set $U(G; G/H \times G/K)$ can be given an intuitively more appealing description. Consider diagrams $G/H \leftarrow G/L \to G/K$. Two such diagrams are called equivalent if there exists an isomorphism $\sigma: G/L \to G/L'$ making the diagram

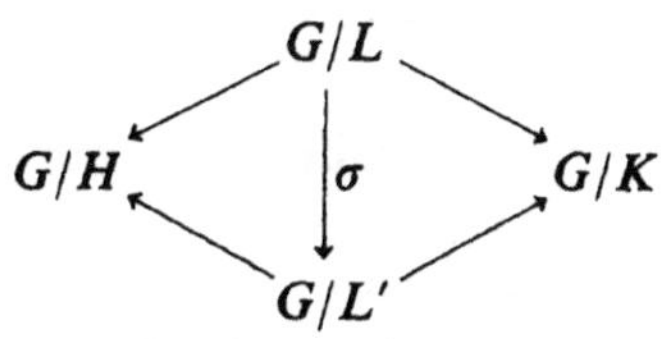

commutative up to G-homotopy.

(8.4) Proposition. *$U(G; G/H \times G/K)$ is the free abelian group on the equivalence classes of diagrams $G/H \leftarrow G/L \to G/K$.*

Proof. The diagram $G/H \leftarrow G/L \to G/K$ is just a map $G/L \to G/H \times G/K$ and therefore represents an element of $U(G; G/H \times G/K)$. By (1.19), equivalent diagrams represent the same element. Now use (1.16). □

The composition of

(8.5) $\quad G/H \xleftarrow[\alpha]{} G/L \xrightarrow[\text{id}]{} G/L \quad$ and

(8.6) $\quad G/L \xleftarrow[\text{id}]{} G/L \xrightarrow[\beta]{} G/K$

and the pullback definition

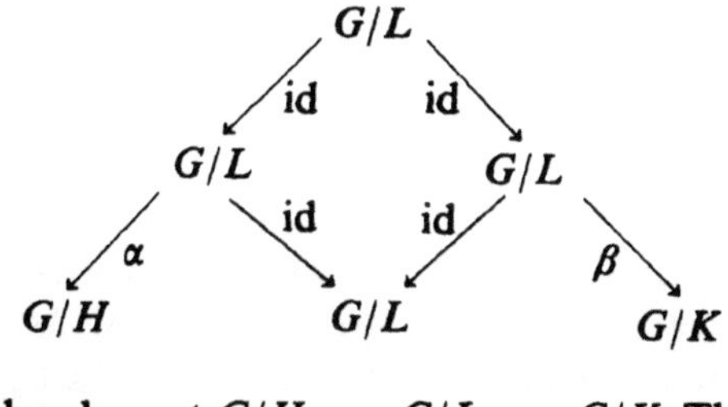

yields the element $G/H \xleftarrow[\alpha]{} G/L \xrightarrow[\beta]{} G/K$. Thus each morphism from G/H to G/K is the composition of two special types of morphisms:

If we map $\beta: G/L \to G/K$ to the morphism (8.6), we obtain a covariant functor from the homotopy category $\pi_0 \operatorname{Or}(G)$ of the orbit category $\operatorname{Or}(G)$ into

$\Omega(G)$. The morphisms of this type are the ordinary morphisms of the category and we write them in the usual way. The morphisms (8.5) from G/H to G/L yield a contravariant functor from $\pi_0 \operatorname{Or}(G)$ into $\Omega(G)$ as is easily checked. They correspond to transfer morphisms, which appear in various places in topology. The morphism (8.5) will also be written as

$$\textbf{(8.7)} \qquad G/H \overset{\alpha}{\rightarrowtail} G/L$$

in order to indicate that the ordinary morphism $G/H \overset{\alpha}{\longleftarrow} G/L$ yields a formal morphism in the category $\Omega(G)$ in the other direction.

(8.8) Remark. The category $\Omega(G)$ has a canonical duality. The morphism set $\operatorname{Mor}(G/H, G/K) = U(G; G/H \times G/K)$ is by its very definition symmetric in H and K. The same set can be considered as the morphism set $\operatorname{Mor}(G/K, G/H)$. Thus we obtain an isomorphism between $\Omega(G)$ and the dual category $\Omega(G)^{\mathrm{op}}$.

The composition of the basic morphisms (8.4) is given by a double coset formula. For finite groups, this can be done in purely algebraic terms as explained below. For compact Lie groups in general, one has to use Euler characteristic methods in order to define and compute the double coset formula.

Suppose G is finite. Consider the diagram (8.2) with $A = G/L_1$ and $B = G/L_2$. Write the finite G-set C as a disjoint union of orbits $C = \coprod C_j$. Let us write $C_j = G/K_j$ and denote the restriction of $\bar{\alpha}$ resp. $\bar{\gamma}$ to C_j by $\bar{\alpha}_j$ resp. $\bar{\gamma}_j$. The composition of

$$G/H_1 \underset{\alpha}{\rightarrowtail} G/L_1 \underset{\beta_1}{\longrightarrow} G/H_2 \qquad \text{and}$$

$$G/H_2 \underset{\beta_2}{\rightarrowtail} G/L_2 \underset{\gamma}{\longrightarrow} G/H_3$$

is then the sum over j of the morphisms

$$G/H_1 \underset{\alpha}{\rightarrowtail} G/L_1 \underset{\bar{\alpha}_j}{\rightarrowtail} G/K_j \underset{\bar{\gamma}_j}{\longrightarrow} G/L_2 \underset{\gamma}{\longrightarrow} G/H_3 .$$

We see that already in this case we need linear combinations of basic morphisms. (Compare also with I(4.5, 4.13).)

For the axiomatic representation theory, certain functors on $\Omega(G)$ are important.

A contravariant additive functor $M : \Omega(G) \to R$-Mod into the category of left R-modules over the commutative ring R is called a **Mackey-functor.** (Instead of R-Mod any abelian category can be used. A functor M is called additive if M is a homomorphism of abelian groups on morphism sets.) Such a functor comprises two types of induced morphisms. Let $\alpha : G/H \to G/K$ be a G-map. If we consider this as an ordinary morphism α_1 of $\Omega(G)$, we have the induced homomorphism

$$M(\alpha_1) =: M^*(\alpha) =: \alpha^* : M(G/K) \to M(G/H).$$

If α is induced from $H \subset K$, i.e. $\alpha(gH) = gK$, then α^* is called a **restriction** homomorphism. If we consider α as a transfer morphism $\alpha^!: G/K - G/H$ in $\Omega(G)$, then we have

$$M(\alpha^!) =: M_*(\alpha) =: \alpha_*: M(G/H) \to M(G/K).$$

If α is induced from $H \subset K$, then α_* is called the **induction** homomorphism associated to α. The fact that the $\alpha_!$ and $\beta^!$ sit inside a larger category implies certain relations between the restriction and the induction process (double coset formula) which are vital for the development of the theory.

Let M, N, and L be Mackey functors on $\Omega(G)$. A **pairing** $M \times N \to L$ is a family of bilinear maps

$$M(S) \times N(S) \to L(S), (x, y) \mapsto x \cdot y,$$

one for each object S of $\Omega(G)$, such that for each G-map $f: G/H = S \to T = G/K$ the following holds

(8.9)
$$\begin{aligned} L^*f(x \cdot y) &= (M^*fx) \cdot (N^*fy), x \in M(T), y \in N(T) \\ x \cdot (N_*fy) &= L_*f((M^*fx) \cdot y), x \in M(T), y \in N(S) \\ (M_*fx) \cdot y &= L_*f(x \cdot (N^*fy)), x \in M(S), y \in N(T). \end{aligned}$$

A **Green** functor $U: \Omega(G) \to R$-Mod is defined to be a Mackey functor together with a pairing $U \times U \to U$ such that for each object S the map $U(S) \times U(S) \to U(S)$ turns $U(S)$ into an associative R-algebra with unit and such that the morphisms f^* preserve the units.

If U is a Green functor, then a **left** U**-module** is a Mackey functor M together with a pairing $U \times M \to M$ which equips each $M(S)$ with a left unital $U(S)$-module structure.

We now describe the universal example for a Green functor $\mathfrak{U}$. The value on G/H is defined to be the ring

$$\mathfrak{U}(G/H) := U(G; G/H) \cong U(H; P) = U(H),$$

P a point space. In order to make this a contravariant functor, we view P as G/G and consider $U(G; G/H) = U(G; G/G \times G/H)$ as a morphism set. Then composition of morphisms

$$U(G; P \times G/K) \times U(G; G/H \times G/K) \to U(G; P \times G/H)$$

defines the action of the functor on morphisms. We describe this more explicitly for the basic morphisms.

Let $G/H \overset{\alpha}{\rightarrowtail} G/L \overset{\beta}{\longrightarrow} G/K$ be given. The induced morphism $\mathfrak{U}(\beta) = \beta^*: U(G; G/K) \to U(G; G/L)$ is obtained as follows: Given $f: X \to G/K$, let $F: Y \to G/L$ be the pullback of f along β; then $\beta^*[f] = [F]$. The morphism $\mathfrak{U}_*(\alpha) = \alpha_*: U(G; G/L) \to U(G; G/H)$ is simply the composition with α.

If one uses $\mathfrak{U}(G/K) = U(K)$, then the induced morphisms are given as follows: $\beta\colon G/L \to G/K$ is determined by $\beta(eL) = gK$ with $g^{-1}Lg \subset K$. A K-space X, representing $[X] \in U(K)$, can be considered as L-space X' via restriction to $g^{-1}Lg$ and conjugation by g; then $\beta^*[X] = [X']$. One can see from this description that β^* is a ring homomorphism. Let $\alpha\colon G/L \to G/K$ be given by $\alpha(eL) = uH$ with $u^{-1}Lu \subset H$. Then an L-space X yields an $u^{-1}Lu$-space X_u via conjugation, namely X with action $u^{-1}Lu \times X \to X$, $(u^{-1}lu, x) \mapsto lu$, and we can extend to H

$$H \times_{u^{-1}Lu} X_u =: X''.$$

Then we have $\alpha_*[X] = [X'']$.

Of course, the pairing $\mathfrak{U} \times \mathfrak{U} \to \mathfrak{U}$ is given by using the ring structures on

$$\mathfrak{U}(G/H) = U(G; G/H) \cong U(H).$$

The multiplication in $U(G; G/H)$ is given by fibre-product of representatives over G/H.

(8.10) Proposition. *$\mathfrak{U}$ is a Green functor with values in abelian groups.*

Proof. One has to verify (8.9). We have already remarked that the β^* are ring homomorphisms, which implies the first relation in (8.9). The equation $x \cdot f_* y = f_*(f^* x \cdot y)$ expresses the transitivity of pullbacks: Let $x = [x\colon X \to G/K]$ and $y = [y\colon Y \to G/H]$. Then one uses the diagram with pullback squares

$$\begin{array}{ccccc} Z & \longrightarrow & X' & \xrightarrow{\;f\;} & X \\ \downarrow & & \downarrow{\scriptstyle f^*x} & & \downarrow{\scriptstyle x} \\ Y & \xrightarrow[\;y\;]{} & G/H & \xrightarrow[\;f\;]{} & G/K. \end{array}$$

The pullback of x along $f \circ y$, composed with $f \circ y$, represents $x \cdot f_* y$. The pullback of x along f represents $f^* x$, the pullback of $f^* x$ along y, composed with y, represents $f^* x \cdot y$; and composing this with f gives a representative for $f_*(f^* x \cdot y)$. □

Let M be a Mackey functor into abelian groups. We want to show that M is a module over $\mathfrak{U}$ in a canonical way. To this end we need the following observation.

Let $\mathrm{Hom}_\Omega(G/H, G/H)$ be the endomorphism ring of the object G/H in $\Omega(G)$. We can define a map

$$\delta\colon \mathfrak{U}(G/H) \to \mathrm{Hom}_\Omega(G/H, G/H)$$

by sending $[f\colon X \to G/H]$ to $[(f, f)\colon X \to G/H \times G/H]$ in $U(G; G/H \times G/H)$.

(8.11) Proposition. *δ is a ring homomorphism.*

Proof. The map is additive since in both cases addition is induced by disjoint union. The product of $f: X \to G/H$ and $h: Y \to G/H$ resp. of (f, f) and (h, h) is defined by the diagrams

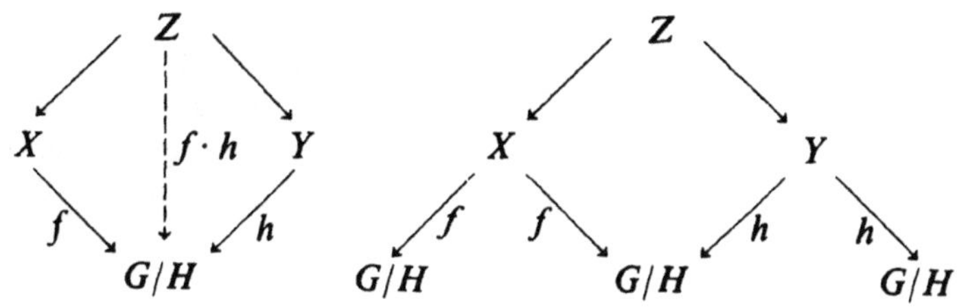

respectively; see (8.2). □

Let M be a Mackey functor. In order to make M a $\mathfrak{U}$-module, we have to define bilinear maps $\mathfrak{U}(G/H) \times M(G/H) \to M(G/H)$ or, equivalently, homomorphisms

$$\mathfrak{U}(G/H) \to \operatorname{Hom}(M(G/H), M(G/H)).$$

Such homomorphisms are obtained by applying the functor M to the ring homomorphisms δ of (8.11).

(8.12) Proposition. *With the pairing defined above, M becomes a $\mathfrak{U}$-module.*

Proof. The verification of (8.9) is left to the reader. □

If U is a Green functor, we can define a map $h_{G/H}: \mathfrak{U}(G/H) \to U(G/H)$ by sending $[f: X \to G/H]$ to $U(\delta(f))(1_{G/H})$ where $1_{G/H} \in U(G/H)$ is the unit. The verification of the next result is again left to the reader.

(8.13) Proposition. *The map $h_{G/H}$ is a ring homomorphism. The maps $h_{G/H}$ constitute a natural transformation of Green functors.* □

In section II.9 we gave a different definition of a Mackey functor, which applied to finite groups. The relation between the two notions is a follows. Let $M = (M_*, M^*): G\text{-Set} \to R\text{-Mod}$ be a bi-functor in the sense of II.9. From this datum we want to define a functor $\tilde{M}: \Omega(G) \to R\text{-Mod}$ by setting $\tilde{M}(G/H) = M_*(G/H) = M^*(G/H)$ on objects; and a morphism $G/H \underset{\alpha}{\leftarrowtail} G/L \underset{\beta}{\longrightarrow} G/K$ shall be mapped to the composition

$$M(G/H) \underset{M_*(\alpha)}{\longleftarrow} M(G/L) \underset{M^*(\beta)}{\longleftarrow} M(G/K).$$

The axioms II(9.2) then imply that $\tilde{M}$ is compatible with composition of morphisms. Therefore, we obtain a contravariant functor $\tilde{M}: \Omega(G) \to R$-Mod. Conversely, if $\tilde{M}$ is given, we have already defined $\tilde{M}^*(\alpha)$ and $\tilde{M}_*(\beta)$ for $\alpha: G/H \to G/K$. One verifies that this can be extended additively to finite G-sets to obtain a Mackey functor in the sense to II.9.

For another notion of a Mackey functor for compact Lie groups, see Lewis-May-McClure [1981].

Actually most of the Mackey functors one meets in pratice, e.g. the representation rings, are defined for all (finite) groups and not just for subgroups of a given group. It is therefore advisable to change the definition of Mackey functor in order to deal with functors on all finite groups. We are not going into this; but see the exercises.

(8.14) Exercises.

1. Give a proof of (8.12) and (8.13).
2. Show that the space C in (8.2) is a compact G-ENR.
3. Define a category with objects G/H and morphisms $G/H \to G/K$ the elements of $\omega_0^G(G/H^+; G/K^+)$, the stable equivariant homotopy classes; see II(8.13), Ex. 8. Show that this category is a quotient category of the induction category. (See Lewis-May-McClure [1981] where Mackey functors are defined on this category.)
4. Let G and H be compact Lie groups. Let $\mathfrak{C}(G, H)$ be the category of finite (G, H)-complexes with left G-action and free right H-action and both actions commuting. Such complexes may be considered as finite $(G \times H)$-complexes. They may also be considered as (G, H)-bundles over finite G-complexes; see I.8. Let $X \mapsto [X] \in \mathrm{U}(G, H)$ be the universal additive invariant for this category. Show that $\mathrm{U}(G, H)$ has a $\mathbb{Z}$-basis consisting of isomorphism classes of (G, H)-bundles over homogeneous spaces. Show that $\mathrm{U}(G, H)$ can be defined using compact (G, H)-ENR's.
5. Show that one obtains a category Ω whose objects are the compact Lie groups and whose morphism groups are $\mathrm{Hom}_\Omega(G, H) = \mathrm{U}(G, H)$; composition

$$\mathrm{U}(G, H) \times \mathrm{U}(H, K) \to \mathrm{U}(G, K)$$

is defined on representatives by $(X, Y) \mapsto X \times_H Y$. Contravariant additive functors $\Omega \to$ Abel may be called **global Mackey functors**. Justify this name.
6. Show that one obtains a contravariant functor $M: \Omega \to$ Abel by defining $M(G) = \mathrm{U}(G)$ on objects and by setting

$$M([E]): \mathrm{U}(H) \to \mathrm{U}(G), [X] \mapsto [E \times_H X].$$

Describe $M([E])$ explicitly in case E is a (G, H)-bundle over a homogeneous space.

7. Show that the complex representation ring leads to a contravariant functor $R: \Omega \to$ Abel in the following way. Set $R(G) = R(G)$ on objects. For $[E] \in U(G, H)$, define a morphism $R[E]: R(H) \to R(G)$ as the composition

$$R(H) \to K_G(E/H) \to R(G);$$

the first map is given by $V \mapsto E \times_H V$ for a representation V and the second map is the transfer in equivariant K_G-theory associated to the map $E/H \to$ Point. (Compare Lewis [1983].)

9. Induction theory.

Induction theory is part of the axiomatic representation theory and has its origin in the investigation of induced representations. The axiomatic theory began with work of Green [1971] and Dress [1973], [1975] and ever since has shown its usefulness in many parts of topology. Methods from the theory of transformation groups can be used to provide a setting for induction theory. We use notations and results from the previous section.

Let M be a Mackey functor for the compact Lie group G. Denote the unique morphism $G/H \to G/G$ by $p(H)$. The main problem of induction theory for a Mackey functor M is the following:

(i) Find a (minimal) family $(G/H_j | j \in J)$ of homogeneous spaces such that

$$\bigoplus_{j \in J} M(G/H_j) \to M(G/G), (x_j) \mapsto \sum_j p(H_j)_* x_j$$

is surjective.

(ii) Compute $M(G/G)$ in terms of the $M(G/H_j)$ and induced homomorphisms between such groups (e.g. describe the kernel of the map in (i)).

One has to think of $M(G/H)$ as being a representation-theoretical object associated to H, e.g. the representation ring $R(H)$ of complex H-representations. Then (i) says that the elements in $M(G/G)$ are integral linear combinations of elements induced up from smaller subgroups H_j.

In order to deal with these problems, we introduce the following terminology. A finite family $\Sigma = (G/H_j | j \in J)$ is called an **induction system** Σ. Any such system leads to two homomorphisms $p(\Sigma)$ and $i(\Sigma)$ (called induction and restriction maps, respectively)

(9.1) $$p(\Sigma): \bigoplus_{j \in J} M(G/H_j) \to M(G/G), (x_j | j \in J) \mapsto \sum_{j \in J} p(H_j)_* x_j$$

(9.2) $$i(\Sigma): M(G/G) \to \bigoplus_{j \in J} M(G/H_j), x \mapsto (p(H_j)^* x | j \in J).$$

We call the induction system **projective** if $p(\Sigma)$ is surjective and **injective** if $i(\Sigma)$ is injective.

Let G/K and G/H be two homogeneous spaces. The identity of $G/K \times G/H$ represents an element in $U(G; G/K \times G/H)$ which can be expanded in terms of the canonical basis

$$[\mathrm{id}] = \sum_{\alpha} n_\alpha [\alpha\colon G/L_\alpha \to G/K \times G/H];$$

see (1.16).

Let $S(K, H)$ denote the set of α over which the summation is taken. Let $\alpha = (\alpha(1), \alpha(2))$ be the two components of α. For each homogeneous space G/H and each induction system $\Sigma = (G/H_j | j \in J)$, we define induction maps

(9.3) $$p(\Sigma, G/H)\colon \bigoplus_{j\in J} \Big(\bigoplus_{\alpha\in S(H_j,H)} M(G/L_\alpha)\Big) \to M(G/H) \quad \text{by}$$

$$(x(j,\alpha)) \mapsto \sum_{j\in J} \Big(\sum_{\alpha\in S(H_j,H)} n_\alpha \alpha(2)_* x(j,\alpha)\Big).$$

Similarly, we have restriction maps

(9.4) $$i(\Sigma, G/H)\colon M(G/H) \to \bigoplus_{j\in J} \Big(\bigoplus_{\alpha\in S(H_j,H)} M(G/L_\alpha)\Big)$$

defined by mapping x to $(\alpha(2)^* x | \alpha \in S(H_j, H), j \in J)$.

Now let $U\colon \Omega(G) \to \mathbb{Z}$-Mod be a Green functor and $M\colon \Omega(G) \to \mathbb{Z}$-Mod a left U-module.

(9.5) Proposition. *Let Σ be an induction system which is projective for U. Then the following holds: For each homogeneous space G/H, the induction map $p(\Sigma, G/H)$ is split surjective and the restriction map $i(\Sigma, G/H)$ is split injective.*

Proof. Consider the case of $p(\Sigma, G/H)$. Since $p(\Sigma)$ is surjective for U, we can choose elements $x_j \in U(G/H_j)$ such that $\sum p(H_j)_* x_j = 1$. Define a homomorphism

$$q(\Sigma, G/H)\colon M(G/H) \to \bigoplus_j \Big(\bigoplus_\alpha M(G/L_\alpha)\Big)$$

by $q(\Sigma, G/H)x = (\alpha(1)^* x_j \cdot \alpha(2)^* x | \alpha \in S(H_j, H), j \in J)$. Here we use the action of $U(G/L_\alpha)$ on $M(G/L_\alpha)$. We claim that $p(\Sigma, G/H) q(\Sigma, G/H) = \mathrm{id}$. This is proved by the following computation which uses (8.9) and the basic identity

$$\sum_{\alpha\in S(H_j,H)} n_\alpha \alpha(2)_* \alpha(1)^* = p(H)^* p(H_j)_*,$$

valid for each Mackey functor.

$$\begin{aligned}
&p(\Sigma, G/H)q(\Sigma, G/H)x \\
&= \sum_j \left(\sum_\alpha n_\alpha \alpha(2)_*(\alpha(1)^* x_j \cdot \alpha(2)^* x)\right) \\
&= \sum_j \left(\sum_\alpha n_\alpha \alpha(2)_* \alpha(1)^* x_j\right) \cdot x \\
&= \sum_j (p(H)^* p(H_j)_* x_j) \cdot x \\
&= p(H)^* \left(\sum_j p(H_j)_* x_j\right) x \\
&= p(H)^*(1) \cdot x \\
&= 1 \cdot x = x.
\end{aligned}$$

Thus $q(\Sigma, G/H)$ is a splitting for $p(\Sigma, G/H)$. A splitting $j(\Sigma, G/H)$ for $i(\Sigma, G/H)$ is defined in a dual fashion

$$\begin{aligned}
j(\Sigma, G/H) \colon \bigoplus_j \left(\bigoplus_\alpha M(G/L_\alpha)\right) &\to M(G/H) \\
(x(j,\alpha)) &\mapsto \sum_j \left(\sum_\alpha n_\alpha \alpha(2)_*(\alpha(1)^* x_j \cdot x(j,\alpha))\right). \quad \square
\end{aligned}$$

The significance of (9.5) is that an induction theorem for U, i.e. a determination of a projective induction system, implies an induction theorem for each Mackey functor which is a U-module. The next result states, roughly, that two projective induction systems for U have a common projective refinement.

More precisely, let $\Sigma = (G/H_j \mid j \in J)$ and $\Sigma' = (G/K_i \mid i \in I)$ be two projective induction systems for the Green functor U. Then we may look at the composition

$$p(\Sigma) \circ \left(\bigoplus_{j \in J} p(\Sigma', G/H_j)\right).$$

Writing out this map, we see that it is again an induction map based on a suitable induction system $(G/L_\alpha \mid \alpha \in A)$. The G/L_α are derived from the expansion of the identity in the $U(G; G/H_j \times G/K_i)$. The significance of this observation is that the L_α are contained in groups of the form $gH_jg^{-1} \cap hK_ih^{-1}$, the isotropy groups of $G/H_j \times G/K_i$, and are therefore not larger than the groups we started with.

If H and K are conjugate subgroups, then $p(H)_*$ and $p(K)_*$ have the same image. Therefore, we call a finite set E of conjugacy classes (H) an **induction set** for U if

$$\bigoplus_{(H) \in E} U(G/H) \to U(G/G), (x(H)) \mapsto \Sigma p(H)_* x(H)$$

is surjective. We define a partial order on induction sets by:

$E \leq F \Leftrightarrow$ for each $(H) \in E$, there exists $(K) \in F$ such that $(H) \leq (K)$.

Recall that $(H) \leq (K)$ means that H is subconjugate to K. Now we can state

(9.6) Proposition. *Each Green functor U possesses a unique minimal induction set $D(U)$.*

Proof. The considerations following the proof of (9.5) show that for any two induction sets F_1, F_2 there exists an induction set E such that $E \leq F_1, E \leq F_2$. By the descending chain property for closed subgroups of G, there can be no infinite proper chain $F_1 \geq F_2 \geq F_3 \geq \dots$. □

The set $D(U)$ is called the **defect set** of the Green functor U.

Suppose now that $\Sigma = (G/H_j | j \in J)$ is an induction system for the Green functor U. Let M be a U-module. We want to describe the kernel of

$$p: \oplus M(G/H_j) \to M(G/G), (x_j) \mapsto \Sigma p(H_j)_* x_j.$$

To this end we use the two homomorphisms

$$p_1, p_2: \bigoplus_{i,j \in J} \bigoplus_{\alpha \in S(i,j)} M(G/L_\alpha) \to \bigoplus_{k \in J} M(G/H_k)$$

defined by

$$p_1(x(i,j,\alpha)) = (\sum_{j \in J} \sum_{\alpha \in S(i,j)} n_\alpha \alpha(1)_* x(i,j,\alpha))$$

$$p_2(x(i,j,\alpha)) = (\sum_{i \in J} \sum_{\alpha \in S(i,j)} n_\alpha \alpha(2)_* x(i,j,\alpha)).$$

Here, $S(i,j) = S(H_i, H_j)$ and the $\alpha \in S(i,j)$ belong to the decomposition of $[\mathrm{id}] \in U(G; G/H_i \times G/H_j)$.

(9.7) Proposition. *The following sequence is exact*

$$\bigoplus_{i,j \in J} \bigoplus_{\alpha \in S(i,j)} M(G/L_\alpha) \xrightarrow{p_2 - p_1} \bigoplus_{k \in J} M(G/H_k) \xrightarrow{p} M(G/G) \to 0.$$

Proof. We have already shown that p is surjective by constructing a splitting homomorphism q such that $pq = \mathrm{id}$. We are going to construct a homomorphism q_1 such that

$$(p_2 - p_1)q_1 + qp = \mathrm{id}.$$

(Think of a contracting chain homotopy.) Then exactness follows easily. Note that p_2 is defined as $\bigoplus_{k \in J} p(\Sigma, G/H_k)$. We define $q_1 = \bigoplus_{k \in J} q(G/H_k)$ and obtain from the proof of (9.5) that $p_2 q_1 = \mathrm{id}$. Thus it remains to show that $p_1 q_1 = qp$. This is done by a computation similar to the one in the proof of (9.5). □

The simplest application of the axiomatic induction theory is the Hyperelementary Induction Theorem (HIT). A proof of the HIT will be given in (9.12). Let N be a set of subgroups of G and $p \in \mathbb{Z}$ a prime. Choose another set $M(p)$ of subgroups with the following property: For each $L \subset H \in N$, the defining group of the prime ideal $q(L, p)$ of the Burnside ring $A(G)$ is a subgroup of a member of $M(p)$. (Recall that L' is a defining group for $q(L,p)$ if $q(L',p) = q(L,p)$ and $G/L' \notin q(L,p)$. Compare with (4.2).)

Let an index (p) denote localization at the prime ideal (p). Let $\mathrm{Ke}(N)$ denote the kernel of the restriction map

$$A(G)_{(p)} \to \prod_{H \in N} A(H)_{(p)}$$

and let $\mathrm{Im}(M_{(p)})$ denote the image of the induction map

$$\bigoplus_{H \in M(p)} A(H)_{(p)} \to A(G)_{(p)}.$$

(9.8) Proposition. $\mathrm{Ke}(N) + \mathrm{Im}(M(p)) = A(G)_{(p)}$.

Proof. $\mathrm{Ke}(N) + \mathrm{Im}(M(p))$ is an ideal of $A(G)_{(p)}$ since $\mathrm{Ke}(N)$, as kernel of a ring homomorphism, is an ideal and the image of the induction map is seen to be an ideal.

If this ideal were different from $A(G)_{(p)}$, then we could find a maximal ideal q of $A(G)_{(p)}$ with $\mathrm{Ke}(N) + \mathrm{Im}(M(p)) \subset q$. The ideal q has the form $q = q(L,p)$; see section 4. The inclusion $A(G)_{(p)}/\mathrm{Ke}(N) \to \prod_{H \in N} A(H)_{(p)}$ is an integral ring extension. By a basic theorem of commutative algebra (Atiyah-MacDonald [1969], 5.10), the ideal q extends to a prime ideal of $\prod_{H \in N} A(H)_{(p)}$. Any such prime ideal is obtained by projecting onto some factor $A(H)$ and lifting one of its prime ideals. Therefore, we can assume that $L \subset H$ for some $H \in N$. Let L' be the defining group of $q(L,p)$ and choose $K \in M(p)$ with $L' \subset K$. Then $G/L' \in A(G)$ is the image of K/L' under induction $A(K) \to A(G)$; thus $G/L' \in \mathrm{Ke}(N) + \mathrm{Im}(M(p)) \subset q = q(L,p)$, a contradiction. □

For the universal rings $\mathfrak{U}(G/G)$, we have similar induction and restriction maps and therefore an analogous ideal $\mathrm{Ke}(N) + \mathrm{Im}(M(p)) \subset \mathfrak{U}(G/G)_{(p)}$. Since the canonical homomorphism $v\colon \mathfrak{U}(G/G) \to A(G)$ has a kernel consisting of nilpotent elements, see (1.14), v induces an isomorphism of prime ideal spectra. From (9.8) we obtain

(9.9) Corollary. $\mathrm{Ke}(N) + \mathrm{Im}(M(p)) = \mathfrak{U}(G/G)_{(p)}$. □

In the next proposition we use the notation of (9.8). Let U be a Green functor into abelian groups.

(9.10) Theorem. *Assume that any torsion element in $U(G/G)$ is nilpotent. Assume that the restriction map*

$$U(G/G) \otimes \mathbb{Q} \to \prod_{H \in N} U(G/H) \otimes \mathbb{Q}$$

is injective. Then the induction map

$$\bigoplus_{K \in M(p)} U(G/K)_{(p)} \to U(G/G)_{(p)}$$

is surjective.

Proof. The injectivity and nilpotency hypotheses imply that each element in the kernel of $U(G/G)_{(p)} \to \prod_{H \in N} U(G/H)_{(p)}$ is nilpotent. By (9.9), we can find $x \in \mathrm{Ke}(N)$ and $y \in \mathrm{Im}(M(p))$ such that $x + y = 1 \in \mathfrak{U}(G/G)_{(p)}$. Apply the natural transformation $h\colon \mathfrak{U}(G/G)_{(p)} \to U(G/G)_{(p)}$; see (8.13). Then $h(x) + h(y) = 1 \in U(G/G)_{(p)}$. The element $h(x)$ is contained in the kernel of $U(G/G)_{(p)} \to \prod_{H \in N} U(G/H)_{(p)}$ and hence nilpotent. Thus $h(y) = 1 - h(x)$ is a unit. But $h(y)$ is contained in the image of

$$\bigoplus_{K \in M(p)} U(G/K)_{(p)} \to U(G)_{(p)}.$$

Since this image is an ideal and contains a unit, it must be all of $U(G)_{(p)}$. □

We apply the foregoing considerations to the set N of cyclic subgroups of G (cyclic = powers of a generator are dense). A subgroup K of G is called p-**hyperelementary** if there exists an exact sequence

$$1 \to S \to K \to P \to 1$$

with a finite p-group P and a cyclic group S such that $|S/S_0|$ is prime to p. It is called **hyperelementary** if it is p-hyperelementary for some prime p.

(9.11) Proposition. *A suitable set $M(p)$ for the set N of cyclic subgroups is the set of p-hyperelementary subgroups.*

Proof. Let C be cyclic. We can find a cyclic subgroup $D \subset C$ such that $q(C,p) = q(D,p)$ and $|D/D_0| \not\equiv 0 \bmod p$. By the proof of (4.5, iii), we find S such that $D \lhd S$, S/D a torus, NS/S finite, S/S_0 a quotient of D/D_0, $q(D,p) = q(S,p)$. By (4.7), the defining group of $q(S,p)$ is p-hyperelementary. □

We now can prove the Hyperelementary Induction Theorem.

(9.12) Theorem. *Let N be the set of cyclic subgroups. Let $U(G/G)$ be torsion free. Let the restriction map $U(G/G) \to \prod_{H \in N} U(G/H)$ be injective. Then the induction*

$$\bigoplus_{H \in M} U(G/H) \to U(G/G)$$

from the set M of hyperelementary subgroups is surjective.

Proof. From (9.10) and (9.11) we see that the induction map is surjective when localized at each prime ideal $(p) \subset \mathbb{Z}$. By commutative algebra, the map itself must be surjective. □

A particular example, where (9.12) applies, is the functor complex representation ring $G \mapsto R(G)$. Since representations are determined by characters, the elements of $R(G)$ are detected by the restrictions to cyclic subgroups. Therefore, $R(G)$ satisfies hyperelementary induction (Segal [1968b], 3.11). Let G be finite and $G \mapsto R(G; \mathbb{Q})$ the Green functor rational representation ring. Then this functor satisfies hyperelementary induction and, by (9.5), any module over this functor has similar induction properties. In some cases, the set of hyperelementary groups can still be reduced. For finite G and the complex representation ring, the induction theorem of Brauer states that it suffices to use the elementary groups $S \times P$, S cyclic, P a p-group (see Serre [1971], 10.).

10. The Burnside ring and localization.

Since stable equivariant homology groups are modules over the Burnside ring, we can localize such groups at prime ideals of the Burnside ring. There are also some unstable theories, e.g. bordism theories, which are modules over the Burnside ring because they carry a Mackey structure in the sense of II.9. Therefore, let us consider an axiomatic situation.

We are given the (unstable) additive G-equivariant homology theory $h_*^G(X, Y)$. We assume that $h_n^G(X, Y)$ is a left $A(G)$-module, natural in (X, Y). For $U \subset G$, we put $h_*^U(X, Y) = h_*^G(G/U \times X, G/U \times Y)$ and assume that $h_n^U(X, Y)$ is an $A(U)$-module, natural in (X, Y). For $K \subset U$, there exists a natural transformation

$$\mathrm{res} = r\colon h_n^U(X, Y) \to h_n^K(X, Y),$$

called restriction, which is assumed to be compatible with the restriction (2.12) $s\colon A(U) \to A(K)$, i.e. we require

(10.1) $r(xy) = s(x)r(y),\ x \in A(U),\ y \in h_n^U(X, Y).$

In stable homology theories, the restriction is induced by a transfer map II.8 associated to $G/K \to G/U$. We also have a natural transformation

$$\mathrm{ind}\colon h_n^K(X, Y) \to h_n^U(X, Y)$$

which is induced by $G/K \to G/U$, called induction, and we assume that

(10.2) $h_*^U(X, Y) \xrightarrow[\mathrm{res}]{} h_*^K(X, Y) \xrightarrow[\mathrm{ind}]{} h_*^U(X, Y)$

is multiplication by $[U/K] \in A(U)$. If res is induced by a transfer map, then this is indeed the case.

We consider the prime ideal $q = q(H, p)$ of $A(G)$ where $H \subset G$, NH/H is finite of order prime to p if $p \neq 0$; see section 4. Let $F_1 \supset F_2$ be open isotropy families of G and consider homology with families $h_*^G[F_1, F_2]$ as explained in II.7. We assume

$$K \in F_1 \backslash F_2 \Rightarrow q(K, p) = q(H, p).$$

Let an index (p) or q denote localization at a prime ideal $(p) \subset \mathbb{Z}$ or $q \subset A(G)$ in the usual sense of commutative algebra, i.e. the elements not contained in the ideal are made invertible.

(10.3) Proposition. *Multiplication with $y \notin q(H, p)$ is an automorphism of the homology theory $h_*^G[F_1, F_2]_{(p)}$. The canonical map $h_*^G[F_1, F_2]_{(p)} \to h_*^G[F_1, F_2]_q$ is an isomorphism. (Recall that $G/H \notin q(H, p)$.)*

Proof. Using exact sequences and exactness of localization, it suffices to consider adjacent $F_1 \supset F_2$, say $F_1 \backslash F_2 = (K)$ with $q(H, p) = q(K, p)$; compare the proof of II(7.1). We write $N = NK$, $W = WK$. By II(7.8), Ex. 4, the space $(G \times_N EW) * EF_2$ is a classifying space EF_1. If A and B are G-spaces, there is a G-homeomorphism

$$A * B \cong (A * P) \times B \cup A \times (B * P),$$

P a point. Using excision, this yields

$$\begin{aligned} &h(A * B, B) \cong \\ &h(A * P) \times B \cup A \times (B * P), P) \cong \\ &h(A * P) \times B \cup A \times (B * P), (A * P) \times B) \cong \\ &h(A \times (B * P), A \times P). \end{aligned}$$

We have written h for h_*^G. If CB denotes the cone over B, we observe that the pair (CB, B) is G-homotopy equivalent to $(B * P, B)$. In our case, we thus have a natural isomorphism

$$h[F_1, F_2](X, A) \cong h(G \times_N EW \times (CEF_2, EF_2) \times (X, A)).$$

As in II.8, we use the spectral sequence coming from the skeleton filtration of the classifying space EW. The E_2-term is the homology of the following chain complex

$$\ldots h(G \times_N W^i \times Z) \xleftarrow[d_i]{} h(G \times_N W^{i+1} \times Z) \leftarrow \ldots$$

with $Z = (CEF_2, EF_2) \times (X, A)$ and $d_i = \sum_{j=0}^{i} (-1)^j p_{j*}$ where p_j omits the $(1+j)$-th factor of W^{i+1}.

Multiplication by y is a natural transformation of homology theories and therefore induces an endomorphism of this spectral sequence. Hence it suffices to show that multiplication with y is an isomorphism on $h(G \times_N W^i \times Z)_{(p)}$ for $i \geq 1$. The group in question is isomorphic to $h(G/K \times W^{i-1} \times Z)_{(p)}$ and therefore, by (10.1), the action of $y \in A(G)$ only depends on its restriction $z = s(y) \in A(K)$. This restriction has the form

$$z = \varphi_K(y)[K/K] + \sum a_j[K/K_j]$$

with $a_i \in \mathbb{Z}$ and $(K_i) < (K)$, $(K_i) \neq (K)$. Since $q(H,p) = q(K,p)$ and $y \notin q(H,p)$ by assumption, we have $\varphi_K(y) \not\equiv 0 \bmod p$. Since we have localized at (p), multiplication by $\varphi_K(y)[K/K]$ is an isomorphism. The proof of (10.3) will be complete if we can show that multiplication with $[K/K_i]$ is zero. By (10.2), this multiplication factors over

$$h(G/K_j \times W^{i-1} \times (CEF_2, EF_2) \times (X, A))_{(p)}$$

and this group is zero by the localization theorem III(3.15), Ex. 5. □

For our equivariant homology theory h, let us now look at the two chain complexes

(10.4) $$h \xleftarrow[d_0]{} h(G/H) \xleftarrow[d_1]{} h(G/H^2) \xleftarrow[d_2]{} \cdots$$

$$h \xrightarrow[d^0]{} h(G/H) \xrightarrow[d^1]{} h(G/H^2) \xrightarrow[d^2]{} \cdots$$

with $d_i = \sum_{j=0}^{i} (-1)^j (p_j)_*$ and $d^i = \sum_{j=0}^{i} (-1)^j p_j^*$. Here, p_j^* is a transfer homomorphism which is assumed to have the property:

(10.5) $(p_j)_* p_j^*$ is multiplication by $[G/H] \in A(G)$. Moreover, p_j^* is natural with respect to pull-backs.

(10.6) Proposition. *The complexes* (10.4) *are exact after localization at* $q = q(H,p)$ *if* $G/H \notin q(H,p)$.

Proof. One can construct a contracting chain homotopy s for the first complex as follows:

$$s = [G/H]^{-1}(p_0)^*: h(G/H^i)_q \to h(G/H^{i+1})_q.$$

One verifies that $ds + sd = \mathrm{id}$ by using (10.5). Similarly for the second complex. □

We apply (10.6) in the following situation. The restriction $h_*^G(X) \to h_*^H(X)$ becomes injective when localized at $q(H,p) = q$ and the image is equal to the kernel of

$$p_0^* - p_1^* : h_*^G(G/H \times X)_q \to h_*^G(G/H^2 \times X)_q.$$

We denote this kernel by $h_*^H(X)_q^{\text{inv}}$ and call it the subgroup of G-invariant elements of $h_*^H(X)_q$. In this notation, we thus have an isomorphism

(10.7) $h_*^G(X)_q \cong h_*^H(X)_q^{\text{inv}}$.

Let $p \neq 0$ be a prime. Let H and K be subgroups of G. The restriction (2.12) $r_H^G : A(G) \to A(H)$ makes $A(H)$ an $A(G)$-module. We localize this $A(G)$-module with respect to $q(K,p) \subset A(G)$. The ring homomorphism $\varphi_L : A(G) \to \mathbb{Z}$ turns $\mathbb{Z}$ into an $A(G)$-module, which we denote $\mathbb{Z}(L)$. If $L \subset H \subset G$, we have the prime ideals $q(L,p)$ in $A(H)$ and $A(G)$; we use the notation $q(L,p;H)$ and $q(L,p;G)$ to distinguish them. With these conventions, the following holds.

(10.8) Proposition.
(i) $\mathbb{Z}(L)_{q(K,p;G)} \neq 0$ *if and only if* $q(L,p;G) = q(K,p;G)$.
(ii) $A(H)_{q(K,p;G)} \neq 0$ *if and only if there exists* $L \subset H$ *such that* $q(L,p;G) = q(K,p;G)$.

Proof. (i) Suppose that $q(L,p;G) = q(K,p;G)$. Then $\varphi_L = \varphi_K : A(G) \to \mathbb{Z}/p$. The elements $x \notin q(K,p;G)$ become invertible in $\mathbb{Z}/p$. Thus $\varphi_L \bmod p$ factors over $A(G)_{q(K,p;G)}$. Consequently, this localization is nonzero. Conversely, if $q(K,p;G) \neq q(L,p;G)$, then the determination of the maximal ideals of $A(G)$ implies the existence of some $x \in q(L,p;G) \backslash q(K,p;G)$. Since this element is mapped to zero under φ_L, the localization $\mathbb{Z}(L)_{q(K,p;G)}$ must be zero.
(ii) If $L \subset H$ and $q(H,p;G) = q(K,p;G)$, then we have a surjective homomorphism $\varphi_L : A(H) \to \mathbb{Z}(L)$ which induces a surjection $B := A(H)_{q(K,p;G)} \to \mathbb{Z}(L)_{q(K,p;G)}$. From (i) we see that $B \neq 0$.

Suppose the localization B is non-zero and let $l : A(H) \to B$ be the natural map. Let q be a maximal ideal of B. Then $l^{-1}q$ is a prime ideal $q(L,p;H)$ of $A(H)$ which does not intersect $r_H^G(A(G) \backslash q(K,p;G))$ (Atiyah-MacDonald [1969], 3.11 (iv)). Thus $(r_H^G)^{-1} l^{-1} q = q(L,p;G)$ is contained in $q(K,p;G)$ and this implies $q(K,p;G) = q(L,p;G)$. □

(10.9) Corollary. $(A(G)_{q(L,p)})_{q(K,p)} \neq 0$ *if and only if* $q(L,p) = q(K,p)$.

Using (10.9) and elementary commutative algebra (Atiyah-MacDonald [1969], 3.9), one obtains

(10.10) Proposition. *Let M be an $A(G)_{(p)}$-module such that the set J of maximal ideals $q \subset A(G)_{(p)}$ with $M_q \neq 0$ is finite. Then the natural map*

$$M \to \prod M_q,\ q \in J$$

is an isomorphism. □

Now let G be a finite group. Let $\phi_p(G)$ be the set of (H) such that $|WH| \not\equiv 0 \bmod p$. For $(H) \in \phi_p(G)$, let $H_p \lhd H$ be unique smallest normal subgroup of H such that H/H_p is a p-group. Then $q(H,p) = q(L,p)$ if and only if $(L_p) = (H_p)$. We consider the families

$$FH = \{L \subset G \mid L \text{ subconjugate } H\}$$
$$F_0 H = \{L \subset G \mid q(L,p) = q(H,p)\}$$
$$F'H = FH \backslash F_0 H.$$

If $F_1 \supset F_2$ are open families of subgroups of G, $A(G;F_1)$ denotes the ideal of $A(G)$ generated by sets (or spaces) X with isotropy groups in F_1 and $A(G;F_1,F_2)$ stands for the ideal $A(G;F_1)$ modulo the subideal $A(G;F_2)$.

By (10.10), the ring $A(G)_{(p)}$ splits into the direct sum of rings $A(G)_{q(H,p)}$, $(H) \in \phi_p(G)$, and there is a corresponding splitting for each $A(G)_{(p)}$-module. In particular, equivariant (co-)homology theories which have properties as stated in the beginning of this section split when localized at (p).

(10.11) Proposition. *Suppose $(H) \in \phi_p(G)$. Taking H_p-fixed points induces an isomorphism*

$$A(H; FH, F'H) \cong A(H/H_p).$$

Proof. For both groups, an additive basis is given by those $[H/K]$ with $(H_p) \le (K) \le (H)$ and $H/K^{H_p} = H/K$. □

(10.12) Proposition. *Suppose $(H) \in \phi_p(G)$. The inclusion $i\colon A(G;FH) \to A(G)$ and the quotient map $j\colon A(G;FH) \to A(G;FH,F'H)$ induce isomorphisms of their $q(H,p)$-localizations.*

Proof. The kernel $A(G;F'H)$ of j is detected by fixed point mappings $\varphi_L\colon A(G;F'H) \to \mathbb{Z}(L)$ for L with $q(L,p) \ne q(H,p)$. Thus, by (10.8, i), $A(G;F'H)_{q(H,p)} = 0$. For a similar reason, the cokernel of i is zero when localized at $q(H,p)$. □

(10.13) Proposition. *The canonical map*

$$A(G;FH,F'H)_{(p)} \to A(G;FH,F'H)_{q(H,p)}$$

is an isomorphism.

Proof. An analogous map for $A(G;F_1,F_2)$ with $F_1 \supset F_2$, $F_1 \backslash F_2 = (K)$, $q(K,p) = q(H,p)$ is an isomorphism since both localizations map under φ_K into $\mathbb{Z}(K)_{q(H,p)}$ injectively with the same image. □

Under the canonical isomorphism

$$A(G)_{(p)} \cong \oplus A(G)_{q(H,p)}, (H) \in \phi_p(G),$$

the factor $A(G)_{q(H,p)}$ corresponds to the subring $e(H)A(G)_{(p)}$ where $e(H) \in A(G)_{(p)}$ is a suitable indecomposable idempotent. The maps i and j of (10.12) induce isomorphisms

(10.14) $e(H)A(G)_{(p)} \leftarrow e(H)A(G; FH)_{(p)} \rightarrow A(G; FH, F'H)_{(p)}.$

By the general theory, we have

$$A(G; FH, F'H)_{(p)} \cong A(H; FH, F'H)_{(p)}^{\text{inv}}.$$

Combining all these facts, we obtain

(10.15) Proposition. *Taking H_p-fixed points for the various $(H) \in \phi_p(G)$ induces a ring isomorphism*

$$A(G)_{(p)} \rightarrow \sum_{(H)\in\phi_p(G)} A(H/H_p)_{(p)}^{\text{inv}}. \quad \square$$

We return to the equivariant homology theory h_*^G with Mackey structure. We have a natural transformation of homology theories

(10.16) $r_H: h_*^G(X)_{(p)} \rightarrow h_*^H(X)_{(p)}^{\text{inv}} \rightarrow h_*^H[FH, F'H](X)_{(p)}^{\text{inv}}$

in which the first map is restriction and the second map comes from the exact homology sequence of the pair FH, $F'H$. (Note that EFH is H-contractible.)

(10.17) Proposition. *Let $(H) \in \phi_p(G)$ and $q = q(H, p)$. Then the following holds:*

(i) $(r_H)_q$ *is an isomorphism.*
(ii) r_H *is split surjective.*
(iii) *The product of the maps* r_H

$$r = (r_H): h_*^G(X)_{(p)} \rightarrow \prod_{(H)\in\phi_p(G)} h_*^H[FH, F'H](X)_p^{\text{inv}}$$

is an isomorphism.

Proof. (i) By (10.3), we have

$$h_*^H[FH, F'H](X)_{(p)}^{\text{inv}} \cong h_*^H[FH, F'H](X)_q^{\text{inv}}$$

and by (10.7), we have $h_*^G(X)_q \cong h_*^H(X)_q^{\text{inv}}$. Thus it remains to be shown that $h_*^G[F'H](X)_q$ is zero. Using the additivity of the theory and skeleton filtration of the classifying space $EF'H$, we see that is suffices to show that groups of the type $h_*^G(G/K \times Y)$, for $K \in F'H$, are zero. But this follows from the fact that the $A(G)$-module structure factors over the restriction $A(G) \rightarrow A(K)$ and $A(K)_{q(H,p)} = 0$ by (10.8). $\square$

We finally look at complex equivariant K-theory $K_G(X)$ for a compact Lie group G. This is part of a stable equivariant cohomology theory (see Segal [1968]) and thus $K_G(X)$ is a module over the Burnside ring. Actually, $K_G(X)$ is naturally a module over the complex representation ring $R(G)$ and the $A(G)$-module structure comes from the canonical ring homomorphism $\chi_G: A(G) \to R(G)$.

(10.18) Definition. A closed subgroup S of G is called a **Cartan subgroup** of G if WS is finite and S is topologically cyclic, i.e. powers of a suitable element are dense. A Cartan subgroup is p-regular if the group of components has order prime to p.

A topologically cyclic group is isomorphic to the product of a torus and a finite cyclic group. Let C be the set of conjugacy classes of Cartan subgroups of G and $C(p)$ the subset of p-regular groups. We refer to Segal [1968] or Bröcker-tom Dieck [1985] for the proof of

(10.19) Proposition. *The set C is finite.* □

For $H \subset G$, let H_p again denote the smallest normal subgroup such that H/H_p is a finite p-group.

(10.20) Proposition. *$R(G)_{q(H,p)} \neq 0$ if and only if H_p is a p-regular Cartan subgroup.*

Using (10.9), (10.20), (10.10) and (4.5) we obtain the

(10.21) Corollary. *The canonical map induced by the restrictions*

$$K_G(X)_{(p)} \to \prod_{(S)\in C(p)} K_G(X)_{q(S,p)}$$

is an isomorphism of rings. □

Proof of (10.20). The map $R(G) \to \prod_{(S)\in C} R(S)$ induced by the restriction is a monomorphism since representations (characters) are detected by cyclic subgroups and each cyclic subgroup is contained in a Cartan subgroup. Thus $R(G)_{q(H,p)} \neq 0$ implies that there exists an $(S) \in C$ such that $R(S)_{q(H,p)} \neq 0$ and this implies $A(S)_{q(H,p)} \neq 0$. By (10.8), there exists a cyclic subgroup $L \subset S$ such that $q(L,p) = q(H,p)$; then $q(L_p, p) = q(H,p)$. If G is finite then $(L_p) = (H_p)$ and we are done. In general, it follows from (4.5) that there exists a group $L_p \lhd K$ such that K/L_p is a torus and WK is finite. The group K must then be a p-regular Cartan subgroup and $q(K,p) = q(L_p, p) = q(H,p)$. By (4.7), $K \lhd H$, up to conjugation, and H/K is a p-group.

Conversely, let S be a p-regular Cartan subgroup. We have to show that

$R(G)_{q(S,p)} \neq 0$. It suffices to show that $R(S)_{q(S,p)} \neq 0$. Let g be a generator of S and let $e_g\colon R(S) \to \mathbb{C}$ be the evaluation of characters at g. The commutative diagram of ring homomorphisms

$$\begin{array}{ccc} A(S) & \xrightarrow{\chi_G} & R(S) \\ \downarrow{\scriptstyle \varphi_S} & & \downarrow{\scriptstyle e_g} \\ \mathbb{Z} & \xrightarrow{\subset} & \mathbb{C} \end{array}$$

allows us to view everything as $A(G)$-module. Now use (10.8). □

(10.22) Exercises.

1. Let G be a finite group. Show that $\varphi_L\colon A(G) \to \mathbb{Z}$ induces an isomorphism $A(G)_{q(L,0)} \cong \mathbb{Z}(L)_{q(L,0)}$.
2. Show that the idempotent $e(H)$ (10.14) is contained in $A(G; FH)_{(p)}$.
3. Establish a canonical isomorphism $A(G; F_1, F_2) \cong \omega_0^G[F_1, F_2]$.
4. Use (10.17) to show $e(H)h_*^G(X)_{(p)} \cong h_*^H[FH, F'H](X)_{(p)}^{\mathrm{inv}}$.
5. Let G be a finite group and suppose that the homology theory h_*^G with Mackey structure takes values in $\mathbb{Z}[|G|^{-1}]$-modules. Show that $h_*^G(X)$ splits naturally into groups $h_*^H[F_\infty, F_e](X)^{\mathrm{inv}}$ where (H) runs through the conjugacy classes of subgroups and F_∞ resp. F_e denotes the family of all resp. all proper subgroups.
6. Show that those $A(G)$-module structures on $K_G(X)$ coincide which come from $\chi_G\colon A(G) \to R(G)$ and from the stability of the cohomology theory $K_G^*(X)$.
7. Let $\lambda_{-1}(V) \in R(G)$ be the Euler class of the representation V; see II.5. Let $S \subset R(G)$ be the set of Euler classes of representations without trivial direct summand. Show that $S^{-1}R(G) \neq 0$ if and only if G is cyclic.
8. Let G be finite cyclic of order m and let x be the standard irreducible representation of G. Then $R(G) \cong \mathbb{Z}[x]/(x^m - 1)$. Let $e_g\colon R(G) \to \mathbb{Z}[u_m]$, $u_m = \exp(2\pi i/m)$, be the evaluation of characters at a generator g of G. Show that e_g induces an isomorphism

   $$S^{-1}R(G) \cong \mathbb{Z}[m^{-1}, u_m]$$

   with S as in exercise 7. If G is a subgroup of some finite group M, show that e_g induces an isomorphism

   $$R(G)_{q(G,p;M)} \cong \mathbb{Z}_{(p)}[u_m].$$

   Study the action of NG/G under this isomorphism.
9. Let P be a finite p-group. Show that $A(P)_{(p)}$ is a local ring.

10. Let $m = |G|$. Show that there exists a canonical isomorphism

$$K_G(X)[m^{-1}] \cong \prod E_C^{-1} R(C) \otimes K(X^C)^{WC}$$

where C runs through the conjugacy classes of cyclic subgroups of G and E_C is the set of Euler classes of representations without trivial summand. X is a finite G-complex.

Bibliography

Adams, J. F., 1984: *Prerequisites (on equivariant stable homotopy) for Carlsson's lecture.* Algebraic topology, Proc. Conf., Aarhus 1982, Lect. Notes Math. **1051**, 483–532.

– Gunawardena, J. H., and H. Miller, 1985: *The Segal conjecture for elementary abelian p-groups.* Topology **24**, 435–460.

Aigner, M., 1975: *Kombinatorik.* Berlin–Heidelberg–New York: Springer.

Allday, C. J., 1976: *The stratification of compact connected Lie group actions by subtori.* Pac. J. Math. **62**, 311–327.

– 1978: *On the rational homotopy of fixed point sets of torus actions.* Topology **17**, 95–100.

– 1979: *Rational homotopy and torus actions.* Houston J. Math. **5**, 1–19.

– and T. Skjelbred, 1974: *The Borel formula and the topological splitting principle for torus actions on a Poincaré duality space.* Ann. of Math. **100**, 322–325.

– and V. Puppe, 1985: *On the localization theorem at the cochain level and free torus actions.* Algebraic, topology, Proc. Conf., Göttingen 1984. Lect. Notes Math. **1172**, 1–16.

Allen, R. J., 1979: *Equivariant embeddings of $\mathbb{Z}_p$-actions in euclidean space.* Fund. Math. **103**, 23–30.

– 1980: *Equivariant embeddings of finite abelian group actions in euclidean space.* Fund. Math. **110**, 25–31.

Arens, R., 1946: *Topologies for homeomorphism groups.* Amer. J. Math. **68**, 593–610.

Atiyah, M. F., 1966: *K-theory and reality.* Quart. J. Math. Oxford (2) **17**, 367–386.

– 1967: *K-theory.* New York–Amsterdam: Benjamin.

– 1968: *Bott periodicity and the index of elliptic operators.* Quart. J. Math. Oxford (2) **19**, 113–140.

– 1974: *Elliptic operators and compact groups.* Berlin–Heidelberg–New York: Springer.

– and R. Bott, 1967: *A Lefschetz fixed point formula for elliptic complexes. I.* Ann. of Math. **86**, 374–407.

– – 1968: *A Lefschetz fixed point formula for elliptic complexes. II. Applications.* Ann. of Math. **88**, 451–491.

– – 1984: *The moment map and equivariant cohomology.* Topology **23**, 1–28.

– and F. Hirzebruch, 1970: *Spin-manifolds and group actions.* „Essays on Topology and Related Topics, Memoires dédiés a Georges de Rham", 18–22. Berlin–New York: Springer.

– and I. G. MacDonald, 1969: *Introduction to commutative algebra.* Reading, Mass.: Addison-Wesley.

– and G. B. Segal, 1968: *The index of elliptic operators II.* Ann. of Math. **87**, 531–545.

– – 1969: *Equivariant K-theory and completion.* J. Differential Geometry **3**, 1–18.

– – 1971: *Exponential isomorphisms for λ-rings.* Quart. J. Math. Oxford (2) **22**, 371–378.

– and I. M. Singer, 1968: *The index of elliptic operators I.* Ann. of Math. **87**, 484–530.

– – 1968a: *The index of elliptic operators III.* Ann. of Math. **87**, 546–604.

– and D. O. Tall, 1969: *Group representations, λ-rings and the J-homomorphism.* Topology **8**, 253–298.

Bass, H., 1968: *Algebraic K-theory.* New York: Benjamin.

Bauer, S., 1986: Dissertation. Göttingen.

Beardon, A. F., 1983: *The geometry of discrete groups.* Berlin–Heidelberg–New York: Springer.

Becker, J. C., 1986: *A relation between equivariant and non-equivariant stable cohomotopy.* Math. Z.

– and R.E. Schultz, 1974: *Equivariant function spaces and stable homotopy theory I.* Comment. math. Helv. **49**, 1–34.
– – 1976: *Equivariant function spaces and stable homotopy theory II.* Indiana Univ. Math. J. **25**, 481–492.
– – 1978: *Fixed point indices and left invariant framings.* Geometric appl. of homotopy theory I. Proc. Conf., Evanston 1977, Lecture Notes Math. **657**, 1–31.
Bierstone, E., 1973: *The equivariant covering homotopy property for differential G-fibre bundles.* J. Differential Geometry **8**, 615–622.
– 1974: *Equivariant Gromov theory.* Topology **13**, 327–345.
– 1975: *Lifting isotopies from orbit spaces.* Topology **14**, 245–252.
Blass, A., 1979: *Natural endomorphisms of Burnside rings.* Trans. Amer. Math. Soc. **253**, 121–137.
Bloomberg, E.M., 1975: *Manifolds with no periodic homeomorphisms.* Trans. Amer. Math. Soc. **202**, 67–78.
Boardman, J.N., and R.M. Vogt, 1973: *Homotopy invariant algebraic structures on topological spaces.* Lect. Notes Math. **347**. Berlin–Heidelberg–New York: Springer.
Bochner, S., 1945: *Compact groups of differentiable transformations.* Ann. of Math. **46**, 372–381.
– and D. Montgomery, 1946: *Locally compact groups of differentiable transformations.* Ann. of Math. **47**, 639–653.
– – 1947: *Groups on analytic manifolds.* Ann. of Math. **48**, 659–669.
Borel, A., 1949: *Some remarks about Lie groups transitive on spheres and tori.* Bull. Amer. Math. Soc. **55**, 580–586.
– 1950: *Le plan projectif des octaves et les sphères comme espaces homogènes.* Comptes Rendue de l'Académie Sciences, Paris **230**, 1378–1380.
– 1955: *Nouvelle démonstration d'un théorème de P.A. Smith.* Comment. Math. Helv. **29**, 27–39.
– 1960: *Seminar on transformation groups.* Princeton: Princeton Univ. Press.
– and J.-P. Serre, 1953: *Sur certain sous groupes des groupes de Lie compacts.* Comment. Math. Helv. **27**, 128–139.
– et J. de Siebenthal, 1949: *Les sous-groupes fermès de rang maximum des groupes de Lie clos.* Comment. Math. Helv. **23**, 200–221.
Bourbaki, N., 1960: *Topologie générale. Chap. 3. Groupes topologiques.* Paris: Hermann 3. éd.
– 1961: *Topologie générale. Chap. 10, Espaces fonctionnels.* Paris: Hermann 2. éd.
– 1961a: *Topologie générale. Ch. I, II.* Paris: Hermann 3. éd.
– 1961b: *Algèbre commutative. Ch. 1.2.* Paris: Hermann.
– 1967: *Variétés différentielles et analytiques. Fascicule de résultats.* Paris: Hermann.
– 1968: *Groupes et algèbres de Lie. Ch. IV–VI.* Paris: Hermann.
Bredon, G.E., 1964: *The cohomology ring structure of a fixed point set.* Ann. of Math. **80**, 524–537.
– 1967: *Equivariant cohomology theories.* Lecture Notes in Math. **34**. Berlin–Heidelberg–New York: Springer.
– 1967a: *Sheaf theory.* New York: McGraw-Hill.
– 1968: *Cohomological aspects of transformation groups.* Transformation groups, Proc. Conf., New Orleans 1967, 245–280.
– 1972: *Introduction to compact transformation groups.* New York: Academic Press.
– 1973: *Fixed point sets of actions on Poincaré duality spaces.* Topology **12**, 159–175.
Bröcker, Th., 1971: *Singuläre Definition der äquivarianten Bredon-Homologie.* Manuscr. math. **5**, 91–102.
– and T. tom Dieck, 1970: *Kobordismentheorie.* Lecture Notes in Math. **178**. Berlin–Heidelberg–New York: Springer.
– – 1985: *Representations of compact Lie groups.* Berlin–Heidelberg–New York: Springer.
– and E.C. Hook, 1972: *Stable equivariant bordism.* Math. Z. **129**, 269–277.

– and K. Jänich, 1973: *Einführung in die Differentialtopologie*. Berlin–Heidelberg–New York: Springer.
Browder, W., 1983: *Cohomology and group actions*. Invent. math. **71**, 599–608.
– and W.-C. Hsiang, 1982: *G-actions and the fundamental group*. Invent. math. **65**, 411–424.
– and G. R. Livesay, 1967: *Fixed point free involutions on homotopy spheres*. Bull. A. M. S. **73**, 242–245.
Brown, K. S., 1974: *Euler characteristics of discrete groups and G-spaces*. Invent. math. **27**, 229–264.
– 1975: *Euler characteristics of groups: The p-fractional part*. Invent. math. **29**, 1–5.
– 1982: *Cohomology of groups*. New York–Heidelberg–Berlin: Springer.
– 1982a: *Complete Euler characteristics and fixed-point theory*. J. Pure Appl. Algebra **24**, 103–121.
Capell, S. E., and J. L. Shaneson, 1979: *Nonlinear similarity of matrices*. Bull. Amer. Math. Soc. (N.S.) **1**, 899–902.
– – 1981: *Non-linear similarity*. Ann. of Math. **113**, 315–355.
– – 1982: *The topological rationality of linear representations*. Publ. Math. Inst. Hautes Etud. Sci. **56**, 101–128.
Carlsson, G., 1983: *On the homology of finite free $(\mathbb{Z}/2)^n$-complexes*. Invent. math. **74**, 139–147.
– 1984: *Equivariant stable homotopy and Segal's Burnside ring conjecture*. Ann. of Math. **120**, 189–224.
Cartan, H., and S. Eilenberg, 1956: *Homological algebra*. Princeton: Princeton Univ. Press.
Cassels, J. W. S., 1978: *Rational quadratic forms*. London: Academic Press.
Chang, T., 1976: *On the number of relations in the cohomology of a fixed point set*. Manuscripta math. **18**, 237–247.
Chang, T., and T. Skjelbred, 1972: *Group actions on Poincaré duality spaces*. Bull. Amer. Math. Soc. **78**, 1024–1026.
– – 1974: *The topological Schur lemma and related results*. Ann. of Math. **100**, 307–321.
– – 1976: *Lie group actions on a Cayley projective plane and a note on homogeneous spaces of prime Euler characteristic*. Amer. J. Math. **98**, 655–678.
Chevalley, C., 1946: *Theory of Lie groups*. Princeton: Princeton University Press.
Conner, P. E., 1960: *Retraction properties of the orbit space of a compact topological transformation group*. Duke Math. J. **27**, 341–357.
– 1979: *Differentiable periodic maps*. (Second edition). Lecture Notes in Math. **738**. Berlin–Heidelberg–New York: Springer.
– and E. E. Floyd, 1964: *Differentiable periodic maps*. Berlin–Göttingen–Heidelberg: Springer.
– – 1966: *Maps of odd period*. Ann. of Math. **84**, 132–156.
– F. Raymond, and P. Weinberger, 1972: *Manifolds with no periodic maps*. Transformation groups, Proc. Conf., Amherst 1971. Lect. Notes Math. **299**, 81–108.
Connolly, F., and T. Koźniewski, 1986: *Finiteness properties of classifying spaces of proper Γ-actions*. J. of Pure and Applied Algebra **41**, 17–36.
Cooke, G., 1978: *Replacing homotopy actions by topological actions*. Trans. Amer. Math. Soc. **237**, 391–406.
Copeland, A. H., Jr., and J. de Groot, 1961: *Linearization of a homeomorphism*. Math. Ann. **144**, 80–92.
Curtis, C. W., and I. R. Reiner, 1962: *Representation theory of finite groups and associative algebras*. New York: Interscience Publishers.
Davis, M., 1978: *Smooth manifolds as collections of fibre bundles*. Pacific J. of Math. **77**, 315–363.
tom Dieck, T., 1969: *Faserbündel mit Gruppenoperation*. Arch. Math. **20**, 136–143.
– 1969a: *Glättung äquivarianter Homotopiemengen*. Arch. Math. **20**, 288–295.
– 1970: *Fixpunkte vertauschbarer Involutionen*. Arch. Math. **21**, 296–298.
– 1970a: *Bordism of G-manifolds and integrality theorems*. Topology **9**, 345–358.
– 1970b: *Actions of finite abelian p-groups without stationary points*. Topology **9**, 359–366.

– 1971: *Characteristic numbers of G-manifolds. I.* Invent. math. **13**, 213–224.
– 1971a: *Lokalisierung äquivarianter Kohomologie-Theorien.* Math. Z. **121**, 253–262.
– 1972: *Orbittypen und äquivariante Homologie. I.* Arch. Math. **23**, 307–317.
– 1972a: *Kobordismentheorie klassifizierender Räume und Transformationsgruppen.* Math. Z. **126**, 31–39.
– 1972b: *Periodische Abbildungen unitärer Mannigfaltigkeiten.* Math. Z. **126**, 275–295.
– 1973: *Equivariant homology and Mackey functors.* Math. Ann. **206**, 67–78.
– 1974: *On the homotopy type of classifying spaces.* Manuscripta math. **11**, 41–45.
– 1974a: *Characteristic numbers of G-manifolds. II.* J. of Pure and Applied Algebra **4**, 31–39.
– 1975: *Orbittypen und äquivariante Homologie. II.* Arch. Math. **26**, 650–662.
– 1975a: *The Burnside ring of a compact Lie group. I.* Math. Ann. **215**, 235–250.
– 1977: *A finiteness theorem for the Burnside ring of a compact Lie group.* Compositio math. **35**, 91–97.
– 1977a: *Idempotent elements in the Burnside ring.* J. of Pure and Applied Algebra **10**, 239–247.
– 1978: *Homotopy-equivalent group representations.* J. f. d. reine angew. Math. **298**, 182–195.
– 1978a: *Homotopy equivalent group representations and Picard groups of the Burnside ring and the character ring.* Manuscripta math. **26**, 179–200.
– 1979: *Transformation groups and representation theory.* Lecture Notes Math. **766**. Berlin–Heidelberg–New York: Springer.
– 1979a: *Semi-linear group actions on spheres: Dimension functions.* Algebraic topology, Conf. Proc. Aarhus 1978, Lect. Notes Math. **763**, 448–456.
– 1981: *Über projektive Moduln und Endlichkeitshindernisse bei Transformationsgruppen.* Manuscripta math. **34**, 135–155.
– 1982: *Homotopiedarstellungen endlicher Gruppen: Dimensionsfunktionen.* Invent. math. **67**, 231–252.
– 1982a: *Homotopy representations of the torus.* Arch. Math. **38**, 459–469.
– 1984: *The singular set of group actions on homotopy spheres.* Arch. Math. 43, 551–558.
– 1984a: *Die Picard-Gruppe des Burnside-Ringes.* Algebraic topology, Conf. Proc., Aarhus 1982, Lect. Notes Math. **1051**, 573–586.
– 1984b: *Über λ-Ringstrukturen auf dem Burnside-Ring.* Math. Z. **187**, 505–510.
– 1985: *The Picard group of the Burnside ring.* J. für die reine angew. Math. **361**, 174–200.
– 1985a: *The homotopy type of group actions on homotopy spheres.* Arch. Math. **45**, 174–179.
– and T. Petrie, 1978: *Geometric modules over the Burnside ring.* Invent. math. **47**, 273–287.
– – 1982: *Homotopy representations of finite groups.* Publ. math. I.H.E.S. **56**, 129–169.
– Kamps, K.H., and D. Puppe, 1970: *Homotopietheorie.* Lecture Notes in Math. **157**. Berlin–Heidelberg–New York: Springer.
– and P. Löffler, 1985: *Verschlingung von Fixpunktmengen in Darstellungsformen. I.* Algebraic topology, Proc. Conf., Göttingen 1984. Lect. Notes Math. **1172**, 167–187.

Dold, A., 1963: *Partitions of unity in the theory of fibrations.* Ann. of Math. **78**, 223–255.
– 1970: *Chern classes in general cohomology.* Istituto Nazionale di Alta Matematica, Symposia Mathematica Vol. V, 385–410.
– 1972: *Lectures on algebraic topology.* Berlin–Heidelberg–New York: Springer.
– 1974: *The fixed point index of fibre-preserving maps.* Inventiones math. **25**, 281–297.
– 1976: *The fixed point transfer of fibre-preserving maps.* Math. Z. **148**, 215–244.
– 1984: *Fixed point indices of iterated maps.* Invent. math. **74**, 419–435.

– and D. Puppe, 1980: *Duality, trace and transfer*. Geometric topology, Proc. Conf., Warszawa 1978, 81–102.

Dotzel, R. M., 1984: *An Artin relation (mod 2) for finite group actions on spheres*. Pacific J. of Math. **114**, 335–343.

– and G. Hamrick, 1981: *p-group actions on homology spheres*. Invent. math. **62**, 437–442.

Dovermann, K. H., and T. Petrie, 1982: *G-Surgery II*. Memoirs Amer. Math. Soc. **260**.

– – 1983: *An induction theorem for equivariant surgery* (G-Surgery III). Amer. J. Math. **105**, 1369–1403.

Dress, A., 1969: *A characterization of solvable groups*. Math. Z. **110**, 213–217.

– 1973: *Contributions to the theory of induced representations*. Algebraic K-theory, Proc. Conf., Seattle 1972, Lect. Notes Math. **342**, 182–240.

– 1975: *Induction and structure theorems for orthogonal representations of finite groups*. Ann. of Math. **102**, 291–325.

Duflot, J., 1981: *Depth and equivariant cohomology*. Comment. Math. Helv. **56**, 617–637.

– 1983: *The associated primes of $H_G^*(X)$*. J. Pure and Appl. Algebra **30**, 137–141.

Dwyer, W., and D. Kan, 1985: *Equivariant homotopy classification*. J. Pure and Applied Algebra **35**, 269–285.

Elmendorf, A., 1983: *Systems of fixed point sets*. Trans. Amer. Math. Soc. **277**, 275–284.

Engelking, R., 1977: *General Topology*. Monografie Matematyczne **60**. Warszawa: Polish Scientific Publishers.

Gay, C. D., G. C. Morris, and I. Morris, 1983: *Computing Adams operations on the Burnside ring of a finite group*. Journal f. d. reine angew. Math. **341**, 87–97.

Gleason, A. M., 1950: *Spaces with a compact Lie group of transformations*. Proc. Amer. Math. Soc. **1**, 35–43.

Gluck, D., 1981: *Idempotent formula for the Burnside algebra with applications to the p-subgroup simplicial complex*. Ill. J. Math. **25**, 63–67.

Gordon, R. A., 1975: *Contributions to the theory of the Burnside ring*. Dissertation, Saarbrücken.

– 1977: *The Burnside ring of a cyclic extension of a torus*. Math. Z. **153**, 149–153.

Green, J. A., 1971: *Axiomatic representation theory for finite groups*. J. of Pure and Applied Algebra **1**, 41–77.

Hattori, A., and T. Yoshida, 1976: *Lifting compact group actions in fibre bundles*. Japanese J. Math. New Ser. **2**, 13–26.

Hauschild, H., 1975: *Allgemeine Lage und äquivariante Homotopie*. Math. Z. **143**, 155–164.

– 1975a: *Äquivariante Transversalität und äquivariante Bordismentheorien*. Arch. Math. **26**, 536–546.

– 1977: *Äquivariante Homotopie I*. Arch. Math. **29**, 158–165.

– 1977a: *Zerspaltung äquivarianter Homotopiemengen*. Math. Ann. **230**, 279–292.

– 1978: *Äquivariante Whiteheadtorsion*. Manuscripta math. **26**., 63–82.

Hauschild, V., 1983: *Deformationen graduierter Artinscher Algebren in der Kohomologietheorie von Transformationsgruppen*. Habilitationsschrift, Konstanz.

Helgason, S., 1962: *Differential geometry and symmetric spaces*. New York–London: Academic Press.

Hewitt, E., and A. K. Ross, 1963: *Abstract harmonic analysis I*. Berlin–Heidelberg–New York: Springer.

Hirsch, M. W., 1976: *Differential topology*. New York–Heidelberg–Berlin: Springer.

Hirzebruch, F., and K. H. Mayer, 1968: *O(n)-Mannigfaltigkeiten, exotische Sphären und Singularitäten*. Lecture Notes in Math. **57**. Berlin–Heidelberg–New York: Springer.

– and D. Zagier, 1974: *The Atiyah-Singer theorem and elementary number theory*. Boston: Publish or Perish.

Hochschild, G., 1965: *The structure of Lie groups*. San Francisco: Holden-Day.

Hsiang, W.-C., and W.-Y. Hsiang, 1967: *On compact subgroups of the diffeomorphism groups of Kervaire spheres*. Ann. of Math. **85**, 359–369.

– and W. Pardon, 1982: *When are topologically equivalent orthogonal trans-*

formations linearly equivalent? Invent. math. **68**, 275–316.

Hsiang, W.-Y., 1967: *On the bound of the dimensions of the isometry groups of all possible Riemannian metrics on an exotic sphere.* Ann. of Math. **85**, 351–358.

– 1975: *Cohomology theory of topological transformation groups.* Berlin–Heidelberg–New York: Springer.

Huppert, B., 1967: *Endliche Gruppen I.* Berlin–Heidelberg–New York: Springer.

Hurewicz, W., and H. Wallman, 1948: *Dimension theory.* Princeton: Princeton Univ. Press.

Illman, S., 1972: *Equivariant algebraic topology.* Thesis. Princeton.

– 1972a: *Equivariant singular homology and cohomology for actions of compact Lie groups.* Transformation groups, Proc. Conf., Amherst 1971, Lect. Notes Math. **298**, 403–415.

– 1973: *Equivariant algebraic topology.* Ann. Inst. Fourier **23**, 87–91.

– 1975: *Equivariant singular homology and cohomology. I.* Mem. Amer. Math. Soc. **156**.

– 1978: *Smooth equivariant triangulations of G-manifolds for G a finite group.* Math. Ann. **233**, 199–220.

– 1980: *Approximation of G-maps by maps in equivariant general position and imbeddings of G-complexes.* Trans. Amer. Math. Soc. **262**, 113–157.

– 1983: *The equivariant triangulation theorem for actions of compact Lie groups.* Math. Ann. **262**, 487–501.

Jackowski, S., 1985: *Families of subgroups and completion.* J. of pure and applied Algebra **37**, 167–179.

Jänich, K., 1968: *Differenzierbare G-Mannigfaltigkeiten.* Lecture Notes in Math. 59. Berlin–Heidelberg–New York: Springer.

James, I. M., and G. B. Segal, 1978: *On equivariant homotopy type.* Topology **17**, 267–272.

Jaworowski, J. W., 1976: *Extensions of G-maps and euclidean G-retracts.* Math. Z. **146**, 143–148.

Jones, L. E., 1971: *A converse to the fixed point theory of P. A. Smith, I.* Ann. of Math. **94**.

– 1972: *The converse to the fixed point theorem of P. A. Smith. II* Indiana Univ. Math. J. **22**, 309–325. Correction, **24**, 1001–1003 (1975).

Karoubi, M., 1978: *K-Theory.* Berlin–Heidelberg–New York: Springer.

Kawakubo, K., 1980: *Weyl group actions and equivariant homotopy equivalence.* Proc. Amer. Math. Soc. **80**, 172–176.

– 1980a: *Equivariant homotopy equivalence of group representations.* J. Math. Soc. Japan **32**, 105–118.

Kelley, J. L., 1955: *General topology.* Princeton: D. van Nostrand.

Kister, J. M., and L. Mann, 1962a: *Equivariant imbeddings of compact abelian Lie groups of transformations.* Math. Ann. **148**, 89–93.

Knutson, D., 1973: *λ-rings and the representation theory of the symmetric group.* Lecture Notes in Math. **308**. Berlin–Heidelberg–New York: Springer.

Kobayashi, S., and K. Nomizu, 1963: *Foundations of differential geometry.* New York–London: Interscience Publ.

Kosniowski, C., 1974: *Equivariant cohomology and stable cohomotopy.* Math. Ann. **210**, 83–104.

– 1978: *Actions of finite abelian groups.* London: Pitman.

Kratzer, Ch., and J. Thévenaz, 1984: *Fonction de Möbius d'un groupe fini et anneau de Burnside.* Comment. Math. Helv. **59**, 425–438.

Kreck, M., 1984: *Bordism of diffeomorphisms and related topics.* Lecture Notes in Math. **1069**, Berlin–Heidelberg–New York: Springer.

Laitinen, E., 1986: *Unstable homotopy theory of homotopy representations.* Transformation groups, Conf. Proc., Poznań, Lect. Notes Math. **1217**, 210–248.

Lang, S., 1965: *Algebra.* Reading, Mass.: Addison-Wesley.

Lashof, R., 1979: *Obstructions to equivariance.* Algebraic topology, Conf. Proc., Aarhus 1978, Lect. Notes Math. **763**, 476–503.

– 1982: *Equivariant bundles*. Illinois J. Math. **26**, 257–271.
– J.P. May, and G.B. Segal, 1983: *Equivariant bundles with abelian structural group*. Homotopy theory, Proc. Conf., Evanston 1982, Contemp. Math. **19**, 167–176.
– and M. Rothenberg, 1978: *G-smoothing theory*. Algebraic and geometric topology. Proc. Conf., Stanford 1976, Proc. Symp. Pure Math. **32**, Part 1, 211–266.
Lee, C.-N., and A.G. Wasserman, 1975: *On the groups JO(G)*. Mem. Amer. Math. Soc. **159**.
Lee, R., 1973: *Semicharacteristic classes*. Topology **12**, 183–199.
Lewis, L.G., 1983: *The uniqueness of bundle transfers*. Math. Proc. Camb. Phil. Soc. **93**, 87–111.
Lewis, L.G., J.P. May, and J.E. McClure, 1981: *Ordinary RO(G)-graded cohomology*. Bull. Amer. Math. Soc. N. S. **4**, 208–212.
– – – 1982: *Classifying G-spaces and the Segal conjecture*. Algebraic topology, Conf. Proc., London Ont. 1981, Canad. Math. Soc. Conf. Proc. **2**, Part 2, 165–179.
Lewis, L.G., J.P. May and M. Steinberger (with contributions by J.E. McClure), 1986: *Equivariant stable homotopy theory*. Lect. Notes Math. Berlin–Heidelberg–New York, Springer.
Liulevicius, A., 1978: *Homotopy rigidity of linear actions: Characters tell all*. Bull. Amer. Math. Soc. **84**, 213–221.
Löffler, P., 1981: *Homotopielineare $\mathbb{Z}_p$-Operationen auf Sphären*. Topology **20**, 291–312.
– 1981a: *$\mathbb{Z}_p$-Operationen auf rationalen Homologiesphären mit vorgeschriebener Fixpunktmenge*. Manuscr. math. 187–221.
– 1980: *Äquivariante Homotopie, äquivarianter Bordismus und freie differenzierbare Involutionen auf Sphären*. Math. Z. **170**, 233–246.
– und M. Raußen, 1985: *Symmetrien von Mannigfaltigkeiten und rationale Homotopietheorie*. Math. Ann. **271**, 549–576.
López de Medrano, S., 1971: *Involutions on Manifolds*. Berlin–Heidelberg–New York: Springer.
Lück, W., 1986: *The geometric finiteness obstruction*. J. London Math. Soc.
Lundell, A.T., and S. Weingram, 1969: *The topology of CW complexes*. New York: Van Nostrand.
Mac Lane, S., 1963: *Homology*. Berlin–Göttingen–Heidelberg. Springer.
Madsen, I., 1977: *Smooth spherical space forms. Geometric Applications of Homotopy Theory I*. Proceedings, Evanston 1977, Springer Lecture Notes Vol. **657**, 303–353.
– 1983: *Reidemeister torsion, surgery invariants and spherical space forms*. Proc. London Math. Soc. (3) **46**, 193–240.
– and M. Raußen, 1985: *Smooth and locally linear G-homotopy representations*. Algebraic topology, Proc. Conf., Göttingen 1984. Lect. Notes Math. **1172**, 130–156.
– and M. Rothenberg, 1985: *On the classification of G-spheres I: Equivariant transversality. II: PL automorphism groups. III: Top automorphism groups*. Aarhus Universitet, Preprint series 1985/86, No. **1**, No. **2**, No. **14**.
– C.B. Thomas, and C.T.C. Wall, 1976: *The toplogical spherical space form problem, II: Existence of free actions*. Topology **15**, 375–382.
– – – 1983: *Topological spherical space form problem III: Dimensional bounds and smoothing*. Pac. J. Math. 106, 135–143.
Mann, L.N., 1966: *Gaps in the dimensions of transformation groups*. Illinois J. Math. **10**, 532–546.
Massey, W.S., 1967: *Algebraic topology: An introduction*. New York: Harcourt Brace & World.
– 1978: *Homology and cohomology theory*. New York–Basel: Marcel Dekker.
Matumoto, T., 1971: *On G-CW-complexes and a theorem of J.H.C. Whitehead*. J. Fac. Sci. Univ. of Tokyo **18**, 363–374.
– 1973: *Equivariant cohomology theories on G-CW-complexes*. Osaka J. Math. **10**, 51–68.

Maunder, C.R.F., 1970: *Algebraic topology*. London: van Nostrand.

May, J.P., 1975: *Classifying spaces and fibrations*. Mem. Amer. Math. Soc. **155**.

– 1982: *Equivariant homotopy and cohomology theory*. Symposium in Honor of J. Adem (Oaxtepec, Mexico, 1981). Contemporary Math. **12**, 209–217.

– 1982a: *Equivariant completion*. Bull. London Math. Soc. **14**, 231–237.

– and J.E. McClure, 1982: *A reduction of the Segal conjecture*. Algebraic topology, Proc. Conf., London Ont. 1981, Canadian Math. Soc. Conf. Proc. Vol. **2**, Part 2, 209–222.

– – and G. Triantafillou, 1982: *Equivariant localization*. Bull. London Math. Soc. **14**, 223–230.

Meyerhoff, A., and T. Petrie, 1976: *Quasi equivalence of G modules*. Topology **15**, 69–75.

Milgram, R.J., 1967: *The bar construction and Abelian H-spaces*. Illinois J. Math. **11**, 242–250.

Milnor, J., 1957: *Groups which act on S^n without fixed points*. Amer. J. Math. **79**, 623–630.

Milnor, J., 1956: *Construction of universal bundles. II*. Ann. of Math. **63**, 430–436.

– 1962: *On axiomatic homology theory*. Pacific J. Math. **12**, 337–341.

– 1964: *Micro bundles*. Topology **3**, Suppl. 1, 53–80.

– and J.D. Stasheff, 1974: *Characteristic classes*. Princeton: Princeton Univ. Press.

Mitchell, B., 1972: *Rings with several objects*. Advances in Math. **8**, 1–161.

Montgomery, D., 1960: *Orbits of highest dimension*. In: Seminar on Transformation groups, Ann. of Math. Studies, No. **46**, Chap. IX. Princeton, Princeton Univ. Press.

– and H. Samelson, 1943: *Transformation groups on spheres*. Ann. of Math. **44**, 454–470.

– – and C.T. Yang, 1956: *Exceptional orbits of highest dimension*. Ann. of Math. **64**, 131–141.

– – and L. Zippin, 1956: *Singular points of a compact transformation group*. Ann. of Math. **63**, 1–9.

– and C.T. Yang, 1957: *The existence of a slice*. Ann. of Math. **65**, 108–116.

– – 1958: *Orbits of highest dimension*. Trans. Amer. Math. Soc. **87**, 284–293.

– and L. Zippin, 1955: *Toplogical transformation groups*. New York: Interscience publishers.

Morimoto, M., 1982: *On the groups $J_G(*)$ for $G = SL(2,p)$*. Osaka J. Math. **19**, 57–78.

Mostow, G.D., 1957: *On a conjecture of Montgomery*. Ann. of Math. **65**, 513–516.

– 1957a: *Equivariant embeddings in euclidean spaces*. Ann. of Math. **65**, 432–446.

Myers, S., and N. Steenrod, 1939: *The group of isometries of a Riemannian manifold*. Ann. of Math. **40**, 400–416.

Nagasaki, I., 1985: *Homotopy representations and spheres of representations*. Osaka J. Math. **22**, 895–905.

Namboodiri, U., 1983: *Equivariant vector fields on spheres*. Trans. Amer. Math. Soc. **278**, 431–460.

Nishida, G., 1978: *The transfer homomorphism in equivariant generalized cohomology theories*. J. Math. Kyoto **18**, 435–451.

– 1984: *Hecke functors and the equivariant Dold-Thom theorem*. Publ. Res. Inst. Math. Sci. **20**, 65–77.

Ochoa, G., 1984: *λ-ring structure on the Burnside ring of hamiltonian groups*. Arch. Math. **43**, 312–321.

Okonek, Chr., 1983: *Bemerkungen zur K-Theorie äquivarianter Endomorphismen*. Arch. Math. **40**, 132–138.

Oliver, R., 1975: *Fixed point sets of group actions on finite acyclic complexes*. Comment. Math. Helv. **50**, 155–177.

– 1976: *Fixed points of disk actions*. Bull. Amer. Math. Soc. **82**, 279–280.

– 1976a: *A proof of the Conner conjecture*. Ann. of Math. **103**, 637–644.

– 1976b: *Smooth compact Lie group actions on disks*. Math. Z. **149**, 79–96.

– 1977: *G-actions on disks and permutation representations. II* Math. Z. **157**, 237–263.

– 1978: *G-actions on disks and permutation representations*. J. Algebra **50**, 44–62.

– 1982: *A transfer homomorphism for compact Lie group actions*. Math. Ann. **260**, 351–374.

– and T. Petrie, 1982: *G-CW-surgery and $K_0(\mathbb{Z}G)$*. Math. Z. **179**, 11–42.

Palais, R., 1957: *Imbedding of compact differentiable transformation groups in orthogonal representations*. J. of Math. and Mech. **6**, 673–678.

– 1957a: *On the differentiability of isometries*. Proc. Amer. Math. Soc. **8**, 805–807.

– 1957b: *A global formulation of the Lie theory of transformation groups*. Mem. Amer. Math. Soc. **22**.

– 1960: *The classification of G-spaces*. Mem. Amer. Math. Soc. **36**.

– 1960a: *Slices and equivariant imbeddings*. In: Seminar on Transformation Groups, Ann. of Math. Studies **46**, Chap. VIII. Princeton: Princeton Univ. Press.

– 1960b: *On the existence of slices for actions of non-compact Lie groups*. Ann. of Math. **73**, 295–323.

Pedersen, E.K., 1985: *Topological $H_0 \times H_1$-actions on spheres and linking numbers*. Algebraic topology, Proc. Conf., Göttingen 1984. Lect. Notes Math. **1172**, 163–166.

Petrie, T., 1971: *Free metacyclic group actions on homotopy spheres*. Ann. of Math. **94**, 108–124.

– 1976: *G maps and the projective class group*. Comment. math. Helv. **51**, 611–626.

– 1978: *Pseudoequivalences of G-manifolds*. Algebraic and geometric topology, Proc. Conf., Stanford 1976, Proc. of Symposia in Pure Math. **32**, Part 1, 169–210.

– and J.D. Randall, 1984: *Transformation groups on manifolds*. New York–Basel: Marcel Dekker.

Pontrjagin, L.S., 1957: *Topologische Gruppen*. Leipzig: Teubner.

Puppe, V., 1974: *On a conjecture of Bredon*. Manuscripta math. **12**, 11–16.

– 1978: *Cohomology of fixed point sets and deformation of algebras*. Manuscripta math. **23**, 343–345.

– 1979: *Deformations of algebras and cohomology of fixed point sets*. Manuscripta math. **30**, 119–136.

– 1984: *P.A. Smith theory via deformations*. Homotopie algébrique et algèbre locale, Journ. Luminy/France 1982, Astérisque 113–114, 278–287.

Quillen, D., 1971: *The spectrum of an equivariant cohomology ring. I* Ann. of Math. **94**, 549–572.

– 1971a: *The spectrum of an equivariant cohomology ring. II* Ann. of Math. **94**, 573–602.

Quinn, F., 1978: *Finite nilpotent group actions on finite complexes*. Geometric appl. of homotopy theory I, Proc. Conf., Evanston 1977, Lect. Notes Math. **657**, 375–407.

– 1979: *Nilpotent classifying spaces and actions of finite groups*. Houston J. Math. **4**, 239–248.

de Rham, G., 1940: *Sur les complexes avec automorphismes*. Comment. Math. Helv. **12**, 191–211.

– 1950: *Complexes à automorphismes et homéomorphie différentiables*. Ann. Inst. Fourier, Grenoble, **2**, 57–67.

– 1964: *Reidemeister's torsion invariants and rotations of S^n*. Differential Analysis (Bombay Colloq., 1963), 27–36. New York, Oxford University Press.

Ritter, J., 1972: *Ein Induktionssatz für rationale Charaktere von nilpotenten Gruppen*. Journal f. d. reine angew. Math. **254**, 133–151.

Rothenberg, M., 1978: *Torsion invariants and finite transformation groups*. Algebraic and geometric topology, Proc. Conf. Stanford 1976. Proc. Symp. Pure Math. **32**, Part 1, 267–311.

Rubinsztein, R.L., 1973: *On the equivariant homotopy of spheres*. Preprint **58**. Polish Academy of Sciences.

Schubert, H., 1964: *Topologie*. Stuttgart: Teubner.

– 1970: *Kategorien I. II*. Berlin–Heidelberg–New York: Springer.

Schultz, R., 1973: *Homotopy decompositions of equivariant function spaces. I*. Math. Z. **131**, 49–75.

– 1973a: *Homotopy decompositions of equivariant function spaces. II* Math. Z. **132**, 69–80.

– 1977: *Equivariant function spaces and equivariant stable homotopy theory.* Transformation groups, Proc. Conf., Newcastle-upon-Tyne 1976, London Math. Soc. Lect. Notes No. **26**, 169–189.

Schwänzl, R., 1977: *Koeffizienten im Burnsidering.* Arch. Math. **29**, 621–622.

– 1979: *On the spectrum of the Burnside ring.* J. of pure and applied Algebra **15**, 181–185.

Schwarz, G., 1980: *Lifting smooth homotopies of orbit spaces.* Publ. Math. Inst. Hautes Études Sci. **51**, 37–136.

Segal, G. B., 1968: *Equivariant K-theory.* Publ. Math. Inst. Hautes Études Sci. **34**, 129–151.

– 1968a: *Classifying spaces and spectral sequences.* Publ. Math. Inst. Hautes Études Sci. **34**, 105–112.

– 1968b: *The representation ring of a compact Lie group.* Publ. Math. Inst. Hautes Études Sci. **34**, 113–128.

– 1971: *Equivariant stable homotopy theory.* Actes Congr. internat. Math. 1970, **2**, 59–63.

– 1972: *Permutation representations of finite p-groups.* Quart. J. Math. Oxford (2), **23**, 375–381.

Serre, J.-P., 1971: *Représentations linéaires des groupes finis.* Paris: Herrman 2. éd.

– 1971a: *Cohomologie des groupes discrets.* Ann. Math. Studies **70**, 77–169. Princeton: Princeton Univ. Press.

– 1980: *Trees.* Berlin–Heidelberg–New York: Springer.

Siebeneicher, C., 1976: *λ-Ringstrukturen auf dem Burnsidering der Permutationsdarstellungen einer endlichen Gruppe.* Math. Z. **146**, 223–238.

Skjelbred, T., 1974: *Integral global weights for torus actions on projective spaces.* Math. Scand. **34**, 249–253.

– 1978: *Cohomology eigenvalues of equivariant mappings.* Comment. Math. Helv. **53**, 634–642.

– 1978a: *Combinatorial geometry and actions of compact Lie groups.* Pacific J. of Math. **79**, 197–205.

Smith, P. A., 1938: *The topology of transformation groups.* Bull. Amer. Math. Soc. **44**, 497–514.

– 1938a: *Transformations of finite period.* Ann. of Math. **39**, 127–164.

– 1939: *Transformations of finite period. II.* Ann. of Math. **40**, 690–711.

– 1941: *Fixed point theorems for periodic transformations.* Amer. J. Math. **63**, 1–8.

– 1941a: *Transformations of finite period. III. Newman's theorem.* Ann. of Math. **42**, 446–458.

– 1941b: *Periodic and nearly periodic transformations.* „Lectures in Topology“ pp. 159–190. Univ. of Michigan Press, Ann Arbor, Michigan.

– 1942: *Fixed points of periodic transformations.* Appendix *B* in: S. Lefschetz, Algebraic Topology, New York, Amer. Math. Soc.

– 1942a: *Stationary points of transformation groups.* Proc. Nat. Acad. Sci. U.S.A. **28**, 293–297.

– 1944: *Permutable periodic transformations.* Proc. Nat. Acad. Sci. U.S.A. **30**, 105–108.

– 1945: *Transformations of finite period, IV. Dimensional parity.* Ann. of Math. **46**, 357–364.

– 1959: *Orbit spaces of abelian p-groups.* Proc. Nat. Acad. Sci. U.S.A. **45**, 1772–1775.

– 1960: *New results and old problems in finite transformation groups.* Bull. Amer. Math. Soc. **66**, 401–415.

– 1961: *Orbit spaces of finite Abelian transformation groups.* Proc. Nat. Acad. Sci. U.S.A. **47**, 1662–1667.

– 1963: *The cohomology of certain orbit spaces.* Bull. Amer. Math. Soc. **69**, 563–568.

– 1965: *Periodic transformations of 3-manifolds.* Illinois J. Math. **9**, 343–348.

– 1967: *Abelian actions on 2-manifolds.* Michigan Math. J. **14**, 257–275.

– and M. Richardson, 1937: *Periodic transformations of complexes.* Ann. of Math. **38**, 611–633.

Snaith, J., 1971: *J-equivalence of group representations.* Proc. Camb. Phil. Soc. **70**, 9–14.

Solomon, L., 1967: *The Burnside algebra of a finite group*. J. Combin. Theory **2**, 603–615.

Spanier, E.H., 1966: *Algebraic topology*. New York: Mc Graw-Hill.

Steenrod, N.E., 1967: *A convenient category of topological spaces*. Michigan Math. J. **14**, 133–152.

– 1968: *Milgram's classifying space of a topological group*. Topology **7**, 349–368.

Stewart, T.E., 1961: *Lifting group actions in fibre bundles*. Ann. of Math. **74**, 192–198.

Stieglitz, A., 1978: *Actions of groups with periodic cohomology on Poincaré-duality spaces*. Math. Ann. **236**, 29–42.

Stong, R.E., 1970: *Unoriented bordism and actions of finite groups*. Mem. Amer. Math. Soc. **103**.

Su, J.C., 1963: *Transformation groups on cohomology projective spaces*. Trans. Amer. Math. Soc. **106**, 305–318.

Swan, R.G., 1960: *Periodic resolutions for finite groups*. Ann. of Math. **72**, 267–291.

– 1960a: *The p-period of a finite group*. Illinois J. Math. **4**, 341–346.

– 1960b: *A new method in fixed point theory*. Comm. Math. Helv. **34**, 1–16.

Switzer, R.M., 1975: *Algebraic topology-homotopy and homology*. Berlin–Heidelberg–New York: Springer.

Tits, J., 1983: *Liesche Gruppen und Algebren*. Berlin–Heidelberg–New York: Springer.

Tornehave, J., 1982: *Equivariant maps of spheres with conjugate orthogonal actions*. Algebraic topology, Proc. Conf., London Ont. 1981, Canadian Math. Soc. Conf. Proc., Vol. **2**, part 2, 275–301.

– 1984: *The unit theorem for the Burnside ring of a 2-group*. Aarhus Universitet. Preprint Series 1983/84, No. **41**.

Traczyk, P., 1978: *On the G-homotopy equivalence of spheres of representations*. Math. Z. **161**, 257–261.

– 1982: *Cancellation law for homotopy equivalent representations of groups of odd order*. Manuscripta math. **40**, 135–154.

Triantafillou, G.V., 1982: *Equivariant minimal models*. Trans. Amer. Math. Soc. **274**, 509–532.

Ulrich, H., 1983: *Der äquivariante Fixpunktindex vertikaler G-Abbildungen*. Dissertation, Heidelberg.

van Dantzig, D., and B.L. van der Waerden, 1928: *Über metrische homogene Räume*. Abh. Math. Sem. Univ. Hamburg **6**, 367–376.

Verona, A., 1980: *Triangulation of stratified bundles*. Manuscripta Math. **30**, 425–445.

Vogt, R.M., 1971: *Convenient categories of topological spaces for homotopy theory*. Arch. der Math. **22**, 545–555.

Wall, C.T.C., 1970: *Surgery on Compact Manifolds*. Academic Press, New York.

Waner, S., 1980: *Equivariant homotopy theory and Milnor's theorem*. Trans. Amer. Math. Soc. **258**, 351–368.

Wasserman, A., 1969: *Equivariant differential topology*. Topology **8**, 127–150.

Weyl, H., 1939: *The classical groups*. Princeton: Princeton Univ. Press.

Whitehead, G.W., 1978: *Elements of homotopy theory*. New York–Heidelberg–Berlin: Springer.

Whitehead, J.H.C., 1949: *Combinatorial homotopy I*. Bull. Amer. Math. Soc. **55**, 213–245.

Willson, S.J., 1975: *Equivariant homology theories on G-complexes*. Trans. Amer. Math. Soc. **212**, 155–171.

Wirthmüller, K., 1974: *Equivariant homology and duality*. Manuscripta math. **11**, 373–390.

– 1974: *Equivariant S-duality*. Arch. Math. **26**, 427–431.

Wolf, J.A., 1967: *Spaces of constant curvature*. New York: McGraw-Hill.

Yoshida, T., 1983: *Idempotents of Burnside rings and Dress induction theorem*. J. Alg. **80**, 90–105.

Further reading

1. Compact transformation groups in general

Above all, the reader should study Bredon's book [1972] as a necessary complement to the present book.

The following seminar or conference reports contain a variety of further results, problems, and viewpoints:

(i) Seminar on transformation groups. Edited by A. Borel. Princeton University Press 1960.

(ii) Proceedings of the Conf. on Transformation Groups. New Orleans, 1967. Edited by P. S. Mostert. Berlin–Heidelberg–New York, Springer 1968.

(iii) Proceedings of the Second Conf. on compact transformation groups. Univ. of Massachusetts, Amherst, 1971. Lect. Notes Math. 298, 299. Berlin–Heidelberg–New York, Springer 1972.

(iv) Group Actions on Manifolds. R. Schultz, Ed. Contemporary Mathematics Vol. 36 (1985). Conf. Proceedings, Boulder 1983.

2. Cobordism and transformation groups

Conner-Floyd [1964], Conner [1979]

3. Equivariant surgery. Manifolds

Dovermann-Petrie [1982], [1983], Petrie-Randall [1984], Schwarz [1980]

4. Equivariant algebraic topology

Lewis, May, Steinberger, and McClure [1986]

5. Analytical methods

Atiyah [1968], [1974], Atiyah-Bott [1967], [1968], Atiyah-Singer [1968], [1968a], Hirzebruch-Zagier [1974]

6. Homotopy theory

Dwyer-Kan [1985], Carlsson [1984]

7. Bundle cohomology

W.-Y. Hsiang [1975], V. Puppe [1978], [1979], V. Hauschild [1983]

Subject index and symbols

More symbols

Sets

X/Y	X with Y identified to a point (similar symbol for orbit spaces!)
$A \backslash B$	set theoretic difference, elements in A but not in B (similar notation for orbit spaces!)
$\amalg$	disjoint union
pr	projection from a product

Groups

$H \lhd K$	H normal subgroup of K
$H \sim K$	H and K conjugate subgroups
G/H	space of left cosets gH, $g \in G$
(H)	conjugacy class of the subgroup H
$(H) < (K)$	H subconjugate to K
$K \backslash (G/H)$	space of double cosets
$e \in G$	unit element of group G (or generic element of a space E)
G_0	component of e in G
$NH = N_G H$	normalizer of the subgroup H in G
$WH = NH/H$	
$R(G)$	complex representation ring of G

Transformation groups

$X/G \quad G \backslash X$	orbit space of G-space X
f/G	orbit map of G-map f
X^H	H-fixed point set of X
G_x	isotropy group at the point x
$X_H = \{x \in X \mid G_x = H\}$	
$X_{(H)} = \{x \in X \mid G_x \sim H\}$	
$X(\mathscr{F})$	46
$G \times_H X$	32
$X \times_H Y$	36
$G^+ \wedge_H X$	34
$X^{>H} = X^H \backslash X_H$	

Topology

$[X, Y]_G$	set of G-homotopy classes of G-maps $X \to Y$
$[X, Y]_G^0$	set of pointed G-homotopy classes of pointed maps
Y^Z	set of continuous maps $Z \to Y$
$\mathrm{Map}(X, Y)$	set of maps $X \to Y$
$C_G(X, Y)$	space of continuous maps $X \to Y$ with compact-open-topology
$C^0(X, Y)$	space of pointed maps $X \to Y$ with compact-open-topology
$\mathrm{Top}_G(X, Y)$	set of continuous G-maps $X \to Y$
G-Top	category of G-spaces and continuous G-maps
$X \vee Y$	wedge = pointed sum
$X \wedge Y = X \times Y / X \wedge Y$	smash-product
$\chi(X)$	Euler characteristic of X
$\chi_c(X)$	Euler characteristic of X, using homology with compact support

Bei Fragen zur Produktsicherheit wenden Sie sich bitte an:
If you have any questions regarding product safety,
please contact:

Walter de Gruyter GmbH
Genthiner Straße 13
10785 Berlin
productsafety@degruyterbrill.com